712.2  Sto

# Interaction by Design

## Bringing People and Plants Together for Health and Well-Being

An International Symposium

# Interaction by Design

## Bringing People and Plants Together for Health and Well-Being

### An International Symposium

Editor-in-Chief

## Candice A. Shoemaker

Associate Editors

Elizabeth R. Messer Diehl
Jack Carman
Nancy Carman
Jane Stoneham
Virginia I. Lohr

**Iowa State Press**
*A Blackwell Publishing Company*

Iowa State Press
2121 State Avenue, Ames, Iowa 50014

Orders:  1-800-862-6657
Office:  1-515-292-0140
Fax:  1-515-292-3348
Web site: www.iowastatepress.com

∞ Printed on acid-free paper in the United States of America

First edition, 2002

Library of Congress Cataloging-in-Publication Data

Interaction by design : bringing people and plants together for health and well-being : an international symposium / editor-in-chief, Candice A. Shoemaker.
        p. cm.
Includes bibliographical references and index.
    ISBN 0-8138-0323-3
    1. Gardening—Therapeutic use.  I.  Shoemaker, Candice A.
    RM735.7.G37 1584 2002
    615.8′51—dc21                                    2002008717


The last digit is the print number:  9  8  7  6  5  4  3  2  1

# Contents

Contributors   xi

Foreword   xix
(Paula Diane Relf)

Preface   xxiii
(Candice A. Shoemaker)

## Section I. Communications with the Horticulture, Healthcare, and Design Communities

Chapter 1.   3
Communicating with the Horticulture Community about Human Issues in Horticulture (Paula Diane Relf)

Chapter 2.   19
Communicating with the Healthcare Community (Roger S. Ulrich)

Chapter 3.   33
Greenlining: An Invitation to Cross the Bridges to Urban Green Space and Understanding (Margaret Ross Bjornson)

Chapter 4.   41
Computer Use among Registered Members of the American Horticultural Therapy Association (Lea Minton Westervelt and Richard H. Mattson)

## Section II. Design for Human Health and Well-Being

Chapter 5.   53
Linking People with Nature by Universal Design (Fusayo Asano Miyake)

Chapter 6.   63

From Vision to Reality: The Chicago Botanic Garden's Buehler Enabling Garden
(Gene Rothert)

Chapter 7.   75

A Holistic Approach to Creating a Therapeutic Garden (Virginia Burt)

Chapter 8.   83

The Medical Center Gardens Project (Royce K. Ragland)

Chapter 9.   91

Participatory Design of a Terrace Garden in an Acute In-Patient Unit
(Shelagh Rae Smith)

Chapter 10.   99

Building an Alzheimer's Garden in a Public Park (Mark Epstein)

Chapter 11.   111

Alzheimer's Garden Project of the American Society of Landscape Architects and
the National Alzheimer's Association (Jack Carman)

Chapter 12.   115

Healing Landscapes: Design Guidelines for Mental Health Facilities
(Myra Kovary)

Chapter 13.   123

A Children's PlayGarden at a Rehabilitation Hospital—A Successful
Collaboration Produces a Successful Outcome (Sonja Johansson and
Nancy Chambers)

Chapter 14.   135

The Role of Perception in the Designing of Outdoor Environments
(Marni Barnes)

Chapter 15.   141

Design Strategies for Integrating Natural Elements with Building Design
(Phillip G. Mead)

Chapter 16.   149

Designing Natural Therapeutic Environments (Katie Johnson)

## Section III. Therapeutic Design Applications

Chapter 17.   157

Implementation of Therapeutic Design Applications (Jane Stoneham)

Chapter 18.   165

Long-Term Memory Response of Severely Cognitive-Impaired Elderly to
Horticultural and Reminiscence Therapies (Emi Kiyota)

Chapter 19.   175

Using Bonsai as a Component of Life Enhancement for Older Adults
(Brian H. Santos, John Tristan, and Leda McKenry)

Chapter 20.   179

Horticultural Intervention as a Stress Management Technique among
University of Massachusetts/Amherst Students
(John Tristan, Mary Anne Bright, Michael Sutherland, and Chantal Duguay)

Chapter 21.   187

Fertile Ground in Long-Term Care Medical Facilities (Carole Staley Collins)

Chapter 22.   195

Sensory Garden Tours at Denver Botanic Gardens (Janet Laminack)

Chapter 23.   203

The Role of Botanical Gardens in Research, Design, and Program Development:
A Case Study of Botanica, the Wichita Gardens (Patricia J. Owen)

Chapter 24.   211

Welfare and Medical Institutions in Kyushu, Japan, Engaged in Horticulture:
A Survey (E. Matsuo, Y. Fujiki, H. Takafuji, H. Kweon, F. A. Miyake,
K. Masuda, and K. Mekaru)

## Section IV. Research

Chapter 25.   219

Postoccupancy Evaluation and the Design of Hospital Gardens (Clare Cooper
Marcus)

Chapter 26.   227

Research Methodologies for Studying Human Responses to Horticulture
(Candice A. Shoemaker)

Chapter 27.   237

The Effect of Cut Flowers in the Japanese Tea Ceremony (Kenji Yamane, Miwa
Umezawa, Seiya Uchida, Nobuaki Fujishige, Masao Yoshida, and
Masayoshi Katagiri)

Chapter 28.    245

Impact of Cut Roses at Different Flower-Opening Stages on Customers'
Perceptions of a Restaurant Environment (Megumi Adachi, Yasushi Takano,
and Anthony D. Kendle)

Chapter 29.    253

The Effect of Flower Color on Respondents' Physical and Psychological
Responses (Chun-Yen Chang)

Chapter 30.    261

Plant Decoration in Front of the House Entrance or Gate
(H. Kweon, E. Matsuo, H. Takafuji, F. A. Miyake, K. Masuda,
and K. Mekaru)

Chapter 31.    267

Childhood Contact with Nature Influences Adult Attitudes and Actions toward
Trees and Gardening (Virginia I. Lohr and Caroline H. Pearson-Mims)

Chapter 32.    279

Exhilarating Fragrances in the Indo-Islamic Garden (S. Ali Akbar Husain)

Chapter 33.    291

Where the Lawn Mower Stops: The Social Construction of Alternative Front Yard
Ideologies (Andrew J. Kaufman and Virginia I. Lohr)

## Section V. Abstracts

Chapter 34.    303

Abstracts: Communications

    Restorative Gardens: Nature's Therapeutic Complement to
        Healthcare Environments (David Kamp)    303

    The Banning of Flower Cultivation in Japan Near the End of World War II
        (Haruo Konoshima)    303

    Assessing the Impact of Urban Forests on Elderly People in Long-Term
        Care Settings: Toward a Multicultural Framework
        (Gowri Betrabet and Susana Alves)    304

    Japanese and European Names of Colors Originating from Plant Names in Japan
        (Eisuke Matsuo, Kinuko Masuda, H. Kweon, K. Mekaru, and F. A. Miyake)    304

    Present Status of Horticultural Therapy and Human Issues in Horticulture in Korea
        (Hye Ran Kwack and Paula Diane Relf)    305

    Infection Control and Therapeutic Gardens: A Survey of Policies and Practice
        (Nancy J. Gerlach-Spriggs)    305

Chapter 35.   307

Abstracts: Design

A Universal Design Approach to Recreation and Learning in
the Landscape (Susan Goltsman)   307

Restorative Gardens—Metaphorically Transcending the Human Experience
of Life (Scott C. Scarfone)   307

Therapeutic Garden Design in a Pediatric Healthcare Setting:
A Developmental Approach (Roberta Hursthouse)   308


Chapter 36.   309

Abstracts: Therapeutic Application

Horticultural Therapy in Rehabilitation (Katherine A. Feuillan and
Alicia Gaca)   309

Horticultural Therapy at the Royal National Orthopedic Hospital NHS Trust
(Linda Exley)   309

Seed Balls (Keiko Murayama and Ritsko Yasue)   310

Interactions between Elderly Adults and Preschool Children in a Horticultural
Therapy Research Program (Mary L. Predny and Paula Diane Relf)   310

The Therapeutic Role of Horticulture in Educational, Recreational,
Therapeutic, and Vocational Programs (Lori Keltner)   311

Engaging Disabled Students in Developing a Flower Garden as a Recreational
Therapy (Karen Midden and Michelle McLernon)   312

Every Community Needs a Providence Farm (Christine Pollard)   312

Incorporating Horticulture Therapy with Psychosocial and Pharmacological
Interventions of Mental Illness in Children and Adolescents
(Catherine Trapani)   313


Chapter 37.   315

Abstracts: Research

Effects of Horticultural Activity on the Level of a Stress Hormone and
Susceptibility to Upper Respiratory Infection (Hyejin Cho and
Richard H. Mattson)   315

Personal, Social, and Cultural Meanings of a Community Garden for Students:
An In-Depth Case Study and Evaluation (Catherine A. Bylinowski and
Richard H. Mattson)   315

Physical and Psychological Responses of Patients to the Hospital Landscape
(Chun-Yen Chang and Tzu-Hui Tseng)   316

Plant Successions and Urban Successions: How Human Beings Have Affected
Them (Hideki Hirano)   316

Air Conditioning and Noise Control Using Vegetation (Peter Costa)    317

Psychophysiological and Emotional Influences of Floral Aroma on Human
Stress (Mingwang Liu and Richard H. Mattson)    317

## Section VI. Appendix

Appendix. Symposium Overview    321

Index    323

# Contributors

## Editor-in-Chief

Candice A. Shoemaker, Associate Professor, Department of Horticulture, Forestry, and Recreation Resources, Kansas State University, Manhattan

## Associate Editors

Elizabeth R. Messer Diehl, Landscape Architect, Austin, TX: Communications with the Horticulture, Healthcare, and Design Communities

Jack Carman, President, Design for Generations, Medford, NJ: Design for Human Health and Well-Being

Nancy Carman, Design for Generations, Medford, NJ: Design for Human Health and Well-Being

Jane Stoneham, The Sensory Trust, St. Austell, Cornwall, England: Therapeutic Design Applications

Virginia I. Lohr, Professor, Horticulture and Landscape Architecture, Washington State University, Pullman: Research

## Manuscript Reviewers

Mark Epstein, Landscape Architect, Parametrix, Inc., Kirkland, WA

Jennifer L. Hart, Graduate Assistant, Department of Horticulture and Landscape Architecture, Washington State University, Pullman

Andrew J. Kaufman, Graduate Assistant, Department of Horticulture and Landscape Architecture, Washington State University, Pullman

Jack Kerrigan, Chair, Cuyahog County Extension Agent, Horticulture, Ohio State University Extension, Cleveland Heights

Vicki L. Newman, Graduate Assistant, Department of Horticulture and Landscape Architecture, Washington State University, Pullman

Caroline Pearson-Mims, Research Technologist, Department of Horticulture and Landscape Architecture, Washington State University, Pullman

Fred Rozumalski, Landscape Ecologist, Minneapolis, MN

Peter Thoday, Principal, Thoday Associates, UK

Phillip S. Waite, Assistant Professor, Department of Horticulture and Landscape Architecture, Washington State University, Pullman

Gina Ingoglia Weiner, Landscape Architect, Brooklyn, NY

## *Production Assistant*

Paula Johnson, Highland Park, IL

## *Authors*

*The number(s) in brackets following each name is a chapter number.*
*A number followed by "a" indicates an abstract.*

Megumi Adachi [28]

Researcher in Floriculture, Laboratory of Horticultural Science, Graduate School of Agricultural Life Sciences, University of Tokyo, 1-1 Yayoi, Bunkyo-ku, Tokyo, 113-8657, Japan
Email: ameg@nodai.ac.jp

Susana Alves [34a]

University of Wisconsin—Milwaukee, WI

Marni Barnes [14]

Landscape Architect, Deva Designs
846 Boyce Avenue, Palo Alto, CA 94301
Email: marnib@mindspring.com

Gowri Betrabet [34a]

2131 E. Hartford Ave., Milwaukee, WI 53211
Email: gulwadi@att.net

Margaret Ross Bjornson [3]

Educator/Researcher, 1807 Grant Street, Evanston, IL 60201
Email: m-bjornson@nwu.edu

Mary Anne Bright [20]

Associate Professor, School of Nursing, University of Massachusetts/Amherst

Virginia Burt [7]

Visionscapes Landscape Architects, Inc., R.R. #3, 5318 Cedar Springs Road, Campbellville, Ontario, L0P 1B0 Canada
Email: visions@interlynx.net

**Catherine A. Bylinowski [37a]**

Graduate Student, Department of Horticulture, Forestry, and Recreation Resources, Kansas State University
1604 Hill Crest Drive, Unit #25, Manhattan, KS 66502

**Jack Carman [11]**

President and Landscape Architect, Design for Generations, 92 Tallowood Drive, Medford, NJ 08055
Email: jpcarman@waterw.com

**Nancy Chambers [13]**

Director and Horticultural Therapist, Enid A. Haupt Glass Garden at the Rusk Institute of Rehabilitation Medicine, New York University Medical Center, New York, NY 10016
Email: nancy.chambers@med.nyu.edu

**Chun-Yen Chang [29, 37a]**

Associate Professor, Department of Horticulture, National Chung Hsing University, 250 Kuo Kuang Road, Taichung, Taiwan 40227
Email: cychang1@nchu.edu.tw

**Hyejin Cho [37a]**

Graduate Student, Department of Horticulture, Forestry, and Recreation Resources, Kansas State University, 2021 Throckmorton, Manhattan, KS 66506
Email: hch5974@ksu.edu

**Carole Staley Collins [21]**

Nurse and Doctoral Graduate Student, University of Maryland—Baltimore, 243 Anchorage Drive, Annapolis, MD 21401
Email: mtngarden@msn.com

**Clare Cooper Marcus [25]**

Professor Emerita, University of California, 2721 Stuart Street, Berkeley, CA 94705

**Peter Costa [37a]**

Senior Engineer, Allied Environmental Consulting, 136 Stirling Highway, Unit 6, N. Fremantle, W. Australia 6159, Australia

**Chantal Duguay [20]**

Staff Nurse, Indian River Memorial Hospital, Vero Beach, FL

**Mark Epstein [10]**

Landscape Architect, Parametrix, Inc., 5808 Lake Washington Blvd. N.E., Suite 200, Kirkland, WA
Email: mepstein@parametrix.com

**Linda Exley [36a]**

The Royal National Orthopedic Hospital NHS Trust, Brockey Hill, Stanmore, Middlesex, UK HA7 4LP

Katherine A. Feuillan [36a]

Recreational Therapist, Resurrection Medical Center, Day Rehab Professional Building, 7435 West Talcott, Chicago, IL 60631
Email: bpadjen@ameritech.net

Y. Fujiki [24]

Hana-Purasu, LTD., Kamiuchi, Oomuta-shi 837-0902, Japan

Nobuaki Fujishige [27]

Faculty of Agriculture, Utsunomiya University, Tochigi, 321-8505, Japan

Alicia Gaca [36a]

Recreational Therapist, Resurrection Medical Center, 7435 West Talcott, Chicago, IL 60631

Nancy J. Gerlach-Spriggs [34a]

Director, Meristem, Inc., 1233 York Avenue 22C, New York, NY 10021
Email: ngspriggs@aol.com

Susan Goltsman [35a]

Principal, Moore Iacofano Goltsman, Inc., 800 Hearst Avenue, Berkeley, CA 94710
Email: susan@migcom.com

Hideki Hirano [37a]

Forest Agency Ministry of Agriculture, Forestry and Fisheries, 1-2-1 Kasumigaseki, Chiyoda-Ku, Tokyo 100-8952, Japan
Email: hideki_hirano@nm.maff.go.jp

Roberta Hursthouse [35a]

Landscape Architect and Horticultural Therapist, Accessible Gardens, Naperville, IL

S. Ali Akbar Husain [32]

Professor of Architecture, 14 Cathedral Road, Brampton, Ontario L6W 2N9 Canada

Sonja Johansson [13]

Landscape Architect, Johansson Design Collaborative, Inc., 9 Old Concord Road, Lincoln, MA 01773
Email: johanssondc@earthlink.net

Katie Johnson [16]

Landscape Architect, 32210 First Lane SW #I-102, Federal Way, WA 98023
Email: katieljohnson@hotmail.com

David Kamp [34a]

Dirtworks, Inc., 111 East 14th Street, Unit #242, New York, NY 10003

Masayoshi Katagiri [27]

Faculty of International Studies, Utsunomiya University, Tochigi, 321-8505, Japan

Andrew J. Kaufman [33]

Doctoral Student, Department of Horticulture and Landscape Architecture, Washington State University, Pullman, WA 99164-6414
Email: ajk@mail.wsu.edu

Lori Keltner [36a]

The Horticulture Connection, 150 School Road, Grayslake, IL

Anthony D. Kendle [28]

Lecturer, Department of Horticulture and Landscape, Faculty of Plant Science, Whiteknights, PO Box 221, Reading, RG6 6AS, U.K.

Emi Kiyota [18]

Graduate Research Assistant, Galichia Center on Aging, 203 Fairchild Hall, Kansas State University, Manhattan, KS 66506-1102
Email: eki2163@ksu.edu

Haruo Konoshima [34a]

Professor, Faculty of Education, Shiga University, 2-5-1 Hiratsu, Otsu, Shiga 520-0862, Japan

Myra Kovary [12]

87 Uptown Road, D108, Ithaca, NY 14850
Email: mkovary@hotmail.com

Hye Ran Kwack [34a]

Postdoctoral Research Associate, Virginia Tech University, Blacksburg, VA

Hyojung Kweon [24, 30, 34a]

Graduate Student, Laboratory of Applied Plant Science, Graduate School of Agriculture, Kyushu University, Hakozaki, Fukuoka 812-8581, Japan
Email: khjune@brs.kyushu-u.ac.jp

Janet Laminack [22]

Horticultural Therapist, 9830 Highway 199W, Poolville, TX 76487
Email: janetlaminack@yahoo.com

Mingwang Liu [37a]

Assistant Professor, Department of Ornamental Horticulture, Delaware Valley College, Doylestown, PA 18901
Email: lium@delval.edu

Virginia I. Lohr [31, 33]

Professor, Department of Horticulture and Landscape Architecture, Washington State University, Pullman, WA 99164
Email: lohr@wsu.edu

Kinuko Masuda [24, 30, 34a]

Graduate Student, Laboratory of Applied Plant Science, Graduate School of Agriculture, Kyushu University, Hakozaki, Fukuoka-shi 812-8581, Japan
Email: kinukom@green.h.chiba-u.jp

Eisuke Matsuo [24, 30, 34a]

Professor, Division of Applied Plant Science, Faculty of Agriculture, Kyushu University, Fukuoka, 812-8581, Japan
Email: eisuke_m@agr.kyushu-u.ac.jp

Richard H. Mattson [4, 37a]

Professor, Department of Horticulture, Forestry, and Recreation Resources, Kansas State University, 2021 Throckmorton Plant Sciences Center, Manhattan, KS 66506-5501
Email: rmattson@oznet.ksu.edu

Leda McKenry [19]

Director, Office of Academic Outreach, School of Nursing, University of Massachusetts/Amherst, MA 01003

Michelle McLernon [36a]

Southern Illinois University—Carbondale, Office of Intramural Sports, Student Recreation Center, Carbondale, IL 62901

Phillip G. Mead [15]

Professor, Texas Tech University, College of Architecture, Box 42091, Lubbock, TX 79409
Email: p.mead@ttu.edu

K. Mekaru [24, 30, 34a]

Laboratory of Applied Plant Science, Graduate School of Agriculture, Kyushu University, Hakozaki, Fukuoka 812-8581, Japan

Karen Midden [36a]

Southern Illinois University, Plant, Soil and General Agriculture, Carbondale, IL 62901
Email: kmidden@sill.edu

Fusayo Asano Miyake [5, 24, 30, 34a]

Director, Sen, Inc., 4-11 Tsuruno-cho, Suite 1106, Kita-ku, Osaka 530-0014, Japan
Email: post@sen-inc.co.jp

Keiko Murayama [36a]

Internship Student, Bryn Mawr Rehab, 115 S. Brandywine Street, West Chester, PA 19382

Patricia J. Owen [23]

The Elizabeth Whitney Evans Chair in Garden Therapy, Cleveland Botanic Garden, Cleveland, OH

Caroline H. Pearson-Mims [31]

Research Technologist, Department of Horticulture and Landscape Architecture, Washington State University, Pullman, WA 99164-6414

Christine Pollard [36a]

Program Director, Providence Farm, 1843 Tzouhalem Road, Duncan, British Columbia, V9L 5L6 Canada

Mary L. Predny [36a]

Research Associate, Virginia Tech University, 106 Airport Dr., Apt. 4, Blacksburg, VA 24060
Email: mpredny@vt.edu

Royce K. Ragland [8]

Georgetown University Medical Center, Washington, DC
Email: royceKR@aol.com

Paula Diane Relf [1, 34a, 36a]

Professor, Department of Horticulture, Virginia Tech University, Blacksburg, VA 24061-0327
Email: pdrelf@vt.edu

Gene Rothert [6]

Manager, Horticultural Therapy Services, Chicago Botanic Garden, 1000 Lake Cook Road, Glencoe, IL 60022
Email: grothert@chicagobotanic.org

Brian H. Santos [19]

Student Nurse, University of Massachusetts School of Nursing, Amherst, MA 01003
Email: yoseve@hotmail.com

Scott C. Scarfone [35a]

Senior Associate, LDR International, 9175 Guicford Rd., Columbia, MD
Email: scarfone@ldr-int.com

Candice A. Shoemaker [26]

Associate Professor, Department of Horticulture, Forestry, and Recreation Resources, 2021 Throckmorton, Kansas State University, Manhattan, KS 66506
Email: cshoemak@oznet.ksu.edu

Shelagh Rae Smith [9]

Heart and Soil Horticultural Therapy, 101-1001 West Broadway, P.O. Box 115, Vancouver, British Columbia V6H 4E4 Canada
Email: heartandsoil@canada.com

Jane Stoneham [17]

Director, The Sensory Trust, c/o The Eden Project, Watering Lane Nursery, Pentewan, St. Austell, Cornwall PL26 6BE England
Email: jstoneham@edenproject.com

Michael Sutherland [20]

Director, University Statistics Center, University of Massachusetts, Amherst, MA 01003

H. Takafuji [24, 30]

Graduate Student, Laboratory of Applied Plant Science, Graduate School of Agriculture, Kyushu University Hakozaki, Fukuoka 812-8581, Japan
Email: hiroyuki_t@brs.kyushu-u.ac.jp

Yasushi Takano [28]

Lecturer, Laboratory of Biometrics, Graduate School of Agricultural Life Sciences, University of Tokyo, 1-1 Yayoi, Bunkyo-ku, Tokyo, 113-8657, Japan

Catherine Trapani [36a]

Sonja Shankman Orthogenic School at the University of Chicago, Chicago, IL

John Tristan [19, 20]

Director, Durfee Conservatory, Department of Plant and Soil Sciences, University of Massachusetts, Amherst MA 01003-2190
Email: jtristan@pssci.umass.edu

Tzu-Hui Tseng [37a]
National Taiwan University, Taipei, Taiwan

Seiya Uchida [27]
Department of Biomedical Engineering, University of Tokyo, Tokyo, Japan

Roger S. Ulrich [2]
Professor, Center for Health Systems and Design, Colleges of Architecture and Medicine, Texas A&M University, College Station, TX 77843
Email: Ulrich@archone.tamu.edu

Miwa Umezawa [27]
Faculty of Agriculture, Utsunomiya University, Tochigi 321-8505, Japan

Lea Minton Westervelt [4]
Research Associate, Department of Horticulture, Forestry, and Recreation Resources, Kansas State University, 2021 Throckmorton Plant Sciences Center, Manhattan, KS 66506-5501
Email: llm@ksu.edu

Kenji Yamane [27]
Faculty of Agriculture, Utsunomiya University, Minemachi 350, Utsunomiya City, 321-0942, Japan
Email: yamane@cc.utsunomiya-u.ac.jp

Ritsko Yasue [36a]
Bryn Mawr Rehab, Malvern, PA

Masao Yoshida [27]
Faculty of Agriculture, Utsunomiya University, Tochigi 321-8505 Japan

# Foreword

## PERSPECTIVES ON THE PEOPLE-PLANT COUNCIL

In celebration of the 10th anniversary of the People-Plant Council (PPC) and acknowledgement of this outstanding international symposium, the sixth symposium sponsored under the auspices of the PPC, I felt it appropriate to revisit some of the history and philosophical basis for the formation of this group. As I wrote in 1992 for the Second PPC symposium, *People-Plant Relationships: Setting Research Priorities,* understanding and quantifying the values placed on plants by users of horticultural products and services and the impact these plants have on human health and well-being has significance to all people. Most importantly, research-based knowledge can be used to enhance the quality of life for a wide range of people, from high-pressured executives to residents of low-income housing. For those involved in the horticulture industry, a benefit of research-based knowledge is its value as a market development/planning tool. Ultimately, research-based information can be used to increase jobs, reduce crime and health costs, and stimulate the economy.

In November 1988, leaders of the U.S. horticulture industry met in Washington DC, with professionals interested in research on the impact of horticulture on human health and well-being to develop a plan for documentation of the benefits plants provide for quality of life. This meeting resulted in the agreement that "this is an idea whose time has come; that this is the kind of long-range planning the horticultural community needs." Because of this meeting, the Horticultural Research Institute granted seed money to Virginia Tech (VT) to begin work.

As the first step to establish research initiatives in this area, the Department of Horticulture at VT, the American Society for Horticultural Science (ASHS), the Association of American Botanic Gardens and Arboreta (AABGA), and the American Horticultural Therapy Association (AHTA) cosponsored a national symposium, "The Role of Horticulture in Human Well-Being and Social Development," from April 19-21, 1990, in Arlington, Virginia. The symposium was endorsed by most of the major horticultural associations including the American Association of Nurserymen, the Society of American Florists, the American Floral Endowment, the Associated Landscape Contractors of America, the American Society of Consulting Arborists, the Professional Grounds Management Association, and the U.S. Botanic Garden.

Organizers of the symposium recognized that current trends in horticultural research are inadequate for the future of horticulture in relation to the economic and social role it should play in this country. It is important that members of the horticulture community become active in designing and conducting research cooperatively with social scientists to understand the interplay of horticulture with various aspects of human development. This symposium provided a forum for such interdisciplinary discussions by bringing together four groups: (1) those currently conducting research related to horticulture and human health; (2) horticulturists and social scientists interested in doing future research; (3) representatives from trade groups, public and private agencies, and educational institutions who can use this research; and (4) representatives from public and private funding sources.

At the conclusion of the symposium, more than 30 prominent members of the horticultural community and allied fields met to discuss the future needs for research in Human Issues in Horticulture (HIH). Led by a facilitator, participants first reached a consensus that the time had arrived for cooperating to address these issues. They established an action plan with a twofold approach to structure and organization: (1) work within existing professional and scientific associations; and (2) establish an HIH Consortium or Council.

To sustain this area of horticulture, it was agreed that it must be integrated into the existing structure of the horticulture community. This integration was to be done by encouraging all horticultural associations to support HIH activities within the context of their existing activities: publishing articles in newsletters, soliciting research articles for journals, conducting workshops/presentations at conferences, supporting research through endowments, etc.

In addition, to maintain momentum and give people a sense of identity and a way to share directly with those of similar interests, the PPC was formed on May 24, 1990. The mission of the PPC is to ensure the documentation and communication of the effect that plants and flowers have on human well-being and improved life-quality. The PPC serves as a link between organizations representing many facets of both the horticulture and the social science communities. The PPC is a network designed by the representatives of interested associations to enhance and focus their efforts toward documenting the human benefits derived from horticulture.

Six international symposia on the various aspects of people-plant interaction have been conducted under the auspices of the PPC, and more than 25 symposia, workshops, or conferences have been conducted as part of other affiliated association such as the International Horticultural Congress (IHC), ASHS, AHTA, and AABGA, through the ongoing efforts and encouragement of PPC. The proceedings from five of the PPC symposia have been published, and their availability has helped in developing significantly more research in this area. In addition, the PPC sponsored the development of two computerized partially annotated bibliographies: one addressing people-plant interaction and related research from a broad perspective (1,542 citations), and one focusing on horticultural therapy (1,183 citations). Research priorities in HIH have been identified through focus group discussions with leading researchers in this field, and an increased number of graduate students are entering this field. Three special issues of the ASHS journal, *HortTechnology*, focusing on Human Issues in Horticulture were published in 1992, July 1995, and January 2000. (The January 2000 issue was the first issue available electronically.) New courses offered at Clemson University, Washington State University, Rutgers University, and the Massey University in New Zealand, among others, draw heavily on the resources developed through the PPC.

The demand created by 15 to 20 letters, telephone calls, or e-mails each week requesting information (often from newspaper and magazine writers) led to the creation of a website, http://www.hort.vt.edu/human/human.html, containing many references and citations on work in this area.

The availability of research-based information through the proceedings of the PPC symposia, annotated bibliographies, and the website developed under the auspices of the PPC has contributed to the interest by the popular press in people's responses to plants in their environments and, thus, valuable exposure for horticulture. Among the publications that have published articles in the last few years are *Modern Maturity, National Geographic, Southern Living, Mirabella,* Family Circle's *Easy Gardening* magazine, and (even) *National Enquirer.*

Howard Frumkin, MD, Ph.D., Chair, Department of Environmental and Occupational Health, Rollins School of Public Health, Emory University, credits the resources obtained from the PPC in influencing his thinking and writing in "Beyond toxicity 1: Human health and the natural environment," *Am J Prev Med* 2001:20(3). His suggestion to environmental health specialists, from researchers to clinicians, was to turn their attention to aspects of the environment that may enhance health.

> We need to open collaborations with a broad range of professionals, such as landscape architects to help identify the salient features of outdoor exposures, interior designers to do the same in micro-environments, veterinarians and ethnologists to help us understand more about human relationships with animals, and urban and regional planners to help link environmental health principles with large-scale environmental design. . . . Finally, as we learn more about the health benefits of particular environments, we need to act on these findings. On the clinical level, this may have implications for patient care. Perhaps we will advise patients to take a few days in the country, to spend time gardening, or to adopt a pet, if clinical evidence offers support for such measures. Perhaps we will build hospitals in scenic locations, or plant gardens in rehabilitation centers. Perhaps the employers and managed care organizations that pay for health care will come to fund such interventions, especially if they prove to rival pharmaceuticals in cost and efficacy.

As leaders in the healthcare community call for recognition of the health-giving aspects of the natural environment and advocate collaboration in research and implementation of findings, the role of horticulture in human health and well-being will grow.

It is my hope that the PPC will continue to contribute to a healthy environment for nurturing the seeds of collaborations and cooperation.

Paula Diane Relf
Chair, PPC 1990-2002
Professor, Department of Horticulture
Virginia Tech University

# Preface

This collection of papers is from the symposium "Interaction by Design: Bringing People and Plants Together for Health and Well-Being," the sixth international people-plant symposium sponsored by the People-Plant Council (PPC). The symposium took place July 20-22, 2000, at the Chicago Botanic Garden. The School of the Chicago Botanic Garden organized the symposium, which was the biennial conference of the PPC and the annual conference of the American Horticultural Therapy Association.

The first people-plant symposium, held in 1990 in Arlington, Virginia, had as one of its goals to bring together researchers and educators in horticulture and the social sciences to promote interdisciplinary discussions on the economic, health, and social roles horticulture has in our society. Each symposium since has followed this model by developing a program that would attract participants from many and varied disciplincs. The sixth symposium was no different. This symposium set as one of its goals to bring together landscape architects and designers, horticultural therapists, and researchers and educators in these disciplines to share information on current research and design solutions that bring people and plants together for health and well-being.

The topics addressed at the meeting included communication, universal design, and therapeutic applications as they relate to the design and therapeutic use of outdoor spaces. The manuscripts have been peer reviewed (by two or three reviewers) and organized by these topic areas into four sections. Section I includes keynote presentations and contributed papers on communications with the horticulture, healthcare, and design communities. Section II features presentations on design for human health and well-being. Section III addresses the topic of therapeutic design applications or more specifically, how these designed spaces are used therapeutically. The final section includes research reports on investigations into the psychological, sociological, biomedical, and cultural roles of people and plants.

The papers present a wide range of information on design processes and guidelines for healing and therapeutic gardens; programming of healing and therapeutic gardens; and research in both of these areas. Communicating the roles plants have in human health and well-being, beyond those for food and shelter, is part of the mission of the

PPC. The papers on communication clearly express the value and necessity for effective communication with all audiences, be they in horticulture, design, or health care.

This sixth proceeding was a collaboration of many people. I would like to give thanks and recognition to the associate editors who coordinated the peer review of all the manuscripts, as well as reviewing many of them. I wish to acknowledge all the authors for their valuable contributions to this proceeding. Recognition and thanks to the School of the Chicago Botanic Garden and the Symposium Organizing Committee for creating a program that attracted participants from 11 countries and 45 states. This proceeding would not have been possible without the efforts of many at the Chicago Botanic Garden and the financial contribution by the Garden to help defray the costs of publishing this book. Last, but not least, I must express my heartfelt gratitude to Paula Diane Relf for her support, advice, and counsel throughout the planning of the symposium and the preparation of this proceeding.

Candice A. Shoemaker

# I

# Communications with the Horticulture, Healthcare, and Design Communities

# 1

# Communicating with the Horticulture Community about Human Issues in Horticulture

Paula Diane Relf

## THE UNIVERSAL LANGUAGE

For better or worse, today the universal language for getting someone's attention and interest is money. Those of us practicing, studying, or conducting research in human issues in horticulture are attempting to open the door on a new area of research, teaching, and programming for horticulturists. This is not an easy task nor will it proceed rapidly. However, understanding what we are trying to accomplish and ways to communicate this to others will certainly expedite the recognition and acceptance of this field.

Early attempts to get the commercial horticulture industry and their trade associations interested in issues related to horticultural therapy and horticulture's role in human health and well-being were not very successful. The overall response was essentially that being a "do-gooder" was fine if you were interested in warm and fuzzy feelings, but there really was no place for concerns about people if you wanted to earn a living growing and selling plants. In other words, their bottom line was dollars, and if the information or actions did not add to the bottom line, it was not needed.

This same message is given out across the horticulture community. Money represents salaries, new buildings, better research, industry growth, more students, more visitors to the gardens, etc. A sure way to be heard is to link the message to the money. With the trade associations, we linked the message to a new source of employees and new marketing tools. These were dollar-based concepts that the industry understood and led to their support of the American Horticultural Therapy Association and the Horticulture Hiring the Disabled National Initiative and later to the formation of the People-Plant Council.

## THE HORTICULTURE COMMUNITY

Frequently people tend to think of the horticulture community as comprising only those individuals who share their own narrow career path—university professors, ornamental growers, vegetable producers, garden writers, or any other distinct group.

In fact, the horticulture community is extremely large and needs to be looked at as a whole to fully understand the impact that it can and does have beyond the economics of production. The community encompasses such varied groups as

- Academicians—educators, researchers, and Extension personnel
- Arboreta and botanic garden personnel
- Human-horticulture professionals such as horticultural therapists, teachers, and community garden staff
- Green industry, including growers, marketers, and landscape design, installation, and maintenance professionals
- Fruit and vegetable producers and their allied commercial professionals
- Urban horticulture and forestry professionals
- Users of horticulture crops and services—amateurs and enthusiasts (70% of American households), facility managers
- Allied professionals such as trade association staff, garden writers, etc.

## WHAT IS THE MESSAGE TO COMMUNICATE?

Our message is that horticulture is more than growing, selling, researching, teaching, and writing about plants. With a new understanding and vision of horticulture, we will open the doors to tremendous possibilities for making our communities better places to live, improving human health and well-being, strengthening our children's education, and reaching many other goals toward improved quality of life. At the same time, we will "grow" the horticulture community, expanding the financial resources for all, thus strengthening the bottom line of increased dollars to meet needs.

Historically horticulture has been defined as the science and art of growing fruits, vegetables, flowers, and ornamental plants. If we go to the dictionary, we see that the word horticulture is derived from the root words *Hortus*, a garden, and *cultura* for which the dictionary refers us to the word culture. Culture is defined as the cultivation of the soil; thus, we have traditionally looked on horticulture as simply the cultivation of the soil for a garden; that is, the production elements of growing plants. However, *cultura* also means the development, improvement, or refinement of the mind, emotions, interests, manners, tastes, etc.; the ideas, customs, skills, arts, etc., of a given people in a given period; civilization. By combining *hortus* with these other definitions of *cultura*, we can begin to understand this new vision of horticulture as

- The role of the "garden" in the development, improvement, or refinement of the mind, emotions, interests, manners, tastes, etc., of individuals
- The influence of the "garden" on the ideas/skills of a given people in a given period; that is, the "garden" as part of human communities and the way they function
- The integration of the "garden" in human culture and civilization

Taking into consideration the traditional definition of horticulture, the above alternatives, and the current research into people-plant interaction, we might consider as a comprehensive definition of *horticulture as the art and science of growing flowers, fruits, vegetables, trees, and shrubs resulting in the development of the minds and emotions of individuals, the enrichment and health of communities, and the integration of the "garden" in the breadth of modern civilization.* By this definition, horticulture encompasses PLANTS, including the multitude of products (food, medicine, oxygen) essential for human survival; and PEOPLE, whose active and passive involvement with "the garden" brings about benefits to them as individuals and to the communities and cultures they comprise.

In limiting the definition of horticulture to a combination of "a garden" with "the cultivation of the soil," we in the horticulture community have severely limited the understanding of what horticulture means in terms of human well-being. In effect, we have put blinders on the study and application of horticulture for human life quality.

## WHY COMMUNICATE?

Why should those of us who are concerned with working with human issues in horticulture concern ourselves with communicating with the remainder of the horticulture community? Without communication with others, we have no education of others, and without the education of others, we will have no action from them. The action that we are trying to elicit is the same as everyone's—financial support to allow us to develop this professional area. We want to garner support to conduct research, teaching, and implementation of programs in the area of human issues in horticulture. We want to create new jobs in the field and to increase the salaries for existing practitioners such as horticultural therapists. With financial support we want to develop a professional area that will improve the quality of life for everyone. By opening the vision of horticulture as it is taught at universities, practiced at arboreta and botanic gardens, and promoted by the industry, we hope to

> Make our cities better places to live,
> Give our children a better future, and
> Do our small part to ensure environmental and life quality.

## LINKING A NEW DEFINITION OF HORTICULTURE TO THE BOTTOM LINE OF DOLLARS

It may sound too cynical or too crass to many but in fact, in our society, money is the engine that runs the system that allows us to reach our diverse goals. Some people may wish to become rich or famous while others just want a small successful business. Some need money to provide for their family, others to travel, read, and relax. Some want power and prestige while others want to nurture and care for the environment, animals, or people.

Although these are examples of an individual's goals, the "bottom line" is the money to reach these goals. So your role in communicating effectively is to shape your message to address the needs of your audience rather than simply your own needs or ideas. In other words, you have to decide whom the most important people are for you to convince of the validity of this new aspect to horticulture. Then you have to determine how to talk to them in their language, in terms they understand and relate to. Finally you must adjust your presentation to explain how addressing the human issues in horticulture will meet their needs.

Take, for example, a researcher who is dedicated to wildflowers and to the expansion of the use of wildflowers. He plans to conduct research on the establishment of wildflowers at the corporate headquarters of a major company. His goal is to ensure success of a planting in this location with the anticipation that they will continue to fund additional research at other locations. However, in a highly visible location, where the landscape influences the perceptions of both the customers and the employees regarding the positive characteristics of a business, it is critical to look beyond the cultivation of the soil in determining the long-term sustainability of a wildflower planting. Certainly it will be important to determine the optimum species for the soil and climate and the best installation and maintenance practices. However, if the site is perceived as weedy and unkempt, it will not be retained for long. Or if the labor for wildflowers exceeds the cost of turf, it will be critical that its visual or environmental value exceeds the financial costs. This then means that the viewers of the wildflowers must be integrated into the total research project and their perceptions used in making decisions about the plants and their care. This type of research will be interdisciplinary with a completely different group of peers than is traditional with horticulturists. Therefore, the horticultural researcher must be educated regarding the need for the psycho/social research, who to conduct interdisciplinary work with, and how to do it.

## WHO DO YOU NEED TO TALK TO IN THE HORTICULTURE COMMUNITY?

It is possible to divide the audience into five major groups to understand whom to talk to, why it is important that they understand and act, and how you carry that message to them. The five groups to be addressed are academicians, other educators and communicators, allied "people" professionals, horticulture industry members, and consumers.

### Academicians

Of particular interest are professionals affiliated with research and formal education in universities and colleges, scientists at museums, botanic gardens, and arboreta, and researchers with the United States Department of Agriculture (USDA) and other agencies. Although their ultimate goals vary, similarities among academicians include a need for funding to support research, publications to get tenure and promotions, and student numbers to maintain academic departments or outreach programs.

## Tapping New Sources of Funding

Despite popular misconceptions at some universities, horticulture is more than biotechnology. Becoming an expert in biotechnology with horticultural crops is not the only route to National Science Foundation or National Institute of Health grants. In addition, horticulture is more than just crop production, and USDA dollars are not limited to reducing cost or environmental protection from production techniques. Horticulture includes the perceptions, the interactions, and the values that people have as related to plants. This type of research has implications for everything from health to marketing practices. Horticulturists need to be partners in motivating and directing research in this area with social scientists, health scientists, urban planners, and researchers from many other areas.

Researchers are often so focused on their narrow field of expertise that they only think in terms of working alone or with experts from the same field. However, interdisciplinary work has to be recognized as the force for change into the future. The key to success in the field of human issues in horticulture is combining plant science knowledge with social and health science knowledge to develop new research tools for understanding how to have both healthy plants and healthy people. Developing partnerships, building teams, and establishing interdisciplinary programs allows the researchers from various disciplines to apply their skills and knowledge to their specific interest while exploring the bigger picture. Horticulturists need to invite scientists from business management, gerontology, sociology, and many other fields to work together to address the research needs of the horticulture community.

Currently, I have a cooperative research project under way with a faculty member in gerontology looking at the impact of horticultural activities on dementia clients. This is part of a larger effort related to my Cooperative Extension program with Master Gardener volunteers and linked to my undergraduate teaching program. Dr. Shannon Jarrott brings with her the knowledge of the limitations and needs of dementia patients, and therefore can define the research goals and procedure from that perspective. I work with the students in identifying the activities and the cultural practices. Together we are developing a research mode that will be useful in obtaining future funding.

A faculty member in business and marketing has asked that we team up to interview avid female gardeners to explore their attitudes toward environmental issues and their consumption patterns. A sociologist was a team member on a graduate committee for youth offenders and horticulture. That study resulted in the funding of a greenhouse for a local alternative school for youth on probation and more work focused on employment in the industry.

What about researchers who are more traditional? Understanding tree roots is critical for successful urban tree culture. However, without understanding what people do to the roots and how to change their behavior to protect roots, little will alter in terms of the life of the tree. Genetically altered food is causing a significant social and political uproar because biotech scientists felt it was beyond their ability or responsibility to address the human ramifications of their efforts. Plant breeders took the taste and texture out of food so that it would ship well and be cheap. Others took the nutrients out of beans so they would preserve better. All of these are unintended consequences resulting from leaving people out of the plant research equation.

Research dollars for understanding human needs and responses are from different agencies and foundations than traditional horticulture research dollars. However, the money for the plant element of this type of research must be included in the funding process, thus opening new funding sources. In addition, research on human needs for plants leads to and justifies additional research to understand the cultivation of plants, particularly for urban areas.

## Publishing Research

Publications are major criteria in academia for tenure, promotions, and pay raises. For many years, there was a cry from those few people interested in such cross-disciplinary work as human issues in horticulture that there was no way to publish this type of research. Neither the social science journals nor the plant science journals wanted this hybridized research. However, that is not at all the case today. Most of the major professional horticulture journals publish articles from this field. *HortScience* and the *Journal of Environmental Horticulture* have research articles, and *HortTechnology* has had four special issues focused on some aspect of human issues in horticulture. the *Journal of Horticulture Therapy*, of course, is specifically for individuals working in this area. The *Journal of Extension* and the *Journal of Vocational Technical Education* are examples of others in closely allied fields that have published relevant work. In addition, journals in the social and medical sciences are increasingly including related articles published cooperatively by experts in the profession represented and horticulturists.

## Increasing Student Numbers

At universities, undergraduate student numbers play important roles in the number of faculty and the number of teaching assistantships that are supported by the administration. Graduate student completion is critical to the maintenance of an advanced degree program. Therefore, significant or sustained changes in enrollment influence the entire department and its future. The expansion of the area of human issues in horticulture holds the potential to increase student enrollment of horticulture nonmajors, horticulture majors, and graduate students.

There are several examples of horticulture courses that could be offered to expand our overall enrollment. An introductory Human Issues in Horticulture course could meet university core curriculum requirements thus opening it to students from all majors. Such a course could have significantly more diverse content than Economic Botany. Although it would certainly cover the uses of plants and the impact that gardening and the search for plants has had on history and exploration, it would also deal with issues as varied as urban planning and the arts. This course could look at the psychological and social responses of people to both the presence of cultivated plants in the urban/peri-urban environment and the actual act of cultivating a garden. It could have direct value in future actions and decisions for majors from fields of business management to engineering, from anthropology to education. Horticulture has played an integral role in art, religion, medicine, language, and human rituals. It is important in corporate image and worker satisfaction. These elements of horticulture need to be brought before a wider audience. In addition, such an undergraduate course can spark interest from students and encourage them to take other nonmajor courses (i.e., Home Horticulture, Indoor Plants, and Floral Design) and even some courses designed for majors.

Horticulture courses targeted to professionals from other fields could be a source of a significant number of graduate or distant education students. For several years, we have offered a course for teachers in cooperation with botanic gardens around the state, "Integrating Horticulture into the Elementary School Curriculum to Meet the Standards of Learning (SOL)." The course has been well received with about 35 people taking it each summer. It was developed based on an undergraduate research project and a graduate dissertation that indicated a significant interest in taking such a course if it could be taught locally. However, teachers indicated they could not teach a separate course in horticulture in addition to the other courses they were teaching. Our goal with this course was to listen to the teachers and meet their needs through horticulture, rather than simply get them to do what we wanted.

A similar course for occupational, recreational, and activity therapists is under development and will include basic techniques in using horticulture with elderly and disabled individuals in treatment and healthcare settings. Introducing these other professionals to horticulture as one of the tools that they can use will expand the application of horticultural therapy and ultimately create the need for more horticultural therapists as specialists in a unique area. It will also increase the need for individuals qualified to be consultants in establishing programs with allied professionals.

Courses with the curriculum for horticulture majors likewise need to include human issues in horticulture elements. Although basic horticultural skills and knowledge are essential to get started in a career in horticulture, basic understanding of the motivation and management of people is essential to advance in the field. College graduates should be targeting their career at management level not introductory or field-hand positions. Thus the need for courses that combine these skills continues to increase.

Thus, teaching is an area that could be significantly expanded to attract more students and in turn get additional funding for teaching assistants (TAs), the TAs who help perform the traditional horticulture research.

## Other Educators and Communicators

Certainly the role of education is not limited to a university's degree and continuing education programs. Among the important educators who we must work with are

- Cooperative Extension agents and Master Gardeners
- Arboreta and botanic gardens education and outreach staff
- Garden writers and radio and television garden program talent
- Professional association staff and officers

As with academicians, these professionals have ultimate goals that vary, yet all of these educators share the common goals of having a meaningful and valuable impact on the learners they work with and increasing the number of people that they reach.

## A Meaningful and Valuable Impact on the Learner

Certainly this goal gives one significant intrinsic rewards and, in the case of volunteers such as Master Gardeners, their primary motivation is a direct indicator of success. However, in addition to the warm feelings garnered from satisfied participants, accomplishment in this area is the basis for funding to continue or expand programs, and to realize pay raises or sales of articles and books. Success in this area is not based simply on having the horticultural fact that is being sought but on understanding the needs of the learner and the learning styles that will allow them to effectively use the information provided. With the advent of the World Wide Web and its ease of accessibility to most gardeners, mere facts will be sought in private, while educators will be expected to provide skills, to change human behavior, and to entertain. This implies a greater understanding of people and their relationship to their garden. It requires new styles of teaching and working with new audiences from children to retirees to disabled individuals who have specific needs to be met. It opens the door to new teaching content that has not traditionally been the focus of garden educators.

## Increasing the Number of People Reached

Changes in ways of educating and communicating along with changes in audiences and their needs means changes in what is taught, and how and where it is presented. Communicating to other educators the opportunities presented by a new vision or definition of horticulture will help them see ways to expand their current audiences. Among cooperative extension agents and Master Gardeners, the audience has expanded from home owners to children, teachers, community gardeners, disabled and disadvantaged individuals, professionals working with the elderly, minority community leaders, county boards of supervisors and other officials, water quality boards, tourism professionals, conservationists, historians, and foods and nutrition groups.

As professionals in botanic gardens and arboreta recognize the importance of larger clientele groups to sustain their vigor and growth, they too are addressing the people who can benefit from horticulture in a diversity of ways, in addition to simply being visitors to the garden. It is perhaps the garden writers and others in the media and entertainment end of horticulture that have greatest potential for expanded audiences through the integration of human elements to their stories. An in-depth look at who gardens, how, and why has the same appeal as an exploration of people's homes and is being seen more and more. Glamour magazines have done full issues focused around the garden. Many magazines for specialized groups including topics such as health, seniors, weight loss, and miniature train buffs are interested in gardening articles. Airline magazines, doctor's office magazines, corporate employee magazines, specific car owners, or insurance owner's magazines are among the ones that should be targeted with articles on the role and benefits of plants in the reader's environment, gardening for the health of it, or beautiful plant places to visit.

Perhaps the most important groups for those of us working in this field to educate and develop positive relationships with are the staff and leaders of professional, trade, and plant associations, and editors and staff of magazines and newsletters. These individuals have a significant role in determining what message will go out to the rest of the horticulture community through their publications, conferences, shows, and public relations pieces.

## Allied "People" Professionals

The people professionals, such as horticultural therapists, community garden staff, and botanic garden education staff, already know that this expanded definition of horticulture does a better job of explaining what horticulture really is and how it changes people's quality of life. Although there are not a large number of people in the horticulture community whose careers are recognized to be directly linked to psychological, social, physiological, physical, and other benefits of plants to their clientele, they are well represented among the people attending this symposium. Among us there certainly are a number of academicians and other educators who have linked their research and teaching efforts to human well-being through direct association with plants. In addition, there are many of you who are what we might call field soldiers, those who work directly with clientele from many different areas to help them reach their specific, individual goals. Included among those whose job it is to meet their client's needs using human issues in horticulture are

- Horticultural therapists
- Horticultural vocational rehabilitation specialists
- Community gardening staff
- School gardening professionals
- Landscape designers specializing in healing and therapeutic landscapes

No matter what their personal goals, these professionals share certain overall goals that can be addressed in trying to communicate with them and alter their behavior to enhance the impact of human issues in horticulture. These goals include

- Improved service to clients
- Increased recognition of their professions
- Expanded number of practitioners

### Improved Service to Clients

To enhance the services provided to clients a research-based body of knowledge is required that identifies specific horticultural activities, landscape configurations, and gardening techniques that can be used in various situations, and the specific benefits to be derived from these activities, settings, and techniques for various clientele groups.

Development of this body of knowledge will eliminate the constant requirement on the part of each practitioner to develop and test their own procedures. It will validate the application of horticulture to these uses in the eyes of health, education, and other professionals. It will facilitate obtaining increased funding for programs for your clients.

The question of course is who will do the research. That, as indicated earlier, is the role of academicians. However, there are so few of them who have started working in this area that it is critical that the practitioners also become involved. Partnerships with researchers from various other disciplines and/or in the research division of the various facilities where programs are being conducted will be the most effective way to develop this essential body of knowledge. I know that many of the people working in the field maintain that they do not have the time or knowledge to conduct such research. However, in fact, each of you must collect the needed data on a regular (if slightly less formal) basis as a standard part of your job. Research can be looked at as simply a more structured approach to keeping client records with specific techniques for analyzing them and reporting their meaning. A collaborator in designing the methods for observing and record keeping will help you understand the nature of the interpretation you can make from your information. If your partner is from academia, then analyzing the data and writing a peer-reviewed paper will be an integral part of the research.

Conducting research and building a body of knowledge is essential to the future funding for programs, for professional recognition, for pay raises, for the very existence of this field of work. Practitioners can not sit back and simply say that it is someone else's job, I can't do that, or I don't know where to start. If you do not do the work and motivate others to do it then you will soon find yourself looking for other kinds of work that pays better but gives you little or no personal satisfaction. I am not speaking of a "probability" but rather an observation that I have made over the last 25 years of working with horticultural therapists and others in this field.

## Increased Recognition of Their Professions

The first and most important message here is that each professional in the field must do their part to develop a research-based body of knowledge. The second is that they must *share* that knowledge—otherwise it is meaningless and lost.

However, if participating in research is still beyond you, there is a great deal of other knowledge that you need to share. I have often talked to horticultural therapists or horticulturists working with school children who are very proud of their work. They know that they are doing a good job that is making a difference in the lives of their clients. Their clients know this too and appreciate it. Often the family recognizes and acknowledges their value. And usually their immediate boss knows it. If they are very lucky, someone further up the administrative ladder knows it, too. However, overall, the techniques, values, and impacts are shared by very few people. Thus, the program rarely spreads beyond its immediate site and may be allowed to fade away in times of tight funding.

Being highly visible is a way of continuing vigor for the program. Being highly visible means shining a light on your knowledge, skills, and achievement. Write articles, give lectures, or write a book. There are currently many more books on peonies than on all aspects of human issues in horticulture. If you can't write, dictate your message and have it transcribed. Or work with a writer who can put the message on paper. However you do it, get the words down on paper. Make it happen. Realize that if you want to continue to help the people who you work with, whether it is school children, community gardeners, individuals in treatment, or anyone who benefits from gardening, then you have an obligation to communicate. Part of your job must be to communicate, to write down what you know, what you have experienced, what you have documented in client records, and to share it with a wider audience. If you have not done that then you have not served your clients. You are not contributing to the recognition of the profession.

## Expanded Number of Practitioners

Without the sharing of practical, applied knowledge by the practitioners, faculty at universities will not have the material they need to teach the courses that are needed to produce more professionals. It is all part of a circle in which everyone must play his or her part.

Other roles for practitioners in expanding the number of people using horticulture for human health and well-being are directly related to their communicating with appropriate audiences. Slides from your program along with success stories and how-to information form the basis for lectures and workshops to be given at local and regional conferences for many different groups including

- Allied professionals, such as recreational therapists, occupational therapists, art therapists, vocational rehabilitation specialists, teachers, etc.
- Civic groups, plant societies, and other sources of support and funding
- Media groups, including garden writers who may pick up the story and spread it
- Industry groups and trade associations
- Local and regional government agencies
- Volunteer groups including Master Gardeners
- Facility owners and managers, such as hospitals, nursing care facilities, etc.

The more the people who are action takers and decision makers learn of your work and its value, the more people who will want to replicate it in another venue.

In short, to meet their goals, professionals in the area of human issues in horticulture must work collaboratively with peers from allied professions to develop a research-based body of knowledge published in peer-reviewed journals, provide university teaching programs with content for their curriculums, and promote the field to professionals and the public.

## *Horticulture Industry Members*

The horticulture industry plays a powerful role in what happens in the horticulture departments at universities. Historically, horticulture departments were established to conduct research and train people to support the growth and development of the industry. Cooperative Extension was established to carry the knowledge from the land-grant universities to the public that could not attend the university. This transfer of knowledge was initially done in the agriculture community including fruit and vegetable producers and expanded to include ornamental growers in recent years.

The horticulture industry itself includes growers, wholesalers, distributors, retailers, landscape designers, contractors, and maintenance firms, and all of the related products and services that allow these businesses to function from synthetic soil to container manufacturers. Other important parts of the horticulture industry are the trade associations, trade shows, magazines, and publication firms that focus on horticulture products and services. This wide and diverse industry has even more clearly defined goals than other members of the horticulture community have. The goal is to increase profitability of the industry, which has many contributing factors including increasing net profit on plants and services, increasing overall market size, and increasing the customer base.

### Increase Net Profit on Plants and Services

Net profit is based on the very simple concept of selling price (gross profit) minus total cost equals net profit (the money you get to keep). There are two ways that net profit can be increased—either increase the price or decrease the cost. The focus of the USDA-sponsored research and the teaching and extension programs of land-grant universities has been to decrease the cost of production. A major contributing method for this has been to increase the volume of production such that most profitable segments of the industry consist of huge operations that focus on selling through retail chains and other large distributors. This has allowed for continually reduced prices on the mass-produced plants. Unfortunately, this has left a perception among the general public that plants, food, and other agricultural products should be cheap, that they are simply not worth paying more for. Unfortunately, as wages continue to rise, pesticides and other chemicals used to produce plants cheaply are pulled from the market, and as energy costs go up, it becomes increasingly difficult to keep lowering the price of plants.

To increase or even maintain the net profit means increasing the price of the product. However, since we have spent the last 100 years devaluing agricultural commodities, this presents significant challenges. Selling price is based on both the seller's and the customer's perceived value of the product or service. Horticulture is an industry that has worked for so long to reduce price that the members do not believe in the value of their products. Several years ago, I read a letter from the president of a major horticultural trade show that started out, "As we all know, we are selling a product that no one needs."

To build a marketing program that opens the door to increasing prices requires educating the industry (both the producer and the retailer) that fruits, vegetables, trees, shrubs, and flowers are more than optional ways to fill up space in either your stomach or your landscape. When selling these products, we are selling health, well-being, and quality of life. These products have physical, psychological, and social value that far exceeds the dollar value currently being put on them.

## Increase in Overall Market Size

This approach relies on selling more plants at the same price or sometimes even lower prices and usually is targeted at the established customers increasing their purchases. Although the unit net profit does not increase, the cumulative net profit does. The obvious drawback to this is that either the current members of the industry must be more productive (usually by having larger and more technology-based systems) or there must be an increase in the number of commercial horticulture operations. The first option, that of larger businesses, forces the producers to deal only with the largest retailers and gives those retailers significant control over all aspects of their operation. The second option of increasing the number of businesses usually means that none of them increases their individual net profit.

Thus, market size must be linked to increased prices in order for the individual members of the industry to consistently increase their profit margin while paying competitive wages and implementing the environmental, safety, health, and other regulations that the government and their consumers expect of them. Increase in prices is linked to an understanding of the consumer, his/her motivations to buy, and the benefits derived from the horticultural products and services.

## Increase in Customer Base

On the other hand, increasing the number and diversity of customers opens opportunities for specialized services and products at a more reasonable price. The horticultural industry is just beginning to recognize the significance of the graying of America in terms of shifts in gardening styles, buying habits, and specialized or niche markets. The Americans with Disabilities Act has brought the realization that there is a large buying public that will willingly respond to accessible gardening, specialized tools, appropriate landscape designs and plants, and services marketed to meet their unique needs. Youth and school gardening appears to be coming back into industry focus as an audience that must be educated now to ensure a buying public in the future. Ethnic diversity is one of the great demographic shifts in this country and will influence horticulture as consumers from throughout the world bring their individual tastes and preferences to the retail store.

Understanding how and why these different groups garden and their needs and preferences will play a role in the success of businesses over the next 10 years. The answers to these questions lie in research based on the newer and fuller definition of horticulture that we are discussing.

Our role in communicating with the horticulture industry is to help them understand that we are selling a product that improves health and quality of life. Plants make our cities better places to live and reduce some of the stress of modern life. We are in the unique position of selling a product that is truly good for people, that people want intrinsically, and that makes life better when they are successful at taking care of it. So, the more high-quality plants we sell, the better off the world is . . . and the more money we make. We are in the ultimate WIN-WIN business.

*Our goal with the horticulture industry is to gain their support in funding research, teaching, and outreach initiatives to document and share the benefits of horticulture to people.*

## Consumers

Ultimately, the most important part of the horticulture community is the consumer who uses the plants and services of the professionals in the community. Although most people think of the consumer as home gardeners and amateur enthusiasts, in fact there is a much larger audience of consumers to be concerned with. These consumers are the individuals who are influenced in their perceptions and actions toward a multitude of daily life activities by the use of plants at businesses, in residential areas, in urban and suburban public areas, at tourist sites, along roadways, and in parks and recreational facilities. Their access to and use of plants is controlled by an intermediary group comprised of such professionals as

- Facility managers for corporations that must negotiate the interior and exterior landscaping
- City managers who make decisions on the use of plants in downtown business areas
- Highway department personnel who look at the impact of plants on traffic management
- Real estate developers who are concerned with the impact of the landscape as it influences the decision to buy and the price that will be paid for a home
- Tourist-site managers

The ultimate consumer of horticultural products and services, therefore, includes the homeowner who buys the plants for personal use, the tourist who enjoys them in their surroundings, and the employee who is a little happier at his job because of the presence of plants. These consumers have similar expectations or goals that can be met by plants:

- Quality of life
- Personal satisfaction from their work and home
- Reduced stress, increased relaxation
- Contact with or feeling of connection with nature

Research is beginning to document how the presence of plants and the act of gardening can directly bring about these benefits. However, a great deal more research is needed. The communication about the benefits must go hand in hand with the research. Sharing what is known is critical to garnering the support for more research, but at the same time, it elicits questions that researchers have not yet answered. Overstatement of the benefits can in fact bring about doubts as to the truth of the claims of benefits, as seen by the backlash created to the message of the Plants for Clean Air Council in academia, in building construction, and in parts of the media.

The research on how plants meet the needs of the public can and should be used effectively by the horticulture industry as a marketing and advertisement tool. In addition, it holds significant potential to be integrated into marketing actions by other product companies, for example Microsoft, to show that they are family and environmentally concerned.

The group that particularly needs to be communicated with, regarding the benefits of plants in their business environment, is the professionals who make the decisions about the use of plants in facilities, hospitals, cities, or tourist sites. There are many ways to carry the message to them, including articles in their trade and professional journals, presentations at their conferences, and direct talks to the individuals. The message that needs to be presented is that the plants in their environment will increase their profitability, their worker productivity and retention, their tax dollars, or their public image. Remember, they have specific goals that drive their decisions about where to use their dollars, and basically, they are looking at return on investment.

Consumers intuitively know that being around plants is good for them. Numerous studies have indicated a high preference for environments with easy access to nature. Consumers are willing to pay more for experiences that include a natural surrounding. But most consumers are not aware of the importance of plants or the impact that they have on daily quality of life. Ultimately, the most important member of the horticulture community is the consumer because the consumer, in the end, determines how and where dollars are spent by business, local and national government agencies, and other decision makers. Therefore, it is important that we communicate with the consumer in the truest sense of the word. We must listen to and understand what plants mean to the consumer. This "listening" takes many forms as research seeks answers to our questions. We must then verify what we have heard by presenting it back to the consumers and confirming from them that plants do indeed play the roles that we have learned from our "listening" research. We then must communicate our findings to all other members of the horticulture community to use to communicate still further to decision makers and influencers to ensure that everyone has access to a natural environment and the opportunity to grow and care for plants.

If we learn to communicate effectively, we will see changes within all of our audiences. Positive changes will lead to an increase in horticultural activities resulting in greater personal, community, and cultural health; well-being; and quality of life.

If we do not communicate effectively, then we may see change in quite the opposite direction. About 12 years ago, the Virginia Tech Arboretum and Research Farm was sold to build a strip mall. Graffiti across the front of the K-Mart store under construction appeared one day, "On the Eighth Day . . . we paved it," across a blue and green globe.

# 2

# Communicating with the Healthcare Community

Roger S. Ulrich

## INTRODUCTION

Considering how to communicate and convince healthcare administrators of plant and garden benefits reflects a real growth and evolution since the first national people-plant symposium 10 years ago in Washington DC. How can we communicate in an influential, convincing way with key audiences such as healthcare administrators and physicians? We have to start by keeping an open mind and try to understand the mind-set of these crucial audiences. Where are they coming from? What do they value? What drives their decisions? What are the implications of their mind-set for making an effective case for gardens and plants?

This paper will begin with a review of current trends and changes in healthcare systems that provide critical background for understanding the medical community's point of view. Against this background, information will be presented regarding how to convince administrators of garden and plant benefits, persuade them of the importance of providing funding for plant programs, of perhaps supporting horticultural therapy, and of designing healthcare buildings with prominent plants.

Certain trends and changes in health care, fortunately, are making many healthcare administrators and medical providers more receptive to the idea of providing plants and gardens. One important development has been the shift in recent years in the scientific or mainstream medical community away from a narrow pathogenic conception (germ theory) of disease and health toward an expanded perspective that includes emphasis on health-promoting experiences. Accordingly, conditions or experiences shown by medical researchers to be healthful, such as social support and pleasant, soothing distractions, now become much more important considerations in creating new healthcare environments (Ulrich 2001). By contrast, the traditional pathogenic perspective implied that the main concern in creating healthcare settings should be interpreted narrowly as the reduction of infection or disease risk exposure. Additionally, decades of impressive advances in medical science conditioned healthcare administrators and designers to concentrate also on creating care environments that succeeded as efficient platforms for new technology (Ulrich 1999). The strong emphasis on functional efficiency, together with the pathogenic conception of disease and health, usually produced healthcare environments that are now considered stark-

19

ly institutional, stressful, and detrimental to care quality (Ulrich 1991, 1992; Horsburgh 1995).

In spite of the major stress caused by illness and traumatizing hospital experiences, little priority has been given to creating surroundings that calm patients, strengthen their stress coping resources, or otherwise address emotional and social needs (Ulrich 2001). The new broader perspective in medicine, however, requires that the emotional and social needs of patients be given high priority along with traditional biomedical and economic concerns, including disease risk exposure and functional efficiency, in governing the design of healthcare buildings and management of care activities (Ulrich 2001).

Stating this somewhat differently, a foundation for the medical community's growing interest in gardens and plants has been the major progress in mind-body medical science. If a researcher had seriously suggested two or three decades ago that gardens could improve patient health in hospitals, the notion would have been received with skepticism by most scientists, and probably would have been dismissed by many physicians and healthcare administrators. A substantial body of scientific research, however, has now demonstrated that psychological and environmental factors can affect emotional well-being, physiological systems, and health status (Ulrich 1999).

Calling attention to the progress in mind-body medicine will not alone be adequate to convince many administrators to allocate funding to gardens or plants. Healthcare administrators and providers everywhere are under very strong pressures to control or reduce costs yet increase care quality. Some administrators and healthcare professionals, faced with pressing demands such as paying for costly new imaging or surgical technology, consider gardens and plants as desirable but nonessential luxuries. Put simply, intuitive or subjective arguments in favor of plants usually carry little weight with decision makers who are compelled to pay close attention to the bottom line.

Convincing administrators to assign priority and resources usually requires providing credible evidence that gardens/plants produce benefits yet are cost-effective compared to alternatives such as not having gardens/plants. Here the point should be underscored that most healthcare administrators and especially physicians consider evidence from *health outcomes* research to provide the most sound basis for evaluating whether a particular medical intervention or treatment (here providing a garden or plants) is beneficial and financially sensible. There can be no question, accordingly, that the future importance of garden and plant programs in healthcare environments will be substantially affected by the extent to which sound studies show that gardens can foster improved outcomes (Ulrich 1999).

## WHAT ARE HEALTH OUTCOMES?

Health outcomes refer to measures of a patient's condition or progress, to indicators of healthcare quality. These measures include observable clinical signs or medical measures, patient-based or subjective measures such as reported satisfaction, and economic measures.

## *Types of Outcomes*

- Clinical indicators that are observable signs and symptoms relating to patients' conditions. (Examples: length of stay, blood pressure, intake of pain drugs)
- Patient-/Staff-/Family-based outcomes. (Examples: patient reports of satisfaction with healthcare services, staff reported satisfaction with working conditions)
- Economic outcomes. (Examples: cost of patient care, recruitment or hiring costs as a result of staff turnover, philanthropy to hospital)

Different combinations of outcomes are used for assessing different types of patients. If a researcher's objective is to document the effects of plants or gardens on patients recovering from surgery, for example, then relevant medical outcomes include recovery indicators such as anxiety, reported pain, intake of pain drugs, how quickly the patient can move or walk, length of hospital stay, and time required to achieve independent functioning. However, relevant outcomes might be quite different for a patient recovering from stroke or brain injury—for example, emotional well-being, or how quickly they can begin to recover speaking ability, independent functioning, or are able to maintain a job. A cluster of different outcomes should ideally be assessed to a broader, more in-depth, and valid picture of the patient's condition relative to plant or garden intervention.

Outcomes studies also are important for helping healthcare administrators, physicians, and researchers to gauge whether medical interventions or treatments are ethical. Deliberations on the ethics of a proposed outcome may not be common practice in public horticulture, but as soon as you move into the arena of medicine, ethics must be at the forefront. After all, the first adage that every medical student learns is "first, do no harm" (part of the Hippocratic oath). Therefore, the obligation professionally and ethically is to provide evidence, or at the very least a chain of plausible reasoning, that your therapy or garden intervention causes little or no harm. (More technically speaking, adverse reactions should be mild and occur at acceptably low levels, because virtually any medical intervention will trigger some adverse reactions.) The plant or garden intervention, hopefully, will have positive effects on the great majority of patients. "Because I think this is a great therapy and absolutely obvious to all of us that this must be good, let's implement it" is not evidence. Physicians are trained to be reflexively skeptical of those kinds of self-evident claims. In the absence of credible evidence, there are always other treatment alternatives available for scarce medical resources, including not having any garden or horticultural therapy program.

## "FIRST, DO NO HARM"—EXAMPLES FROM ART IN HEALTH CARE

An analogy relevant to measuring outcomes of a plant or garden intervention in health care is the use of art in health care. Some administrators, artists, and designers assume that nearly any type of visual art or painting will have positive influences on patients. Given that the content of artwork varies enormously, and that the content and styles of much art are strongly emotional or challenging, it is reasonable to expect that certain types of art will be positive for patients whereas other types of art might be stressful and worsen outcomes (Ulrich 1991, 1992).

A few years ago, I assessed the effects of wall art in a small-scale study of patients in a psychiatric ward at a Swedish hospital (Ulrich 1991, 1992, 1999). The ward was extensively furnished with paintings and prints paid for by public funds from a national arts set-aside program in a Scandinavian country. The art was apparently considered critically acceptable and psychologically appropriate or positive for patients. I found that patients consistently reported positive feelings and reactions about paintings and prints dominated by nature. However, many patients had negative reactions to abstract artwork where the content was ambiguous and could be interpreted in multiple ways. An example was a large format print with ambiguous content that produced negative reactions in several patients (Ulrich 1991). By contrast, staff experienced the same print as positive. (For a picture of this print, see Ulrich 1999, p. 67.)

Representative responses from patients (with anxiety and/or depression) follow:

- "Charred skulls. Drops of blood are flying."
- "Wounded people. They're in pain and crying out."

Representative staff comments were

- "I think it's fun. Whimsical. I'd like to have it in my home."
- "Funny little talking apple cores. Maybe some are sweating."

Was this print an example of effective, healthful use of art? Perhaps for the staff lounge, a home, perhaps a museum, but not for the healthcare surroundings for these patients.

A second example of adverse patient reactions occurred when a major university medical center—with one of the finest healthcare arts programs in the United States—conducted a juried search for an art/sculpture installation intended to be restorative or therapeutic (Ulrich 1999). This large-scale series of artwork was to form a "bird garden" in a rooftop space surrounded on all sides by rooms for cancer patients. (Although called a "garden," the space contained no greenery, flowers, or other nature.) Soon after this sculpture garden was installed and dedicated, the hospital administration began to receive anecdotal reports of strong negative reactions by some patients. Because these concerns continued to mount, a group of researchers conducted a questionnaire study of patient reactions to the artwork (Hefferman et al. 1995). Forty-six patients were surveyed; more than 20% reported having a negative emotional or psychological reaction to the "garden." Certain patients had strongly negative responses, interpreting some of the metal bird sculptures, for instance, as predatory, frightening animals. The administration and medical staff decided that the rate and intensity of negative reactions was too high, so the art installation was removed for medical and ethical reasons (Ulrich 1999). In the best traditions of medical research, the arts program at the hospital has widely disseminated information about this episode, thus making it possible for others to gain important new knowledge (McLaughlin et al. 1996).

## TRENDS IN HEALTH CARE

### *Consumer-/Service-Oriented Health Care*

As pointed out earlier, healthcare systems everywhere are under strong pressure to reduce costs yet maintain or somehow increase quality. Administrators consequently are influenced by evidence that proposed treatments or services are going to be medically effective in terms of improving outcomes, and cost effective compared to treatment alternatives. Additionally, administrators and healthcare providers increasingly accord considerable importance to evidence regarding effects of treatments or services on patient satisfaction. This reflects the fact that in the United States, and to a lesser degree in European countries, healthcare providers are facing mounting pressure to become much more patient or consumer oriented. Although managed care in the United States has been associated with a number of problems, it deserves credit for shifting the healthcare system toward a more consumer-oriented direction. One manifestation of this shift is that hospitals and healthcare providers now routinely survey patients/consumers to assess their levels of satisfaction with different aspects of the healthcare experience.

A very large amount of data from patient satisfaction surveys has made it clear that major factors that are important to patients and families include the responsiveness and sensitivity of the nursing staff, the need for good communication from physicians and other staff, and support from family. Additionally, it should be emphasized that the physical environments patients experience can significantly affect satisfaction with the healthcare provider. Factors linked directly to the designed environment that typically are measured in patient satisfaction questionnaires include, for instance, noise conditions in patients' rooms, visual and auditory privacy, whether waiting areas and patient rooms are comfortable and pleasant, and the degree to which finding one's way in large healthcare buildings is easy or difficult.

Further, the point should be highlighted that there is growing evidence that the presence of nature—gardens, plants, window views of nature, atriums with vegetation—increases patient and family satisfaction (Cooper Marcus and Barnes 1995; Whitehouse et al. 2001; Center for Health Design and Picker Institute 1999). Accordingly, the potential for gardens and plants to heighten satisfaction, as well as improve health and economic outcomes, is attracting considerable attention from administrators facing strong pressures in a highly competitive marketplace to increase quality and differentiate their services and market identities (Sadler 2001; Whitehouse et al. 2001). From these comments, it follows that proponents of healthcare gardens and plant programs often should be able to build support and possibly garner resources from administrators by demonstrating that garden/plant interventions increase patient satisfaction. (See Sadler 2001 for an excellent discussion from the perspective of a prominent healthcare CEO of the advantages of providing restorative gardens—for enhancing patient, family, and staff satisfaction, and improving health and economic outcomes.)

The use by a healthcare provider of environmental design or quality to promote a positive consumer image and differentiate its market identity is illustrated by the case of the arts program at the new Northwestern Memorial Hospital (NMH) in Chicago, a $650 million building on the clinical campus of Northwestern University Medical School. Based on the growing scientific evidence that psychologically appropriate art improves patient medical outcomes, the administration at Northwestern Memorial decided to allocate $4 million to acquire art for the new building. I was asked by NMH to develop evidence-based guidelines for selecting art that would have beneficial effects on the great majority of patients and produce relatively few negative reactions. The research-based guidelines were then used by Kathy Hathorn of American Art Resources to acquire and commission 1,700 prints and original artworks.

The NMH administration decided to make the art collection the focus of an extensive media marketing campaign for the new hospital. Full-page advertisements featuring the art appeared in major media such as the *Chicago Tribune* during 1999. One advertisement in this newspaper, for example, showed a color picture of a nature painting accompanied by the text:

> The new Northwestern Memorial Hospital is an amazing building. Right down to the paintings hanging on the walls. Because they form part of an innovative recovery program. Research has found that certain pictures—like landscapes and pastoral settings—can be very calming to patients and can actually help them get better. So every one of the nearly 1,700 works of art that adorn our walls are designed to be pleasing to the eye as well as soothing to the soul. Now if a hospital takes that much care with its decor, just imagine how remarkable its patient care must be.

It seems very likely that healthcare-provider marketing campaigns could be developed that similarly capitalize, for example, on evidence-based programs for providing healing gardens or plants.

## *Changes in Demand/Utilization Patterns*

Communicating effectively with the healthcare community requires being aware of other key trends that shape the priorities and challenges foremost in administrators' minds. Most of the trends hold across "modern" healthcare systems internationally, irrespective of whether systems provide universal care to all residents through government insurance (Canada and Sweden, for example), or whether care is funded by a mix of private and government means and is not universally provided (as in the United States) (Ulrich 2001). The following healthcare trends have played a key role in producing major shifts in patterns of demand and utilization for categories of medical care/services and for new types of healthcare environments.

- Growing numbers of elderly.

  *Implications for communicating effectively with the healthcare community.* To convince administrators and bolster support for gardens and plants, evidence is especially needed showing that gardens and plants improve satisfaction and medical outcomes for the growing numbers of elderly in rehabilitation facilities and long-term care units, including Alzheimer's facilities. Evidence is also needed that garden and plant programs in independent living or "aging in place" residential communities improve quality of life and health for the elderly.

- Sharply increasing demand for ambulatory or out-patient healthcare services. The demand for traditional in-patient acute care beds is declining, and average lengths of hospital in-patient stays are shorter. (In many areas, more than 50% of all patient visits to hospitals occur on an out-patient basis. Advances in medical science and technology have made it possible to treat mainly on an out-patient basis many categories of patients, such as persons with cancer or AIDS, who a few years ago would have been admitted to hospitals as inpatients.)

  *Implications for communicating effectively with the healthcare community.* To bolster support and help garner resources, evidence is especially needed showing that gardens and plants improve satisfaction and medical outcomes for the very large numbers of patients treated in out-patient or ambulatory settings. *Anxiety* (fear, tension) and *pain* are experienced by many outpatients, and it is likely that gardens and plants can effectively alleviate these salient problems.

- Although smaller percentages of patients are admitted to hospitals than a few years ago, hospitalized patients today are significantly sicker on average and require more intensive and costly care. Accordingly, demand and utilization are rising markedly for hospital beds in the highest cost, most technologically intensive units such as neonatal intensive care, general intensive care, and coronary critical care. The stakes are extraordinarily high in monetary as well as human terms—currently, about 1.2% of the entire gross national product (GNP) of the United States is spent on healthcare provided only in intensive or critical care units. (That is far more than all of the botanic garden budgets in this country combined.)

  *Implications for communicating effectively with the healthcare community.* To convince administrators to allocate greater priority and resources to gardens, evidence is especially needed that views or other contacts with nature will improve satisfaction, enhance clinical outcomes, and help control the costs of care for patients in intensive or critical care. Research is also needed for demonstrating that gardens and plants can reduce stress and increase satisfaction for families of patients in intensive care, who often experience much stress and make repeated visits or maintain long vigils.

## Example of Evidence-Based Critical Care
## Unit Design for Providing Contact with Nature

A typical intensive care unit is an often busy place with a high level of work demands and pressures for highly trained staff who are costly, in short supply, and difficult to replace. The environments of patient rooms and staff work areas are quite noisy (75–86 decibel [dB] levels are common), often dominated by the presence of medical equipment and technology, and are usually starkly or harshly clinical in appearance and atmosphere. Patients have a window, but beds are rarely positioned to permit viewing out. They tend to be conscious or awake during much of their stay but usually are forced to view a white wall, ceiling tiles, or a wall-mounted television. Patients in intensive care often experience the deleterious combination of an excess of environmental stressors (noise, lack of control, for instance), together with deprivation, such as lack of positive environmental distractions and stimulation.

A few years ago, Legacy Healthcare decided to mitigate many of these problems by renovating an existing 24-bed critical care unit (CCU) in Legacy Good Samaritan Hospital, Portland, Oregon. This effort produced the Kern Critical Care Unit, which represented an innovative and radical departure from previous CCUs in its explicit emphasis on providing patients, families, and staff with exposure to gardens and nature (Willette 2000).

The spacious private patient rooms in the Kern Unit are notable for having a series of large floor-to-ceiling windows (Willette 2000). The large rooms, in combination with ceiling equipment mounts that permit flexibility in arranging apparatus and monitors, enable beds to be turned so that patients can look directly out their windows. The foreground portion of the view out consists of a patio or large balcony with flowers and small gardens. The visual backdrop to the balcony/patio gardens is the leafy canopy of street trees. (The CCU floor is about three stories above street level.) The staff also can easily view the gardens and trees either while they are working in patient rooms or seated at the decentralized nursing stations. Importantly, the soothing outdoor garden spaces are accessible to family members, who use them as waiting and vigil spaces.

## Growing Stress, Work Demands on Staff

Staffing problems are a huge issue in the United States and most other countries. Much research has documented that healthcare occupations such as nursing traditionally are stressful because they usually involve overload from work demands, lack of control and authority, and stress from rotating shifts (Ulrich 1991). But pressures and workloads have increased even further as healthcare providers have been forced to reduce or control costs yet increase care quality. Mounting demands and stresses lower job satisfaction, increase absenteeism and turnover, and contribute to widespread shortages of qualified personnel. This has been accompanied internationally by growing labor unrest and numerous strikes by healthcare workers.

Staff turnover is notorious for eroding the quality of care that patients receive and increasing the providers' operating costs. As an example of the negative effect on costs, consider that the expense of replacing one nurse who leaves an intensive care unit ranges from about $35,000 to more than $100,000 in the United States, depending on the area of the country (these figures include costs for recruitment and orientation but not salary).

*Implications for communicating effectively with the healthcare community.* The serious staff-related problems outlined above mean that most administrators will be influenced and impressed if they are presented with evidence that gardens and plants reduce employee stress, increase job satisfaction, may reduce turnover, and aid in attracting and hiring qualified personnel. In this regard, it is noteworthy that research by Cooper Marcus and Barnes and others is beginning to show that healthcare gardens are used heavily by staff for restoration and positive escape from workplace stresses (Cooper Marcus and Barnes 1995, 1999; Whitehouse et al. 2001; Sadler 2001).

## PREVIOUS RESEARCH ON EFFECTS OF VIEWING NATURE ON HEALTH OUTCOMES

A key factor contributing to the growing interest in healthcare gardens has been a small number of credible scientific studies showing that simply viewing plants and garden-like settings can measurably improve health outcomes. There are also studies on the benefits of horticultural therapy, but a great majority do not measure outcomes or they have methodological shortcomings that weaken the impact of the findings (such as lack of a control group).

Concerning research on viewing nature, a reliable finding in upwards of a dozen scientific studies has been that certain scenes with prominent plants and nature are effective in producing recovery from stress within 3 to 5 minutes (studies reviewed in Ulrich 1999). Even acutely stressed healthcare patients can experience measurable restoration (reduced blood pressure, for example) after only a few minutes of viewing settings dominated by greenery, flowers, and/or water. As well, three or four rigorous studies suggest that viewing gardenlike scenes can reduce pain, as indicated both by patient reports of subjective pain and recorded intake of pain drugs (studies reviewed in Ulrich 1999).

An example of this line of research is a study I did some years ago that focused on patients recovering from gallbladder surgery (Ulrich 1984). The patients were assigned in a semi-random manner to rooms that were identical except that one half of the individuals had a bedside window view of trees, whereas the others looked out at a wall of a brick building. In order to keep other factors constant that could affect outcomes, the methods ensured that the two groups were equivalent with respect to age, gender, weight, smoking habits, general medical history, and season of surgery. A variety of outcomes information was coded from patient records by a medical professional who was blind in the sense that she did not know which type of view—either trees or a brick wall—was visible from an individual's window.

The outcomes data indicated that persons with the nature view, compared to those who overlooked the wall, spent less time in the hospital and developed fewer minor postsurgical complications (such as persistent headache or nausea requiring medication) (Ulrich 1984). Further, patients with the nature view more frequently received positive written comments about their conditions from doctors and nurses in their medical records. Examples of such positive evaluative comments were

- Patient Smith is in good spirits.
- He does his breathing exercises well, moves well, does not need encouragement.

The patients with the brick wall window view, however, received about four times as many negative comments about their conditions, such as

- Patient Smith is upset and crying.
- He refuses to cooperate in doing his breathing exercises.

In the same study (Ulrich 1984), another major difference in outcomes was that the patients with the nature view needed far fewer doses of strong narcotic pain drugs, but took more weak oral doses. This finding was obtained by extracting from patient records the number of voluntary pain doses taken by each person—and classifying each dose as strong, moderate, or weak on the basis of the type of drug, dose size, weight of the patient, and method of administration (injection or pill). Strong pain doses for these patients were injections of synthetic morphine; moderate strength doses were injections of drugs such as Demerol; and weak analgesic doses were acetaminophen pills (Tylenol).

The combination of outcomes evident in this study made a convincing case that the nature view had positive effects on patient health compared to the wall view. Medical administrators and clinicians looked at the findings and realized that providing windows with views of nature in hospitals could also improve economic outcomes—that is, cost savings would possibly be realized because length of in-patient stays might be shortened, and patients would likely need fewer costly injections of narcotic pain drugs. Hence, the study has proved to be influential among healthcare administrators and physicians.

The above research, together with findings from a small number of other outcome-based studies, affected the proposed revised national criteria for accrediting healthcare facilities. Traditionally, the environmental quality criteria for hospital accreditation have been narrowly focused on considerations such as fire safety (where to locate fire extinguishers, for example). The Joint Commission on Accreditation of Healthcare Organizations has proposed revised environmental quality standards that are being circulated for review and comment. If implemented, the revised criteria may require hospitals to provide "orientation and access to nature and the outside" and availability of the "calming and restorative powers of nature" (Young et al. 1998).

## METHODS FOR STRENGTHENING NATURE/OUTCOMES RESEARCH

### *Is There a Single Best Outcome Measure for Evaluating Garden/Plant Interventions?*

Simply put, no. Different categories of patients or diagnostic groups vary with respect to which outcomes are relevant and appropriate. In my view, a key first stage for a nature/health study is to become knowledgeable regarding the disease or injury characteristics of the patient group of interest, in order that the outcome measures chosen will be relevant and effective (Ulrich 1997). Further, a mix of at least two outcome measures—preferably three or more—is usually stronger in terms of validity or research quality considerations than using only one type of measure. A pattern of findings that suggests agreement or convergence across different measures with respect to the effects of a garden, for instance, will convey greater scientific credibility and have more influence (Ulrich et al. 1991; Ulrich 1997).

## *Plan Research to Obtain Outcome Data Efficiently and Inexpensively*

Certain types of health outcomes data are difficult, costly, and very time consuming to obtain. Other outcomes can be recorded or evaluated only by medical professionals with advanced expertise. These obstacles can often be avoided by concentrating on outcomes that are collected and recorded as a matter of routine procedure in the care process or management of the patients you are interested in studying. For most categories of patients, such outcomes or clinical indicators are well established and known to medical professionals, and usually will be found in patients' records. Familiar examples include blood pressure, pulse, and temperature. Extracting data from patient records, however, necessitates that the investigator negotiates a rather lengthy process for obtaining approval from the human subjects or research ethics panel (Institutional Review Board) of the hospital, university, or other institution responsible for the patients and/or investigator.

An earlier section emphasized that work demands and stresses have increased for healthcare staff. Accordingly, in most cases, researchers should not burden nurses or other staff further by asking them to collect and record additional outcomes data. (Exceptions often occur, especially when funding for a research project permits staff to be compensated.) It may be appropriate and necessary, however, to request staff to standardize or systemize their recordings of outcomes information. This can make it possible for the nature/health researcher to harness the ongoing care process or clinical setting as an efficient engine to generate large amounts of quality outcomes data.

There are a few categories of patients for which outcome measures, unfortunately, are not well established or widely standardized—for example, patients undergoing rehabilitation after brain injury. In a doctoral dissertation study, Patrick Williams examined this problem as part of his effort to improve the quality and impact of research on the effectiveness of horticultural therapy for patients with brain injury (Williams 2000). One problem identified by Williams is that different rehabilitation providers or care facilities vary in the measures or outcomes used for assessing the condition or progress of brain injury patients.

Another approach for obtaining data in an efficient and exceptionally cost-effective manner is to take advantage of information from patient satisfaction surveys that are administered routinely by virtually all hospitals and healthcare providers. The vast majority of hospitals in the United States contract with the four largest satisfaction survey firms (Press Ganey, Gallup, for example). Contracts for survey services with these firms typically include provisions that enable the hospital or administrator to add specific questions tailored to a particular unit or group of patients. This could make it possible for a gardens/plants researcher to evaluate at little or no cost the effects on patient satisfaction, for instance, of installing a restorative garden adjacent to a high-stress treatment area such as an emergency room or chemotherapy infusion area. This scenario requires the researcher to persuade the administration or quality improvement office to direct the survey firm to include a couple of questions relating specifically to the garden. Even if no new questions were added, however, it would still be rather easy to determine whether the addition of the garden improves satisfaction scores for overall quality of care.

## *Use a Study Design that Includes a Control or Comparison Group*

A major pitfall to avoid in nature/outcomes research is using the intuitively appealing but flawed pre-post or before-after design with no comparison group or condition (Ulrich 1997). As Campbell and Stanley emphasized in their classic book on research design, "Basic to any credible or scientific evidence is the process of comparison or contrast . . . of effects" (Campbell and Stanley 1963). Garden or plant researchers who aim for credible and influential findings, therefore, should devise a research plan that makes possible a comparison of outcomes for at least two different study conditions—that is, one situation with the garden/plant intervention and one without (Parsons et al. 1994). It is very important that the different conditions vary only in terms of the garden/plant aspect, and must not differ in other ways that might affect outcomes and confound the findings (Parsons et al. 1994; Ulrich 1997). When there is no comparison or contrast condition, there is no sound basis, strictly speaking, for inferring that the garden or plant intervention *per se,* rather than some other uncontrolled or unknown factor, was responsible for any outcome differences that might be measured (Campbell and Stanley 1963).

Several strategies can be used to include a comparison or control condition in a nature/outcomes study. The strongest approach scientifically—and the one favored by medical researchers—is to randomly assign patients to two or more treatment situations. In a nature/outcomes study, this would mean assigning patients randomly to at least two situations that differ solely in terms of the presence versus absence of the garden/plant exposure or intervention. Another strong research approach is to assign the *same group* of persons to a series of different situations (that vary with respect to the nature intervention), thereby making it possible to compare each patient with himself/herself (Sommer and Sommer 1997). Another widely used but less scientifically strong approach is to assign two comparable groups of patients to different garden or plant interventions—making sure that other things are constant or the same—and then comparing outcomes for the two groups (Ulrich 1997). (Detailed discussions of these and many other procedures are available in several useful books on research methods: see, for example, Campbell and Stanley 1963; Sommer and Sommer 1997).

## CONCLUSION

Against the background of major trends and changes in healthcare systems, there can be no question that the future importance and resources given to garden or plant interventions by the medical community will be strongly shaped by the extent to which sound research demonstrates that these interventions improve outcomes—including medical outcomes such as pain and anxiety, economic outcomes such as the costs of delivering care and staff turnover, and patient and family satisfaction with the healthcare experience. Producing credible evidence that gardens or plants enhance outcomes will pave the way for effective, influential communication with the healthcare community.

## LITERATURE CITED

Campbell, D. T., and J. C. Stanley. 1963. *Experimental and quasi-experimental designs for research.* Boston: Houghton Mifflin.

Center for Health Design and Picker Institute. 1999. *Assessing the built environment from the patient and family perspective: Health care design action kit.* Walnut Creek, CA: The Center for Health Design (http://www.healthdesign.org/).

Cooper Marcus, C., and M. Barnes. 1995. *Gardens in healthcare facilities: Uses, therapeutic benefits, and design recommendations.* Martinez, CA: The Center for Health Design.

Cooper Marcus, C., and M. Barnes. 1999. *Healing gardens: Therapeutic benefits and design recommendations.* New York: John Wiley.

Hefferman, M. L., M. Morstatt, K. Saltzman, and L. Strunc. 1995. A room with a view art survey: The bird garden at Duke University Hospital. Unpublished research report, Cultural Services Program and Management Fellows Program, Duke University Medical Center, Durham, NC.

Horsburgh, C. R. 1995. Healing by design. *New England Journal of Medicine* 333: 735-740.

McLaughlin, J., J. Beebe, J. Hirshfield, P. Lindia, and D. Gubanc. 1996. Duke University's bird garden. In *Proceedings of the 1996 Annual Conference of the Society for the Arts in Healthcare.* Durham, NC: Durham Arts Council and Duke University Medical Center, pp 49-63.

Parsons, R., R. S. Ulrich, and L. G. Tassinary. 1994. Experimental approaches to the study of people-plant relationships. *Journal of Consumer Horticulture* 1(4): 347-372. [Reprinted in J. Flagler and R. P. Poincelot (Eds.). 1994. *People-Plant Relationships: Setting Research Priorities.* Haworth Press.]

Sadler, B. L. 2001. Design to compete in managed healthcare. *Facilities Design and Management,* March, 38-41.

Sommer, B., and R. Sommer. 1997. *A practical guide to behavioral research* (Fourth Ed.). New York: Oxford University Press.

Ulrich, R. S. 1984. View through a window may influence recovery from surgery. *Science* 224:420-421.

Ulrich, R. S. 1991. Effects of health facility interior design on wellness: Theory and recent scientific research. *Journal of Health Care Design* 3:97-109. [Reprinted in Marberry, S.O. (Ed.). 1995. *Innovations in Healthcare Design.* New York: Van Nostrand Reinhold, pp. 88-104.]

Ulrich, R. S. 1992. How design impacts wellness. *Healthcare Forum Journal*, September/October, 20-25.

Ulrich, R. S. 1997. Methods for strengthening arts/health research. Proceedings of the 1997 Meeting of the Society for the Arts in Healthcare, pp. 115-123. Duke University Medical Center, Durham, NC.

Ulrich, R. S. 1999. Effects of gardens on health outcomes: Theory and research. In *Healing gardens: Therapeutic benefits and design recommendations*, eds. Cooper Marcus, C., and M. Barnes. New York: John Wiley, pp. 27-86.

Ulrich, R. S. 2001. Effects of healthcare environmental design on medical outcomes. In *Design and health: Proceedings of the second international conference on health and design*, ed. A Dilani. Stockholm, Sweden: Svensk Byggtjanst, pp. 49-59.

Ulrich, R. S., R. F. Simons, B. D. Losito, E. Fiorito, M. A. Miles, and M. Zelson. 1991. Stress recovery during exposure to natural and urban environments. *Journal of Environmental Psychology* 11: 201-230.

Whitehouse, S., J. W. Varni, M. Seid, C. Cooper Marcus, M. J. Ensberg, J. J. Jacobs, and R. S. Mehlenbeck. 2001. Evaluating a children's hospital garden environment: Utilization and consumer satisfaction. *Journal of Environmental Psychology* 21:301-314.

Willette, S. 2000. Case study: Legacy Good Samaritan Hospital, Portland, OR. Chapter in *ICU 2010*, ed. K. Hamilton. Houston: Center for Innovation in Health Facilities and American Institute of Architects Academy of Architecture for Health, pp. 307-317.

Williams, P. N. 2000. An evaluation of horticultural therapy in brain injury rehabilitation. Doctoral dissertation, Department of Horticultural Sciences, Texas A&M University, College Station, TX.

Young, R. L., A. Salvatore, J. E. Fishbeck, and C. H. Patterson. 1998. The enhanced standards of the Joint Commission on Accreditation of Healthcare Organizations. Proceedings of the 1998 Symposium on Healthcare Design. IMARK and The Center for Health Design, pp. 246-262.

# 3

# Greenlining: An Invitation to Cross the Bridges to Urban Green Space and Understanding

Margaret Ross Bjornson

Visits to urban America's community green spaces introduce the horticultural beauty of unfamiliar neighborhoods and the individuals who create both the gardens and the communities. Urban community gardens can both green the streets and "greenline" the communities formerly "redlined"—barricaded by socioeconomic, cultural, and political agendas that separate those in place from those in power. "Greenlining" visits can remove the gatekeepers of ignorance, prejudice, and economic opportunism.

Recent research around North America demonstrates the possible impact when "outsiders" visit the "greenlined" oases in forgotten urban neighborhoods (Bjornson 1996a; Malakoff 1995). Research and field work focus on how to highlight urban greening efforts.

## GREENLINING

Green and growing networks of advocacy and activism are taking root in urban neighborhoods throughout Chicago and other cities in the United States. These complex greening networks make it possible for formerly marginalized urban residents to gain access to public policy, economic resources, and social interaction. Through the agency of political savvy, already empowered urban gardening advocates confront barriers built by political expediency, economic opportunism, and social inaccessibility. These dynamic interactions and the community gardens that result from them "give the green light" to access and generate the positive, organic energy of a process I call "Greenlining."

Individuals who garden—and according to the National Gardening Association (Sommers 2000), gardening is America's most popular leisure time activity—understand how powerful a garden can be. Gardens inspire. Gardens educate. Gardens are certainly cause to celebrate.

The universal nature of gardening offers the most critical rationale for gardens becoming bridges between and among cultures and communities. This universality provides common ground for exploration and conversation among culturally diverse gardeners and garden lovers. America's current, revitalized urban community gardening movement has deep roots in our nation's history. European immigrants in diverse centuries carried seeds, roots, and plant stocks to carry on the agricultural practices of their home countries. Many discovered existing communal agricultural practices of Native Americans. Sharing of agricultural knowledge between Native Americans and immigrants perhaps constituted the first examples in the colonies of gardening as a cultural bridge (Dorrance 1945; Haughton 1978; Josephy 1992; Nabokov 1992).

In 20th and 21st century America, the possibilities are plentiful for building and crossing "green bridges" that provide access to neighborhoods, communities, ideas, and cultures. These bridges are, to mix metaphors, two-way streets, creating possibilities for cultivating more than the garden grows. Every garden has a story. Those who can tell the story and those who will listen are major players in simple, yet complex, dramas performed every day in cities and towns throughout the United States. Opportunities for telling and hearing these stories and for visiting the sites—the stages on which the dramas take place—constitute the focus of this paper.

The impact and effect of community gardens on individuals and communities may be social, economic, psychological, or political (Bjornson 1996a, 1996b; Malakoff 1995). The active involvement in establishing and nurturing a community garden typically exhibits change (by its very nature), demonstrates growth of individual leaders and community awareness, and usually is reported as therapeutic—for individuals and for the community (Lewis 1996; Suarez 1999).

However, field research demonstrates that benefits extend beyond individuals or groups of gardeners and beyond the greened neighborhoods or communities where gardens are located (Bjornson 1996a; Lewis 1996). Visitors to these gardens, often not from the garden neighborhood, learn and benefit from ventures to see each garden and hear its story. Moreover, the host gardeners may gain a new sense of pride and insight about the work they do. They receive recognition for their leadership and for their community development and environmental efforts. The encounters are extraordinary because of the dynamic and surprising teacher/learner role activity. Both host gardeners and garden visitors experience a connection—and learn how much they have in common with others so unlike themselves.

Community garden visits can be a fascinating adventure. New horizons stretch before the garden visitor—new opportunities as student, educator, writer, researcher, developer, investor, and future neighbor. Every homemade sign and fence, every creative addition of found garden art, every unique growing method opens a window to discovery. Moreover, host gardeners seem to thrive in the presence of interested visitors. They gain new stature and rise to previously unexplored heights when revealing their garden's secrets or responding to questions. No matter how modest the garden or the neighborhood, community gardeners tend to offer hospitality and receive a welcome response to their gifts of garden produce, flowers, seeds, or freshly prepared tastes of their gardens.

Who are the potential visitors to these urban green spaces? Moreover, why do they visit? Potential visitors may be tourists, gardeners from around the area, social service professionals, media professionals, corporate development and philanthropy professionals, educators, students, government agency employees or researchers, developers and investors, or garden club members. The list is limited only by the imagination. Their reasons for visiting are equally varied.

The following are examples of experiences visitors to community gardens have had.

## THE PENNSYLVANIA HORTICULTURE SOCIETY'S GREENE COUNTRIE [SIC] TOWNE PROGRAM, PHILADELPHIA, PENNSYLVANIA

This grant-funded program began in the mid-1980s with the intention of providing the maximum amount of greening education and resources to specified communities or "townes" within Philadelphia. The goal was to introduce projects that would result both in residual and subsequent greening and community development projects. The attention gained by the program brought positive responses, one of which was the return of city services to neighborhoods, which had been ignored before recognition as a Greene Countrie Towne.

A Puerto Rican woman in Philadelphia contacted the Pennsylvania Horticulture Society's (PHS) Greene Countrie Towne Program to help regain her neighborhood from powerful, intimidating street gangs. With the help of PHS volunteers, neighbors in the Puerto Rican community reclaimed their local park and created gardens that became places of respite and beauty.

In 1999, the American Community Gardening Association hosted its 20th annual conference in Philadelphia, and tours to the Puerto Rican neighborhood revealed gardens created along once neglected streets. They visited both the traditional *casita* (little house) with its *jardin* (garden) of Puerto Rico and a garden that honors and memorializes a son of the community lost in gang warfare. The *casita*, in the center of the largest community garden in the neighborhood, is a community-gathering site—a place to teach ancestral culture among generations with the Puerto Rican families. Additionally, a visit to this community garden provides a window to a Puerto Rican neighborhood in North America.

## MARY'S KIDS GARDEN, "BACK OF THE YARDS," CHICAGO, ILLINOIS

In Chicago's "Back of the Yards" neighborhood, Mary Gonzales and her husband, German, struggle to offer welcome, safety, and solace to children overlooked and underattended by their families. Formerly, the neighborhood was home to many who were employed at Chicago's world famous Union stockyards. First, a hearty "stew pot" mix of eastern European, German, Italian, and Irish families resided there; diversity increased with the introduction of Mexican and African-American families (Suarez, 1999).

When the stockyards closed in the late 1960s and early 1970s, most families of skilled, able-bodied employees left; those who remained were largely less skilled workers and their families. Property values plummeted. City services diminished. Many property owners sold and/or became absentee property owners. The population decreased. New residents, attracted by low rents and the low-profile neighborhood (a neighborhood ignored by the media, political decision makers, the business community), introduced drugs; local school dropout rates increased. According to Mary Gonzales, large numbers of the neighborhood's children (largely Latino and African-American) fell into the abyss of neglect.

The Gonzales family decided to stay in the changing neighborhood, and Mary welcomed and nurtured the children. With help from a grant, the Back of the Yards Council acquired land for Mary's Kids Garden. The Chicago Botanic Garden and the University of Illinois Extension both established partnerships with Mary's Kids Garden, which took root on a vacant lot near Mary's home. Each provided resources—monetary, educational, and horticultural—to help Mary and the neighborhood children. It is a green oasis and a place of beauty and homegrown food—greens, peas, *frijoles* (beans), *chiles* (peppers), tomatoes, *tomatillos,* and *cilantro*—for neighborhood tables.

Homemade *quesadillas, tortas,* and *frijoles refritos*, prepared in Mary's kitchen, satisfy the appetites of those who come to visit and to help. High school, college, and corporate volunteers from near and far have visited; given of their time, energy, and expertise; and have learned from their visit—not only about Mary and her "Kids," but also about the neighborhood, its economic and social situation, and its place in Chicago's political hierarchy. Stories about Mary's Kids came to the attention of the *Chicago Tribune*, and a resulting feature story brought reader contributions of food, clothing, games, toys, and monetary gifts to allow Mary to continue her work.

## SHERIFF'S GARDEN AT THE COOK COUNTY CORRECTIONAL FACILITY, CHICAGO, ILLINOIS

In 1994, according to Ron Wolford, University of Illinois Extension Educator, the Cook County Program of Community Supervision and Intervention, for short-term detainees, wanted to begin a vocational training project and asked him to assist.

German Gonzales is a gardener and educator who teaches gardening to detainees at the Cook County Correctional Facility. Under the auspices of the University of Illinois Extension, he has helped the Sheriff's Garden Program thrive. In 2000, selected detainees participated in the first University of Illinois Extension Master Gardener Program offered at the correctional center.

Entering the Sheriff's Garden through locked and guarded gates, a first-time visitor cannot often imagine the experience he/she is about to have. Past the barking guard dogs, beyond chain-linked barbed wire fences that surround it, and within the confines of the brick structures that dominate the compound, an abundant garden of squash, beans, tomatoes, herbs, and chard, surrounded by annual blossoms, is spread out for more than an acre.

In addition to gardening within the confines of the correctional institution, selected detainees participate in an outreach program where they perform horticultural work at specified sites, such as schools or public agencies, within the city. This puts the men in an interactive role with the communities to which they will return.

The men who participate in the garden project do so in lieu of other outdoor recreation and learn marketable skills. Their backgrounds vary, as do their skills. The garden provides more than fresh air and exercise. Many describe earlier gardening experiences, most commonly mentioning their grandmother or their father (Bjornson 2002).

The tons of food produced annually are delivered to a Women, Infants, and Children (WIC) Center, to food pantries, or to homeless centers in the city. Detainees in a trusted status accompany the Sheriff's officers to distribute the produce. To be sure, the detainees have done more than meet a time commitment. Stories of how they grow and distribute food give greater meaning and relevance to their task. Visitors to the correctional facility garden may acquire different insights into correctional programs and criminal offenders.

## CHICAGO'S 13TH STREET GARDEN, CHICAGO, ILLINOIS

Chicago's 13th Street Garden grows in the shadows of the Dan Ryan Expressway with Chicago's impressive urban skyline rising on the horizon as a backdrop. Community garden and recycling guru Ken Dunn, a former philosophy major at the University of Chicago, regularly welcomes visitors to the colorful, art-filled, ecologically based, productive garden. Within a corridor of rapid change, prompted by the expansion plans of the University of Illinois at Chicago, the garden and its adjacent Creative Reuse Warehouse abut what remains of the city's Maxwell Street Market. (The Warehouse offers recycled materials for sale to teachers, youth leaders, and others, with Dunn as director.) Gardeners here include a spectrum of individuals from diverse socioeconomic backgrounds—an advertising executive whose chosen bike path from work to home takes him past the garden and a variety of Chicagoans who work or live near enough to the site to commit themselves to the garden's work and philosophy.

Dunn welcomes visitors to an open space within the garden meant for community gathering. There he reveals the historical, current, and possible future stories of the garden. He describes how he successfully turned a philosophical question about resource utilization into a viable livelihood for himself and others. He describes the political and economic history being made by encounters with the University of Illinois. He welcomes a discussion about agriculture as a dimension of culture and a garden's role in fostering civility. He relates his idea of permanence and mobility of gardens in urban settings: that virtually every important component of a garden can be transported to another site if/when it is forced from its existing location. Motorists speeding past the 13th Street Garden at 60 miles an hour have no idea what they are missing.

## GARFIELD NEIGHBORHOOD GARDEN, CHICAGO, ILLINOIS

In contrast to Dunn, George Clark, an experienced community gardener in Chicago's south side Englewood neighborhood, is a man with little formal education. Yet on a recent visit to his Garfield Neighborhood Garden, college-educated interns from the

Chicago Botanic Garden learned a great deal. They learned about community gardening in a large urban city. They learned about coalition because this garden joined Garfield neighbors, the Archdiocese of Chicago, the Chicago Botanic Garden, and others to create the garden. Their lesson focused on urban skills such as how to get access to land; how to get the attention of and action from one's alderman or commissioner; how to get regular garbage pickup; how to rid a site of rats; and how to get access to water.

## CHICAGO'S SCHOOL GARDEN INITIATIVE, CHICAGO, ILLINOIS

Schools throughout the United States are planting seeds of educational opportunity for children and their families. In the late 1990s, the California legislature mandated a "Garden in Every School." Chicago has established a School Garden Initiative. Individual schools and school districts throughout the nation are removing asphalt, digging deep, and creating garden classrooms. The gardens are used across the curriculum in art, writing, humanities, and social and lab sciences. Visitors and volunteers are welcome.

The Chicago School Garden Initiative (SGI) is a collaboration of the Chicago Botanic Garden, Chicago Park District, Chicago Public Schools, City of Chicago Department of Planning and Development, and Garfield Park Conservatory Alliance that began in the mid-1990s.

With school gardens as its basis, SGI "takes a comprehensive approach to issues of education, community, and environment for Chicago's public school children and the greening of Chicago neighborhoods." Its stated goals and objectives follow:

> . . . to develop a site leadership team consisting of administrators, teachers, students, parents, neighbors, and volunteers; a working garden for use by the school and community; teachers trained to use the garden as an educational resource; a network of school garden projects and staff throughout the city; parents who actively participate in the program; a student body that uses the garden as a basis for . . . multidisciplinary learning.

Everyone learns and everyone can teach in a garden. Visitors learn from teachers and students alike. Teachers and students can learn from visitors and may even benefit in unexpected ways; for example, when the United States Department of Agriculture (USDA) professionals and educators pitched in and helped students with plant identification and weeding when visiting a Chicago high school garden. The students and teachers told the USDA visitors about their garden's beginning and offered ideas for starting and sustaining one. A Chicago Botanic Garden (CBG) professional identified this school yard as a worthy site for excess CBG plants, which the school's new prairie garden sorely needed, and offered professional advice and networking to access horticultural job training programs available through professional organizations. This visit is a model metaphor of cross-pollination.

## CONCLUSION

Garden successes depend in part on the leadership of a few visionary, dedicated, committed, and persistent individuals and those who work alongside them. Among these are advocate agency professionals. However, the ultimate success occurs when the story is shared, the details become known, and knowledge moves both ways across the green bridges.

The outcomes of these extraordinary efforts need to be documented. Field research needs to be conducted to answer questions about sustainability of the gardens, and the roles the gardens have in urban renewal and gentrification, rehabilitation, education, and celebration. Field research about the practical application and quantification of outcomes, results, benefits, and their resulting questions and implications is also important.

Ken Dunn reminds us that agriculture is a form of culture. What we sow in our gardens and what we harvest from our gardens are not the only things we cultivate. Community gardens and visits to these gardens clearly cultivate relationships among individuals of diverse experience—within communities of diverse realities. These visits break down stereotypes, expose misconceptions, and open doors to new sights.

## LITERATURE CITED

Bjornson, Margaret Ross. 1996a. Greenlining: Chicago's urban community gardens build bridges of social, political, and economic access. In *Proceedings People-Plant Interactions in Urban Areas: A Research and Education Symposium*, eds. Williams, P., and J. Zajicek. Department of Horticulture, Texas A&M University, College Station, TX, pp. 49–52.

Bjornson, Margaret Ross. 1996b. Immigrant gardens: Seeds of home in times of change. In *Proceedings People-Plant Interactions in Urban Areas: A Research and Education Symposium*, eds. Williams, P., and J. Zajicek. Department of Horticulture, Texas A&M University, College Station, TX, pp. 123–126.

Bjornson, M. R. 2002. Author's interactions over 5 to 6 years, with the detainees of Cook County Correctional Facility, Chicago, IL.

Dorrance, Anne. 1945. *Green cargoes*. Garden City, N.Y.: Doubleday, Doran & Co., Inc. p. 140.

Haughton, Clare Shaver. 1978. *Green immigrants: The plants that transformed America*. New York and London: Harcourt Brace Jovanovich, pp. iv–xii.

Josephy, Alvin M. Jr. 1992. The center of the universe. In *America in 1492: The world of the Indian peoples before the arrival of Columbus*, eds. Josephy, Alvin M., Jr., and Frederick E. Hoxie. New York: Knopf, pp. 6–7.

Lewis, Charles A. 1996. *Green nature human nature: The meaning of plants in our lives*. Urbana: University of Illinois Press.

Malakoff, David. 1995. *What good is community greening?* Philadelphia: American Community Gardening Association.

Nabokov, Peter with Dean Snow. 1992. Farmers of the woodlands. In *America in 1492: The world of the Indian peoples before the arrival of Columbus*, eds. Josephy, Alvin M., Jr., and Frederick E. Hoxie. New York: Knopf, p. 118.

Sommers, Larry. 2000. Personal communication with vice president of advertising and marketing, National Gardening Association.

Suarez, Ray. 1999. *The old neighborhood: What we lost in the great suburban migration, 1966-1999*. New York: Free Press.

## RESOURCES

Urban community gardens often have a relationship with one or more green advocate organizations found throughout the United States. For more information, contact one of the following:

- American Community Gardening Association
  100 North 20th Street, 5th floor
  Philadelphia, PA 19103-1495
  http://communitygarden.org

  The oldest and only national organization devoted entirely to urban community gardens and related issues. Annual conferences and workshops.

- National Gardening Association
  180 Flynn Avenue
  Burlington, VT 05401
  800.538.7476

  A non-profit member-supported organization, established in 1972, dedicated to helping people be successful gardeners.

- GreenNet: Chicago's Greening Network
  300 North Central Park
  Chicago, IL 60624
  GreenNet Hotline: 773.638.1766 x25

  A coalition of organizations and public agencies committed to sharing information and resources and to developing joint projects to improve the quality, amount, and wide geographic distribution of sustainable green open space in Chicago.

- Chicago Neighborhood Tours/City of Chicago Department of Cultural Affairs
  Chicago Cultural Center
  78 East Washington Blvd.
  Chicago, IL 60602

# 4

# Computer Use among Registered Members of the American Horticultural Therapy Association

**Lea Minton Westervelt**
**Richard H. Mattson**

## INTRODUCTION

This study was conducted to evaluate the type and extent of computer technology use among registered members of the American Horticultural Therapy Association (AHTA). This organization promotes the use of horticulture as a therapy to improve human health. The general membership information provided by AHTA in May 2000 included 957 individuals and organizations of which 264 were registered horticultural therapists. AHTA membership surveys have shown that better communication is a top concern. The use of computers as a communication tool is possible through electronic mail (E-mail) and information exchange through websites (home pages, list serves, and chat rooms) on the Internet. This research surveyed the AHTA registered membership, which lists only 9% of the members as having an E-mail address.

## THE SAMPLE

The population sample was individuals currently registered with AHTA. The AHTA registration process is the only system available for professional recognition of horticultural therapists. It is a voluntary program conducted as a peer review of academic and professional training, work experience, professional activities, and other accomplishments. There are three registration levels: the Horticultural Therapist Technician (HTT) level intended for persons transitioning into the field; the Horticultural Therapist Registered (HTR) level designed as the primary level of registration within AHTA's system; and the Master Horticultural Therapist (HTM) designation, which indicates extensive educational and professional achievements in horticultural therapy. The population sample of 264 was lowered to a test sample of 161 due to invalid telephone numbers. Sixty-three or 41% of the test sample were considered valid of 65 completed interviews. The study response of 13% HTT, 73% HTR, and 14% HTM was closely aligned with the ratio of all registered AHTA members at 17% HTT, 70% HTR, and 13% HTM. The geographical distribution of the respondents was defined by the time zone of the state in which they were located, which follows:

- Eastern: 40%
- Central: 38%
- Pacific: 13%
- Mountain: 9%

## THE SURVEY METHODOLOGY

A telephone interview was the survey method selected. Advantages of telephone surveys include quality control, cost efficiency, and speed of data collection (Lavarakas 1989). All requests for an interview were granted. An average of four calls was placed before an actual personal contact was made to complete each interview. Internet phone service was used to a limited degree. It was available at no charge, but the quality of the voice transmission was poor. The use of computer technology was the primary focus of this survey with additional questions about the utilization of various information resources used for horticultural therapy. The survey included questions designed to develop a profile of the respondent's professional characteristics as well as basic demographic information. (See Appendix 4.1.)

## RESULTS

### Computer Use

The basic question for the foundation of this survey was one that asked if respondents used a computer at work, at home, or at any other location. The total percentage of respondents using a computer in some location was 97%. Horticultural therapists use computers at work more than the average American worker does. In this survey, 75% of the respondents used a computer at work, and the national percentage of workers who used computers on the job was 49.8% (U.S. Census Bureau 1999). The percentage of individuals who used computers at home (84%) was higher than the percentage who used computers at work. Everyone that used a computer in an "other" location also used a computer at work and/or at home. Two respondents did not use a computer, but they indicated an expectation to have one in the future.

The primary computer uses were for sending and receiving E-mail (84%), word processing (84%), and accessing the Internet (82%), followed by preparing spreadsheets (35%), creating and using databases (30%), and preparing graphics (30%). When respondents were questioned about the necessity of a computer to do their job, 62% (n=38) indicated computer use was required for their work, 20% (n=12) indicated a computer was helpful, and 18% (n=11) felt that a computer was not necessary in their work.

In response to the question of how many total hours a week a computer was used, a range of 1 to 40 hours per week was given. The median response was 10 hours. Computer use was similar among all AHTA registration levels. Respondents reported that 84% used E-mail frequently, which was defined as at least weekly. Daily E-mail use was reported by 66% of the respondents.

## Information Resources

Questions on information resources were divided by what sources of information were used and through what channels of communication information was accessed to obtain information on horticultural therapy. Options chosen as their most frequently used source of information were peers, AHTA, their state university, other universities, and regional AHTA chapters (Table 4.1).

Traditional channels of communication such as books, magazines, and newsletters were the most preferred, the next highest preferences were for classes, seminars, and meetings, then websites on the Internet (Table 4.2). Less than half of the respondents chose videotapes, E-mail, television or radio, CD-ROMs, or a list-serve as their preferred information source. Although E-mail was the highest rated computer use, it was not a highly preferred method of professional information exchange. List-serves are E-mail exchanges between a group of individuals (subscribers) where messages and responses can be posted to all subscribers at one time. At the time of this writing, three list-serves are dedicated to horticultural therapy issues:

- One is supported by the Horticultural Therapy program in the Department of Horticulture, Forestry & Recreation Resources at Kansas State University.
- One is supported by the People-Plant Council through Virginia Tech University.
- One is supported by the Human Issues in Horticulture working group of the American Society of Horticultural Sciences.

Table 4.1.  Horticultural therapy information sources most frequently used by registered horticultural therapists

| Information Source | Frequency (%) | Number of Responses |
| --- | --- | --- |
| Peers | 82.5 | 32 |
| AHTA | 74.6 | 30 |
| State University | 49.2 | 17 |
| Other University | 46.0 | 16 |
| AHTA Chapter | 36.5 | 14 |

Table 4.2.  Preferred horticultural therapy information channels most frequently used by registered horticultural therapists

| Information Channel | Frequency (%) | Number of Responses |
| --- | --- | --- |
| Books, magazines, newsletters | 92 | 58 |
| Classes, meetings, seminars | 81 | 51 |
| Internet Websites | 56 | 35 |
| Videotapes | 44 | 28 |
| E-mail | 43 | 27 |
| Radio/TV | 40 | 25 |
| CD-ROM | 24 | 15 |
| List-serves | 18 | 11 |

None of these list-serves was well utilized by the registered horticultural therapists interviewed.

## Professional and Demographic Information

Registered horticultural therapists are highly educated. All respondents held a minimum of an associate's degree, and 94% had a bachelor's, master's, or doctorate degree. Categories of employment such as coordinator, director, or educator were listed by 51% of the respondents (Figure 4.1). The horticultural therapy category represents job titles of horticultural therapist, activity therapist, and horticulturist. "Other" responses included volunteers, a retiree, and a consultant.

The average number of years worked as a horticultural therapist was 13. The mean difference between number of years worked as a horticultural therapist and the registration level was significant at the 0.05 level (Table 4.3). The number of years in the profession was not significantly correlated to the number of hours a computer was used or the number of hours worked weekly.

The majority (85%) indicated that they worked more than 20 hours a week. Eight respondents indicated that they worked 50 or more hours each week. Individuals registered at the HTM level worked significantly more hours a week than did the HTR or HTT respondents (p<.05). The HTR respondents worked significantly more hours a week than did the HTT respondents (p<.05).

The majority (87%) of the registered horticultural therapists interviewed are women. The age range of those interviewed was from 29 to 77 years of age, with a median age of 47.

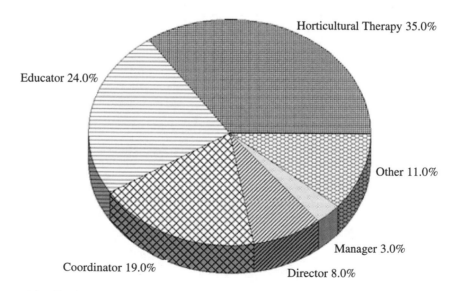

Figure 4.1.  Employment categories of registered horticultural therapists.

Table 4.3.   Years as a horticultural therapist compared to registration level

| Registration Level | Mean Number of Years | Number of Subjects |
|---|---|---|
| HTM | 20.0 a | 9 |
| HTR | 12.7 b | 46 |
| HTT | 10.0 b | 8 |
| Average | 13.4 | |

Means followed by the same letter are not significantly different at the 0.05 level.

## CONCLUSION

An unexpectedly high number of registered horticultural therapists used computers. The total percentage of respondents using a computer was 97%. The computer was certainly considered to be a communication tool with E-mail used daily or at least weekly. However, the use of E-mail as an information source for horticultural thera py was preferred by only 47% of the respondents, only 18% belonged to a list-serve, and only 9% of the members listed in the AHTA Membership Directory showed an E-mail address. This suggests that additional studies should be conducted to understand why electronic communication is a major computer use of horticultural therapists, but is not used in relation to information exchange on horticulture therapy. AHTA might consider experimental use of electronic transmission of information to members. This would increase the actual use of E-mail as an information source and could be a potential source of savings in the future for AHTA. The number of respondents who considered use of a computer as required or helpful to do their job was also a higher percentage than expected.

Internet use was preferred more as a communication tool than as an information source. There was a strong reliance on peers and AHTA as information resources. The lowest response reported was use of universities and local AHTA chapters for information. Although there is a high percentage of computer use, written information sources such as books, magazines, and newsletters are preferred over computer applications such as websites, E-mail, list-serves, and CD-ROMs. The majority of respondents never or rarely use the AHTA website on the Internet. This should also be the source of future study. AHTA could consider ways to improve and promote the use of their Internet website. The acceptance and integration of computer technology could be a resource for increased communication among these highly dedicated and professional registered horticultural therapists.

## LITERATURE CITED

Lavarakas, Paul J. 1989. *Telephone survey methods*. Sage Publ.: Newbury Park, CA.
U.S. Census Bureau. 1999. *Statistical abstract of the United States* (119th edition). Washington DC.

**APPENDIX 4.1. COMPUTER USE AMONG REGISTERED MEMBERS OF THE AMERICAN HORTICULTURAL THERAPY ASSOCIATION**

Respondent #_____

## Part 1:

Q 1.1   **Hello, is this _____?**
If the answer is **YES**, go here:

Q 1.2.a   **I am calling as part of a research project at Kansas State University to study the use of computers by members of the American Horticultural Therapy Association. Your participation is strictly voluntary and your information will be confidential. The questions I am going to ask can be answered in less than five minutes. You may stop at any time or choose whether to answer any question. May I continue?**
If the answer is YES, go to the next page.
If the answer is NO, ask the following:
**When would be a better time to call you?**_____
Surveyor says thank you, and ends call.
If the answer is **NO**, go here:

Q 1.2.b   **Can I reach _____ at this number?**
If the answer is YES, go here:
**When would be the best time to call and reach_____?**
Day and time_____
Surveyor says thank you and ends call.

If the answer is NO, that person is not at that number, go here:
**Do you know how I can reach_____?**
Phone:_____
Address:_____
Surveyor says thank you and ends call.

## Part 2:

Q 2.1   **Do you use horticulture as therapy in your current work?**
1. Yes
2. No
If the answer is YES, go to Part 3.
If the answer is NO, go here:

Q 2.2   **Since your answer is no, why are you a member of the American Horticultural Therapy Association?**

If there is past, current, or future interest in horticulture therapy, go to Part 3 on next page.

If there is NO current or future interest in horticultural therapy, proceed as follows. **Thank you for agreeing to participate in this research project. If you would like further information, please contact Lea Minton or Dr. Richard Mattson at Kansas State University. The phone number is 785-532-6170. For information regarding the rights of human subjects, please contact Clive Fullagar, Chair, Research Involving Human Subjects, 1 Fairchild Hall, (785) 532-3224. Thank you again for your time.**


## *Part 3:*

Q 3.1   **Do you use a computer at work?**
   1. Yes
   2. No

Q 3.2   **Do you use a computer at home?**
   1. Yes
   2. No

Q 3.3   **Do you use a computer at any other location?**
   1. Yes   Where?_____
   2. No

If the answer is YES to any of these questions, go here:

Q 3.4.a   **What do you use the computer for?**
Surveyor does not list the following, but circles all the responses indicated.
   1. For E-mail
   2. Word processing
   3. Spreadsheet
   4. Database
   5. Graphics
   6. Internet
   7. Other, please specify_____

Q 3.5.a   **Which of the following sources do you use for information on horticultural therapy?**
   1. AHTA
   2. AHTA Chapter
   3. Talking to peers
   4. Land Grant University in your state
   5. Other university
   6. Other, please specify_____

**Q 3.6.a   What channel for information do you use most often?**
Surveyor circles all responses indicated.
1. E-mail
2. List-serve
3. Websites
4. CD-ROM
5. Books, magazines, newsletters
6. Videotapes
7. Television/radio
8. Classes, seminars, meetings
9. Other, please specify_____

**Q 3.7.a   Is the use of a computer in your work:**
1. Not necessary
2. Helpful
3. Required

**Q 3.8.a   About how many hours a week do you use the computer?**
_____hours
Questions for E-mail and Internet Use:

**Q 3.9.a   How frequently do you access the AHTA homepage? _____**

**Q 3.10.a   What are your favorite websites for horticultural therapy information and why do you like them?**
_____     _____

**Q 3.11.a   How often do you use E-mail? _____**
If the answer is NO that you do not use a computer at all, go here:

**Q 3.4.b   What are the reasons why you do not use a computer?**
Surveyor circles the responses indicated:
1. No computer
2. Computer cannot operate Internet
3. Have no training on how to operate the computer or to use Internet
4. Unfamiliar with Internet
5. Cost of Internet access
6. Other, please specify_____

**Q 3.5.b   Which of the following sources do you use for information on horticultural therapy?**
1. AHTA
2. AHTA Chapter

3. Talking to peers
4. Land Grant University in your state
5. Other university
6. Other, please specify_____

**Q 3.6.b   What channel for information do you use most often?**
Surveyor circles all responses indicated.
1. E-mail
2. List-serve
3. Websites
4. CD-ROM
5. Books, magazines, newsletters
6. Videotapes
7. Television/radio
8. Classes, seminars, meetings
9. Other, please specify_____

**Q 3.7.b   Is the use of a computer in your work:**
1. Not necessary
2. Would be helpful

**Q 3.8.b   How soon do you expect to start using a computer?**
1. Within 6 months
2. Sometime in the future
3. Never

## Part 4:

**The next questions are being asked so your responses can be compared with other members of AHTA:**

**Q 4.1   What is your current job title?**
1. _____
2. No response

**Q 4.2   About how many hours a week do you work in this position?**
1. _____hours
2. No response

**Q 4.3   How many years have you been or were you a horticultural therapist or used horticulture as therapy?**
1. _____years
2. No response

Q 4.4   **Are you registered with AHTA, and if so at what level?**
1. HTT
2. HTR
3. HTM
4. Not registered
5. No response or does not know

## *Part 5:*

**These last questions are being asked so your responses can be compared with other research on Internet use:**

Q 5.1   **Your gender:**
1. Male
2. Female
3. No response

Q 5.2   **In what year were you born?**
1. _____
2. No response

Q 5.3   **What is the highest level of education you have completed?**
1. _____
2. No response

**That is the final question. Thank you for agreeing to participate in this research project. If you would like further information, please contact Lea Minton or Dr. Richard Mattson at Kansas State University. The phone number is 785-532-6170. For information regarding the rights of human subjects, please contact Clive Fullagar, Chair, Research Involving Human Subjects, 1 Fairchild Hall, (785)532-3224. Thank you again for your time to answer these questions.**

# II

# Design for Human Health and Well-Being

# 5

# Linking People with Nature by Universal Design

Fusayo Asano Miyake

Universal design of outdoor spaces was introduced in Japan with the publication of *Creating People-Friendly Parks* in 1996 (Miyake et al. 1996). Just 4 years later, the concept of universal design has spread to various disciplines such as product development, transportation, architecture, urban planning, and of course, parks. This has occurred jointly with the rising average age of Japanese people at large.

Universal design is based on the principle of designing for people of all abilities (Miyake et al. 1996). The use of many products, for instance, chairs, teacups, or structures such as buildings, is more obvious than the use of outdoor spaces. Outdoor spaces, such as parks, rich in flowers and greenery offer a wide diversity of purposes for people to enjoy. Although some people may want to have a picnic, others may want a place to read a book, or perhaps to go bird watching. Can universal design be followed in designing outdoor spaces with such a wide diversity of needs? This paper will present examples of gardens designed using principles of universal design.

Consider the analogy of "the goose with the golden egg" of Aesop's Fables—the story of a man who cut open the belly of the goose because he wanted more golden eggs, ultimately killing the bird. Similarly, a park may lose its scenic value when trees are cut down to provide an accessible path for people using wheelchairs.

Sighted people certainly think they enjoy the *sight* of nature, be it through the other senses of smell, touch, hearing, and taste (Miyake and Miyake 1999). If any one of the five senses were to be disabled, the remaining senses would still enable you to enjoy flowers or greenery.

By creating spaces where flowers and greenery can be enjoyed by the five senses, people who have different types of disabilities and elderly people can enjoy nature with their remaining senses without being conscious of their own disabilities. Universal design of outdoor spaces is the act of creating each element to be beautiful and usable by all people.

## THE SENSORY GARDEN (FUREAI-NO-NIWA) IN OSAKA

The Sensory Garden in Osaka (Figure 5.1) was designed with two principles in mind. First, to design for all five senses. Second, to follow universal design principles. In the process of developing the design concept of the Sensory Garden, more than 300 peo-

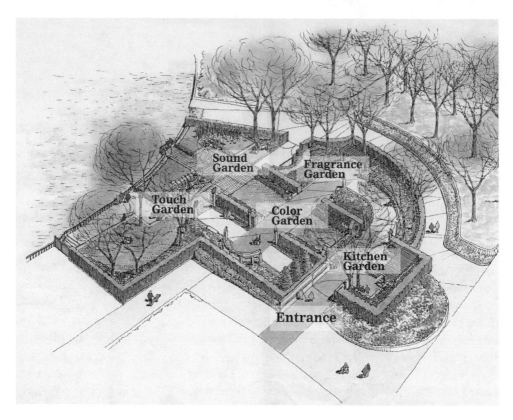

Figure 5.1.   The Sensory Garden (Fureai-no-niwa) in Osaka, Japan.

ple with disabilities, elderly people, and their attendants were interviewed, and asked
what kind of parks they wanted. Most of them replied they did not want special facilities, but rather a place where they could experience flowers, greenery, breeze, and
sunlight just like everybody else.

   The Sensory Garden is a half-acre space in a 200-acre park located in Osaka,
Japan. The word "Sensory" in Japanese is "Fureai." It means "coming in contact with
nature and warm hearts of other people," and was created with that as its goal. The
Sensory Garden was inspired by the main ideas in the story of "The Secret Garden"
by Frances Burnett. It is a story about a flower garden that heals everyone's troubled
soul. The inscription, located in front of the Sensory Garden, says, "Coming in contact with flowers, greenery, wind, light, and warm hearts of other people," and the garden gate is designed to reflect the inner atmosphere even when it is closed. It is a tactile gate so people with visual impairments can touch and feel the garden atmosphere.

   The Sensory Garden is divided into five different zones according to the five senses. The sense of taste is evoked in the Kitchen Garden where edible flowers are planted and cared for without the use of chemicals. The Color Garden shows a beautiful
harmony of colors of flowers. The Fragrance Garden features herbs. Raised flower

beds at different heights are provided so people who are not able to bend can still enjoy the fragrance of flowers.

In the Touch Garden, people can touch water, or lie down on the lawn. The pond is raised so wheelchair users can touch the water. Water lilies planted in the raised pond provide an opportunity for wheelchair users to touch aquatic plants. The raised pond is also a favorite of children. A bench placed in a nook created in the raised pond provides a private place. The raised lawn allows wheelchair users to transfer from the wheelchair to the lawn easily.

The Sound Garden includes a fountain with a Suikinkutsu. Suikinkutsu is a traditional Japanese echoing system. It is like "a water harp" in which water drops make delicate echoes in a jar buried underground. Suikinkutsu was often used in the Japanese garden for the tearoom many years ago. Because most Japanese gardens are not accessible, most wheelchair users or people with visual impairments have never heard the sound of a Suikinkutsu, even if they know about it.

In the Sensory Garden, universal design can be seen everywhere, although it is not easy to recognize. The wall in the entrance zone has a handrail to guide people with visual impairments. Twelve pieces of ceramic relief tiles are installed in the entrance wall. Each of the relief tiles is shaped in the form of a plant that can be found in the Sensory Garden, and the back face of the handrail under the relief tiles has each plant name in Braille. At the end of the entrance wall, there is a tactile information display with a push-button audio system for people with visual impairments who cannot read Braille. The 1-minute audiotape explains the outline of the Sensory Garden. The tactile information display is made of a single panel with an embossed texture. The display has letters, Braille, and a relief map for people with and without visual impairments.

During planning, interviews were conducted to find out the needs of people with visual impairments for parks. From the interviews, we learned that people with visual impairments feel the atmosphere of where they are by listening to a shout of joy or the noise of people around them. The hedge picture window is next to the information display. The Color Garden can be seen through the hedge picture window. Many people give a shout of joy when they see the view through the hedge window of all the beautiful colors. A shout of joy from sighted people becomes of great help in appreciating the scenic beauty for people with visual impairments. The hedge picture window will also make people with visual impairments stop because they feel a gentle breeze from the window.

A pair of pillars is provided where zones change. Each pair of pillars is crowned with bright-colored ornaments shaped like a cloud, wind, the sun, or the moon. The middle portion of the pillars has a different texture for people with visual impairments to use for spatial orientation. There are also double rows of stainless steel railings installed along the path for directional guidance.

All of the benches in the Sensory Garden are set back from the path so that they do not hinder the circulation of visitors. Planting around the bench is carefully considered to provide people sitting on the benches with sensory experiences of scent and feel, the smell of herbs, or flowers with a soft texture. The level of the bench seats and the shape of the armrests are slightly different for each bench. The difference in design will enable different users to select the one best suited for their needs.

The Sensory Garden has a number of design considerations that are not obvious to the visitor but are enjoyed and used by more than the elderly and people with disabilities.

## HEALING LANDSCAPE

My observation is that universal design naturally links people with plants, and that it allows for designs that are people friendly. Additionally, design elements used so people of all abilities can enjoy the outdoor space are unnoticeable.

Figure 5.2 is a concept drawing that is based on my theory of how people relate with a healing landscape. As Matsuo (1998) explained, people encounter nature in two ways. One way of feeling nature is by the five senses. The Sensory Garden explained earlier exemplifies this way of encountering nature. In this setting, people gain healing effects by sensing nature. The other way people encounter nature is through activities, such as gardening. This includes horticultural therapy, community gardens, and school gardens. In this case, people gain healing effects through the actual activities.

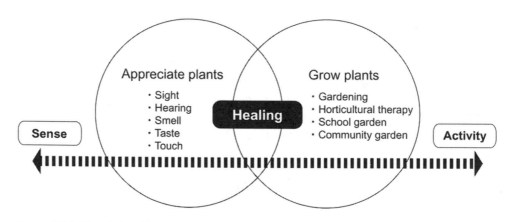

Figure 5.2.   The relationship between people and healing landscapes.

Human beings experience holistic healing of their souls by relating to nature, sometimes through activities while they are conscious, and sometimes by sensing nature while they are unconscious. This sort of environment may be referred to as a "healing landscape" (Miyake and Miyake 1999) (Figure 5.3). I therefore believe that healing landscapes should not just exist in medical facilities or institutions for people with disabilities, but should also be an important consideration for public spaces as well. The next section presents an example of this theory.

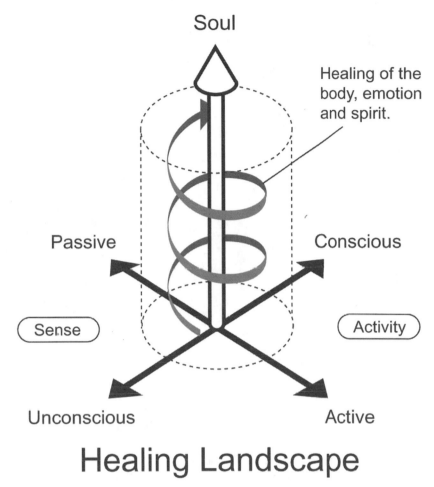

Figure 5.3. Holistic healing by relating to nature.

## AWAJI LANDSCAPE PLANNING AND HORTICULTURE ACADEMY IN HYOGO

The Awaji Landscape Planning and Horticulture Academy (ALPHA) is an example of two gardens adjoining each other that facilitate people's encounters with nature through their five senses and through programmed activities (Figure 5.4). One garden was designed as a sensory garden and the other garden was designed for gardening activities. These two gardens are located at ALPHA in Hyogo prefecture where the great earthquake happened in early 1995. The academy was just beside the epicenter of the earthquake, with damage so severe that a number of contributions were received to restore it, not only from Japan but also from all over the world. To express the gratitude for the generous support, the Sensory Garden and Learning Garden were created in the hope that the academy would convey useful information about the interaction between people and plants.

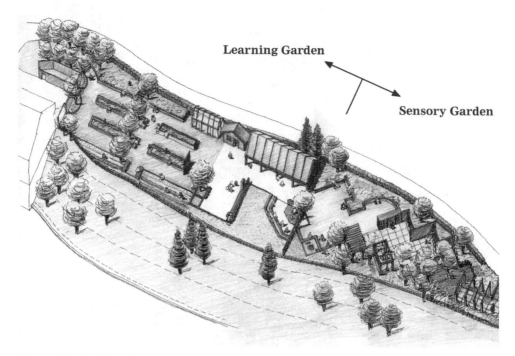

Figure 5.4.   Awaji Landscape Planning and Horticulture Academy (ALPHA) in Hyogo, Japan.

The academy's Sensory Garden provides sensory experiences of fragrance, color, touch, and sound. The garden for activities is called the Learning Garden. Here the experience of taste is provided. In addition, the Learning Garden is where academy students can do research on the relationship between people and plants.

These gardens contain many features of universal design. A tactile map is at the entrance of the Sensory Garden. Wooden deck boards were used for a portion of the walking surface so people with visual impairments are able to know where they are by the sound of footsteps. The flowerbeds include flowers and plants that move (and rustle) in the wind to provide more than a visual experience. A small pond with a tiny fountain provides a gentle sound.

An ordinary tree gate does not allow a wheelchair user access to the tree, so the wooden deck boards were installed as close as possible to the trees, and the tree gates were planted with thyme. Walking or wheeling on it will fill the air with the fragrance of thyme. A trellis was used to display vines.

The Learning Garden is accessible from the Sensory Garden and another entrance in the back, which is provided for lift buses from nursing homes or other institutions. There is an outdoor classroom with stairs for students to sit on. The floor of this classroom is level and accessible by wheelchair users. A workplace is protected from rain and sunlight with a roof and trellises, and individual worktables are provided along the trellises. The height of the worktables is adjustable to the sitting height of clients. Tables in the center are provided for group activities. There is also a barbecue grill for use during the harvest festival.

Some of the raised beds are also designed as benches where elderly people can rest during their work. Trees planted in the raised beds provide shade for the benches. Sinks are provided to wash the dirt from gardening tools. Some of the sinks are accessible by wheelchair users from either side. In the sink, a net is provided to enable wheelchair users to wash scoops or other things without assistance. Elderly clients may be using a cane or a wheelchair, so each raised bed is designed with different heights, ranging from 1.5 to 2.7 feet. The bars on the side of the flower beds are to protect clients' knees from direct contact with concrete. A space is provided under the flower bed for toes or the footrest of a wheelchair. A greenhouse and a cabin are also provided. The cabin is air-conditioned for clients who have difficulty with temperature control, and the cabin has a kitchen to prepare for a tea break.

## "HEALING GARDENERS"

People relate to plants in a variety of ways. The sense of "hospitality: friendly behavior to others" is essential to link people with plants (Miyake and Miyake 1999). Universal design acts as a physical design of hospitality, and providing programs acts as a nonphysical design of hospitality (Figure 5.5). This section explains park volunteers and horticultural therapy assistants who help in providing programs to facilitate the nature experience.

To facilitate the nature experience in parks, a training program for park volunteers was initiated. The park volunteers invite elderly people to come to the park, plant flowers in flower beds, help people who use wheelchairs, and lead people with visu-

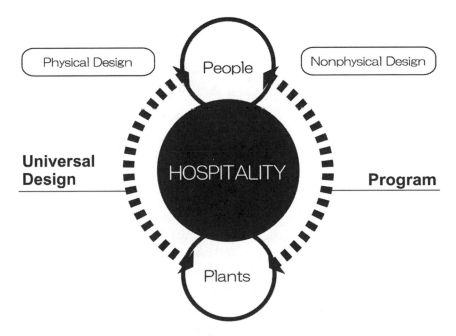

Figure 5.5. Hospitality in universal design and programming.

al impairments through the parks. In the past 5 years, 100 volunteers have been trained to be "Healing Gardeners."

Healing Gardeners are "volunteers in parks with a natural environment designed to heal." The training program for park volunteers includes a 6-month class and 1 year of practical training in the field. Many of the volunteers have come to think that they "are actually healed themselves by healing others." The volunteers get a great deal of fulfillment from their work. One person says, "My mother lives far away, so I am happy to be of service to other elderly people," and another says, "I am continuously moved each time." The best thing is that close to 65% of the volunteers have stayed with the program for 4 years or longer.

In the first class session, the trainees are divided into small groups and have a group discussion on the role of Healing Gardeners during a horticultural therapy workshop. Then they discuss their thoughts in the whole group. They experience how important it is to have good communication with each other. Another training component is shaking hands to understand the feelings of others. Role-playing is also part of the training. For example, trainees describe a scene to each other to practice guiding a person with visual impairments. The trainees describe the scene as if they are painting a picture using as many colors as possible.

In the session on nature observation, they feel the texture and smell the fragrance of leaves while closing their eyes, and learn that their senses of smell and touch are more sensitive when their eyes are closed. They also learn how to give instructions on planting in a flower bed to people with disabilities. In the training, people with different abilities are invited, and the trainees learn how to assist them.

The Healing Gardeners wear a yellow jacket with a Healing Gardener logo on the back of the jacket to identify them as a park volunteer.

## HORTICULTURAL THERAPY TRAINING AT ALPHA

Two gardens at the ALPHA are used for training horticultural therapy assistants.

They invite people from nursing homes and other institutions to assist in the training by being clients. The instructor and students become familiar with the clients' physical conditions and activity plans before they arrive. A nature "feeling program" is started in the Sensory Garden. For example, the clients and students may sit by grass that is waving in the wind, and the students explain how to visualize the sound of wind to the clients.

At the Learning Garden, the clients can enjoy planting in assorted flowerpots at the height-adjustable worktables. The flowerpots are then exhibited on the low shelves near the entrance. After work, they have a tea break with homegrown herbal tea. Students have a wrap-up meeting after the clients have left to discuss if the horticulture activity plan was appropriate, if safety was secured during the work, if the activity proceeded as planned, and how to make a better plan for next time.

## CONCLUSION

Japan is changing into an aged society more than any other country has ever experienced. In another 15 years, 25% of the people in Japan will be classified as elderly. It is furthermore estimated that the number of people 75 and older living alone will triple (Anonymous 2000). Disabilities or depression as a result of advanced age is likely to gradually reduce the opportunity for elderly people to get out of the house. If this situation becomes serious, spaces designed to be enjoyed by the elderly will not be used even if they are created with universal design.

However, a society with a high percentage of elderly people should not mean that everything is negative. If people can develop warm hearts to welcome elderly people and people with disabilities into society, Japan will be a great country where they will be happy to live. Universal design and programs that link people with nature will play an important role in our society in the future.

## LITERATURE CITED

Anonymous. 2000. White Paper on Aged Society. Management and Coordination Agency of Japan. Tokyo: Printing Bureau of Ministry of Finance.

Matsuo, Eisuke. 1998. *Exploring horticultural therapy: To seek healing and humanity.* Nagoya, Japan: Green Information.

Miyake, Fusayo A., Hajime Kameyama, and Yoshisuke Miyake. 1996. *Creating people-friendly parks: From barrier-free to universal design.* Tokyo: Kajima Institute Publishing.

Miyake, Fusayo A., and Yoshisuke Miyake. 1999. *Creating verdant and peaceful parks: Healing landscape and hospitality.* Tokyo: Kajima Institute Publishing.

# 6

# From Vision to Reality: The Chicago Botanic Garden's Buehler Enabling Garden

Gene Rothert

## INTRODUCTION

The aim of this paper is to emphasize two principles that are critical for creating a garden intended to be a safe, comfortable, passive or active gardening experience for the widest audience possible. First, when possible and appropriate, adopt universal design principles as the guiding philosophy for the design process. By following this principle, building features into the Buehler Enabling Garden at the Chicago Botanic Garden (CBG) that eliminated barriers for one group while creating barriers for others was avoided. Second, use a planning process that is as comprehensive as possible, gathering input from the broadest variety of sources that your schedule and budget allow. This principle helped build an institutional and community profile for the Buehler Enabling Garden. It also created an incredible amount of intellectual resources, which ultimately resulted in a better installation.

This paper provides an overview of the planning process and tools used to develop the Buehler Enabling Garden. Creating a garden for the public may be different from a garden designed for an agency serving a narrow population group, such as a school for the blind. The Buehler Enabling Garden needed to be capable of being visited and experienced by nearly half a million people a year from all walks of life, of differing abilities, non-English speaking people, etc. Even with the unique aspects of designing for the public rather than a specific population, essentially the same planning process can be used.

An increasing number of gardens are being created at health and human service sites of care, as observed by this author. These gardens are used for active therapy, education, training, and recreation as well as passive healing and restorative experiences. The therapeutic benefits of plants, gardening, and nature-related experiences are becoming more clearly understood as evidenced by the increased body of knowledge in horticultural therapy and design, as well as research symposia such as this one. In some settings, passive or therapeutic landscapes are being designed to create a restorative place used by patients and visiting family members, as well as staff. Active indoor and outdoor gardening experiences are used in clinical settings to advance therapeutic goals, improve vocational/educational skills, or simply to offer a safe, comfortable gardening opportunity.

This increase in the number of gardens at healthcare facilities is inspiring, yet by personal observation, too many fail due to poor planning. Designers are generally unaware of programming opportunities and do not talk to the program staff enough. Program staff members are not sure of the design possibilities so are ill equipped to talk to the designers. Neither talks to the end users enough. Typically, the garden fails for one or more of the following reasons:

- It is too much work for the agency staff and the program participants to manage.
- It generates an insufficient activity base to engage the intended number of users frequently enough to achieve the desired therapeutic outcomes.
- It is not fully accessible to the intended users.
- The therapeutic programs are not compatible with the garden's design.
- It requires a maintenance regime that is beyond the horticultural skills of those responsible for its upkeep; therefore, the plants rapidly decline.
- It is incongruent with the agency mission and programs.
- It lacks an institutional profile and support.
- It was never supported due to a lack of marketing and/or a fund-raising campaign.

The horticultural therapy staff at the CBG receives at least 20 calls per year from agencies asking for help to fix a failed, therapeutically programmed garden, greenhouse, or nursery project. In addition, at least twice that many requests for funding suggestions for their well-intended idea for a garden are received. However, most, if not all, have no plan available to show a potential funder. Having a well-thought-out plan to present to prospective donors makes funding much more likely.

The planning process for the Buehler Enabling Garden was dynamic. After a clear vision for the project is determined along with an action plan and budget, individual project components such as the hardscape (i.e., walks, walls, fences, etc.), plant selection and design, programs, marketing and public relations, opening events, and postoccupancy evaluations should evolve together. Resources of time and money may limit or expand the resources allocated to a particular step or component; for example, resources will determine how many focus group sessions you may or may not have time to organize and conduct that react to your concept plans. A more successful plan outcome is assured when evaluative measures are built into the planning process. Using evaluative measures throughout the process will help avoid potentially costly mistakes and help meet the expectations of all stakeholders in the project. In planning the Buehler Enabling Garden, focus groups representing a cross-section of people with disabilities, older adults, healthcare and human service professionals, designers, and horticultural therapists provided invaluable feedback from the concept stages through final plans. Finally, if possible, strive for some flexibility in your plans so that if, subsequent to the installation, uses change or new materials, techniques, and technologies emerge, you can incorporate them into the garden at some point in the future.

## THE BUEHLER ENABLING GARDEN

The Buehler Enabling Garden was opened on July 16, 1999. The Enabling Garden encompasses approximately 11,000 square feet. It is strategically positioned within the CBG to maximize the number of visitors who experience it (approximately 400,000 annually). Its location was carefully selected for high visibility and accessibility to audiences and visitors with mobility impairments. It is located within the main cluster of display gardens along a major pathway that forms an east/west central axis through the CBG. It is also adjacent to the Sensory Garden that displays plants in a landscape setting that have unique abilities to reach our senses, an example of passive therapeutic landscaping. The assets of the Sensory Garden are used in programs emanating from the Enabling Garden.

The mission of the Enabling Garden is to be a preeminent exhibit of enabling garden design, tools, techniques, and plant materials and to encourage anyone of any ability to enjoy a lifetime of safe, comfortable gardening. This garden is designed and programmed to present three key messages. First, anyone can access and enjoy safe, comfortable gardening. Second, participation in horticulture, gardening, and nature is therapeutic in many ways. And third, the design and techniques demonstrated benefit everyone who gardens.

These messages are communicated through a variety of methods. Access for visitors to get to, from, and around the garden is made as easy as possible. Paving, paths, and walkways have minimal grades; are firm, well drained, with good traction; and are large enough to provide complete freedom of movement. The garden has several water features to stimulate the senses and serve as auditory orientation points for people who are blind or have low vision. Barrier-free garden design elements are demonstrated throughout the garden through a collection of containers, raised beds, and vertical gardening techniques that position the soil and plants at comfortable working heights.

Special features that may be important to installations at healthcare and human service agencies include nearby parking and a covered teaching pavilion with a universally designed kitchen that includes powered cabinets that rise up and down as needed. This area is also equipped with adjustable height tables that can be used with lower functioning groups or when the activity (like making a salad from the garden) is best performed at tables.

The garden has two mobile exhibit carts designed to mirror the vegetable carts that are used to serve urban neighborhoods. Volunteers staff the carts and demonstrate specially designed gardening tools and unique sensory qualities of plants. The garden also includes universally designed restrooms and a small resource center where visitors can access our printed information as well as purchase enabling garden tools.

## PLANNING AND DESIGN

The guiding philosophy throughout the planning process was the universal design principles. The seven principles of universal design encourage access to places, products, and communications systems by everyone regardless of functional abilities (http://www.design.ncsu.edu/cud/). Using universal design principles to guide project planning will help avoid planning for any one particular user group at the expense of another. For example, a garden designed exclusively for ambulatory individuals may be too challenging for those who prefer to garden while seated; and signage that relies on high color contrast, symbols, and illustrations is more universally understood than signage using English text alone.

The first step when planning the Enabling Garden was to assemble the design team or project steering committee. An interdepartmental team of CBG staff included staff members from virtually all departments: maintenance, security, visitor programs, development, marketing, administration, horticulture, education, etc.

As specialized tasks emerged from the project steering committee, subteams were created for specific components like plant displays, interpretation, media campaign, opening events, etc. Many of the subteams included additional staff members beyond those from the project steering committee. This enriched discussion, ideas, and the final product because of the range of expertise that was present. It also helped build support for the Enabling Garden and foster greater awareness among CBG staff for the project itself and the issues related to accessibility of museums and their programs.

To maintain consistency of architectural and design style throughout the CBG and because of their long history of managing our master site and individual display garden planning, the firm of Marshall, Tyler, and Rausch of Pittsburgh, Pennsylvania, was engaged to articulate the design. The general contractor was a construction management firm that executes all new CBG projects. Someone from this firm served on the steering committee.

From the beginning, focus groups of target audiences and user groups that included a cross section of people with disabilities, older adults, designers, medical and social work professionals, the public, and volunteers participated. As areas of expertise or special projects beyond our scope emerged, consultants were engaged. In addition, a great deal of product research was conducted, allowing the latest innovations to be incorporated into the installation.

The first task of the project steering committee was to collect background material for the project. Strategic and master site planning implications for the Enabling Garden were reviewed along with pertinent donor understandings of the project budget and their expectations for the garden, as well as their own ideas. The vision for the Enabling Garden was then drafted to assure its purpose was congruent with the institutional mission and the strategic and master site plans.

Primary and secondary audiences were identified. Initially, the primary target audience was people with disabilities and older adults with sensory or mobility impairments. Although these groups certainly are among the primary audiences, the primary audience was defined as a cross section of the public—the majority of whom would not experience disability. Healthcare and human service professionals also factored into the primary audience list because this garden was intended to be a model for incorporating gardening-based programming at agencies serving people with disabilities and older adults. Secondary audiences included the architecture and landscape architecture professionals, particularly those designing sites of care, retirement communities, and other human service agencies. The secondary audience also included the CBG membership, donor community, and local elected officials.

## PROJECT ASPIRATIONS

A list of architectural, interpretive, and program aspirations that would have design implications was developed after the mission and audiences were defined. Specific architectural features to fulfill the mission and serve the target audiences were identified. These features included nearby parking, sheltered teaching space, restrooms, containers and raised beds at various accessible heights, vertical gardening methods, irrigation systems, hanging baskets on pulleys, tool collections, and a resource center. The interpretive list included three-dimensional tactile maps, signage, exhibits, audio descript tour, volunteers, and large print visitor map and brochure as well as other printed materials made available in alternative formats (large print, Braille, audiotapes). Collections of accessible garden design features, enabling tools, and plants were to be developed.

A slate of free and fee-based program plans was conceptualized based on input from the target audiences. A range of design and program development tours, internships, and study tours for professional audiences was outlined. A complete range of general visitor tours and demonstrations along with therapeutic sessions targeting groups of individuals from healthcare and human service agencies was also developed. Operating budgets were determined for the Enabling Garden that included staffing, program expenses, utilities, hardscape maintenance, and plant replacements (all opportunities for potential endowment).

Articulation of the final plans and construction drawings (bid documents) followed a standard process. Additional drawings that positioned the garden in context with its surroundings were produced as well as artists' renderings of key components such as the resource center, teaching space, and other unique features. A three-dimensional model of the proposed garden plan was produced, which, along with the artists' renderings, were invaluable tools when seeking feedback from our focus groups and others. The three-dimensional models were particularly useful as we gathered feedback from people who were blind.

## INTERPRETATION AND EXHIBITS

A successful interpretation program in a museum setting must clearly communicate the intended messages and educational concepts to the visitor. The museum visitor may be from anywhere in the world; may be deaf, blind, or use a wheelchair; may have a limited attention span; or many other unique and distinguishing characteristics—hence the need for a more universal approach. We learned through interaction with museum exhibit designers that the universal design of museum exhibits is very much an evolving science. Decisions on interpretation and exhibits for the Enabling Garden were based on universal design principles and advice from interpretive consultants to the museum community knowledgeable about accessible/universal exhibit design.

The interpretation program for the Enabling Garden reaches the broadest audience possible by using a hierarchy of delivery/formats.

- Three-dimensional tactile maps were positioned at the two entrances/exits of the garden. They are constructed with powdered bronze and resin that result in a durable service that remains cool (enough) to touch in sun. These are useful orientation tools for people who are blind or have low vision, but also help the sighted visitor understand the basic layout of the garden.
- Signs are used throughout the garden to describe key features and concepts. They use simple terms, large lettering, symbols, and illustration to convey messages. A high contrast (light on dark) color was chosen for visibility.
- Plant labels use large lettering on a high contrast background (white letters on black background). This decision was a difficult one to make. It is unclear whether Braille or raised letters is more universally understood by the visitor who is blind or has low vision. Given that fewer than 10% of people who are blind are able to read Braille, it would seem that raised letters and embossed graphics should be adopted as the standard. Currently experiments with several plant labeling methods, including black lettering on a white background, raised letters, Braille, and a combination of all three, are being conducted. Based on our own observations and continued research by interpretive specialists, we hope to create and deploy a new standard throughout the garden.
- A visitor map/brochure has been published that relies on large print and illustrations to both orient and convey key concepts and messages to the visitor.
- An audio descript tour of the garden has been developed and tested in partnership with the Center on Accessibility at Indiana University, Chicago's Lighthouse for People who are Blind or Visually Impaired, and consultants on audio description to the theater industry.
- Display panels were constructed with large, high-contrast lettering; drawings and photographs described the principles of universal design and considerations for building a barrier-free garden.
- Two mobile, hands-on exhibit carts staffed by volunteers encourage visitors to try out enabling tools as well as experience the sensory qualities of plants.
- A small resource center called "The Potting Shed" is where printed materials and a small selection of enabling tools are available for sale.
- Specially trained volunteers help maintain the garden and help visitors access the information and resources. Many of the volunteers experience a disability and are older, so while performing routine horticultural maintenance, a powerful and unmistakable message that gardening can and should be a vital part of life is communicated.

- Finally, numerous resources on barrier-free gardening and horticultural therapy are produced. These are produced in alternative formats including large print, Braille, and audiotape. Plans to make all of the materials developed at CBG available on CD-ROM and to produce a closed-captioned video on barrier-free gardening targeting the mass-market audience are being considered. All resources are available on the CBG website (http://www.chicagobotanic.org).

When guided by universal design principles, particularly those pertaining to communications, it is crucial to present information in alternative formats if it is to reach a universal audience. By using a "hierarchy of delivery formats" up to and including, when possible, people who are specially trained to communicate with diverse audiences, you can be assured that your messages and information will be accessible and understood.

## VOLUNTEERS

The baseline assumptions as the Enabling Garden was planned included relying on volunteer support to execute programs as well as maintain the horticultural displays. The CBG is fortunate to have a department that manages the volunteer program—now well over 800 people. The volunteers in that department were members of the project steering committee.

Approximately 50 individuals volunteer in the Enabling Garden. They received the same baseline orientation and training regarding general CBG policies and procedures as all the CBG volunteers, as well as specific training about the Enabling Garden. The specific training was planned and conducted by the full-time horticultural therapist responsible for all operations and programs in the Enabling Garden. Training included information on basic disabling conditions, how these conditions impaired function, and communication tips. Extremely useful in this endeavor was the access resource guide published by the Boston Museum of Fine Arts called *"Be Yourself, Say Hello! Communicating with People Who Have Disabilities."*

Most areas of the country now have centers for independent living that serve as resources for people with disabilities. Many of these offer free or low-cost disability awareness materials and training. Additional content for the training program included accessible features of the Enabling Garden, identification, use of adaptive tools and techniques, and the identification and rationale for the plant materials found there. Volunteers with particular abilities and interests in assisting with therapeutic sessions with agency groups were provided additional training for those roles. Advanced training opportunities are provided during the off-season for seasoned veterans, and "basic training" is repeated annually for new volunteers.

## MARKETING, PUBLICITY, AND OPENING CELEBRATIONS

The marketing and public relations campaign for the Enabling Garden was part of the planning process, and implementation began approximately 1 year in advance of the opening. The CBG's External Affairs Department coordinated a media campaign designed to elevate the profile and position the programs of this unique new display garden. Established marketing strategies were employed to position the Enabling Garden with the local and national print and broadcast media. Our focus groups of people with disabilities and older adults provided insights into reaching this large but very fragmented audience that is difficult to reach. Some of the strategies used included the following:

- A press kit was sent to approximately 500 local, regional, and national print and broadcast media contacts. Many local and national horticultural mass-market and trade publications featured articles on the Enabling Garden. Many publications' websites and organizations serving people with disabilities and older adults featured the garden. More than 11 million media impressions were received from July 1 through October 31, 1999.
- A menu of sponsorship opportunities was compiled ranging from benches and underwriting publications to sponsorship for the entire season. This included a range of benefits to the sponsoring agency including positioning of their marketing materials in the resource center, acknowledgment on entrance signs, and free therapeutic sessions for groups from their agency. Parkside Senior Services, a regional developer of assisted-living communities, sponsored the opening season at the $25,000 level.
- Several VIP/media tours were scheduled for donors, key media representatives, and our political representatives. As a result, the Enabling Garden has garnered continued interest from individual donors and foundations that wish to support it and its activities.
- Approximately 1,000 regional healthcare/human service professionals representing agencies serving people with disabilities and older adults were invited to a special evening of cocktails and tours of the new garden. This event was primarily intended to position the garden and its programs as a resource for agencies looking for something new and different for their clientele.
- Banners announcing the garden's opening were hung throughout the CBG grounds and strategic locations in Chicago.
- The CBG celebrated the opening of the Buehler Enabling Garden on a very warm July 16 evening with cocktails and dinner served to approximately 300 guests.
- The following weekend, special programs were available in the garden for visitors including plant and seed giveaways, special tours, and demonstrations. More than 20,000 visitors experienced the Enabling Garden during the weekend.
- The following week, an opening celebration was available to all staff and volunteers of the CBG that included breakfast, special tours, and demonstrations.

Regardless of the size of your garden project, consider its uniqueness in your community as well as within your agency and develop an appropriate marketing campaign that positions your project and its donor opportunities. Unless your destination is known as being accessible, people with disabilities will generally not visit for fear that once they get there they will not be able to enjoy the experience. This may not be as important for a garden developed to serve patients and their families at a local hospital. Clearly, though, if your project requires support, potential donors will need to know about it if they are going to support it. A carefully planned marketing and media campaign that increases community profile of your garden is necessary.

## PLANT DISPLAYS

Critical to the success of any garden are the plant materials. As a model for home and agency, the goals for the plant collections in the Enabling Garden included

- Exhibiting year-round interest, i.e., interesting winter bark and stem character, persistent fruit, unique growth habit, evergreen foliage, fall color, etc.
- Possessing unique sensory characteristics including fragrance, flavor, texture (visual and tactile), movement, and sound, and where appropriate, bold color and texture contrasts in combinations that are more readily experienced by people with visual impairments.
- Attracting birds, butterflies, and other harmless "wildlife."
- Avoiding highly toxic plants—oral or dermal.
- Selecting plants that have one or more of these features:
  — Are readily available to the gardener—not too exotic
  — Are pest- and disease-tolerant
  — Have effective season-long display
  — Require little or no maintenance including dwarf cultivars of trees and shrubs that would be appropriate for positioning in areas of the garden that are not accessible to the gardener or in a garden designed for aging adults
- Advancing horticultural therapy activities. In other words, plants used for active gardening experiences should generate a sufficient activity base to keep the planned number of participants actively gardening. For example, plants that meet some of these criteria:
  — Require frequent grooming to remain healthy
  — Require frequent harvesting
  — Produce large quantities of fresh cut flowers and/or flowers that dry well
  — Are good for pressing
  — Generate activity in their care and harvest but can also be used in a myriad of craft and cooking activities later such as herbs
- Advancing the CBG's documented plant collections where appropriate. If we wanted a fragrant narcissus, for instance, the Collections Department might recommend a species or cultivar not currently found in the collection.

The final list of selected plants includes more than 1,200 varieties that have resulted in one of the more beautiful display gardens on the grounds.

## PROGRAMMING

Public display gardens and museum exhibits that are intended to be programmed (serve as a venue for face-to-face programs and special events) oftentimes are not planned for that eventuality until after construction. The education and program staff must react to the design because they frequently are not participants during the planning phases. Fortunately, program planning for the Enabling Garden evolved at the same time as the design planning.

In the Enabling Garden, offering a slate of free and fee-based programs of interest to our general visitor as well as scheduled groups from regional healthcare and human service agencies, retirement communities, and schools serving special education students was the goal. In addition, it was very important that the marketing plan communicate the program opportunities to that sector. Programming was very much audience-driven. Input from focus groups of health and human service professionals and people with disabilities was used. Content for therapeutic sessions emanated from the past 20 years adapting horticultural activities to achieve therapeutic outcomes for a wide range of individual/group abilities.

Public programs for the casual or drop-in visitor includes very informal interactions with staff and volunteers. Visitors can also access sensory plant and design tours that illustrate the unique aspects of the Enabling Garden. Volunteers also lead ongoing demonstrations of enabling tools and adaptive gardening techniques, which visitors are welcome to try for themselves. Fee-based design tours are available to architects, landscape architects, developers, and healthcare professionals by appointment.

Fee-based therapeutic sessions are available to groups from regional health and human service agencies. Depending on the weather and functional levels of the participants, groups engage in a variety of carefully planned activities such as maintaining the garden as needed that day, creating a project such as a fresh flower arrangement, using materials harvested from the garden to make a salad, or taking sensory plant tours that explore our senses in the garden.

The Enabling Garden was planned with secondary program uses in mind too. This is an attempt to fully use its extraordinary setting for a range of other special events. The garden often serves as a venue for evening cocktail parties and other events requiring a beautiful and intimate setting. The garden is wired for sound, has "mood" lighting throughout, has sheltered places to set up refreshment and hors d'oeuvres tables, and provides plenty of places to sit and enjoy.

## EVALUATION

Including evaluative measures in the planning process can help avoid costly mistakes. Artist's perspectives and models of the proposed plans were extremely useful when gathering feedback on the layout of the garden, particularly for people who may not understand blueprints and construction drawings. Essential feedback was received from focus groups of health and human service professionals, people with disabilities and older adults, and design professionals. Interpretive tools were pretested with general garden visitors. The audio descript tour was pretested at various states of development with people who are blind.

Planning for postoccupancy evaluation is important too. A postoccupancy evaluation of the Enabling Garden is planned. The evaluation will have four major components. One component is to evaluate the accessible design features while working with user groups of people with disabilities, older adults, and those who are blind or have low vision. The results will help us to refine design protocols for enabling gardens and begin to characterize design prescriptors for various user groups.

The other components will evaluate the interpretive aspects of the garden. Plant displays will be evaluated by a team of horticulturists and horticultural therapists to help develop a list of best plants possible for therapeutic landscapes and horticultural therapy programs. Finally, the program opportunities available in the garden will be assessed to determine if the audiences served are as broad as possible and if the Enabling Garden is being fully used.

## CONCLUSION

The intent in building the Buehler Enabling Garden was to create an excellent model of barrier-free garden design, tools, and techniques as well as plants that successfully engage anyone of any ability in active, lifelong gardening experiences. In addition, the intent was to send clear messages that gardening should be an important part of everyone's life. The planning process was grounded in universal design principles, relied upon by input from diverse resources both within and external to the institution. This input clearly resulted in an improved installation. The Chicago Botanic Garden's Buehler Enabling Garden will hopefully inspire untold thousands of people to continue to garden with age or disability. With any luck, it will inspire many others to pick up a trowel for the first time. This author hopes that it will help inspire change in the character of health care and human services in this country and around the world. Only time will tell.

# 7

# A Holistic Approach to Creating a Therapeutic Garden

**Virginia Burt**

## INTRODUCTION

Therapeutic gardens and the level of care taken on the exterior of any facility are reflective of the level of care given inside and the overall health of the organization. Oftentimes therapeutic gardens created by an individual representing the organization, rather than "stakeholders," who might include representatives of the facility such as board members, staff, patients, and contributors to the facility—both financial and volunteers—are often diminished by low-quality, one-dimensional information gathered for the effort.

Visionscapes Landscape Architects takes a holistic approach in their design work, encompassing physical, mental, emotional, and spiritual aspects. Attention is given to the physical "story" of the land: its topography, geology, history, biology of flora and fauna, and its physical character and how they are affected by (or have affected) human use and movement. These are the technical and tangible aspects of Earth in all of her outward manifestations.

In addressing the emotional aspects of the holistic approach, consider the use of color, contrasts of shadow and light, textures, scent, and people's relationship to space. In addressing the mental aspects, apply sacred geometry (or golden mean, the perfect division of proportion found in nature), use recognizable form(s) that are not abstract and elements that stimulate memory, and attend to the details of completing the project in a timely and creative manner. Implementation is considered during the design process with an awareness of possible costs. Consider how a project could be implemented in several phases or in a single project as related to funds available or fundraising to be done.

The spiritual aspect of the holistic approach is recognized in the belief that we all have a cross-cultural connection to one source. For example, honoring four cardinal directions (north, south, east, and west) is universal to most cultures and often represented by a medicine wheel. A connection to one source teaches us that life is a process of change and attends to archetypes that awaken and invoke our inner knowing and wisdom of relationship to Spirit. This is used in the design process through intention, ritual, and working with Spirit.

As part of developing a holistic approach to designing a therapeutic garden, it should be recognized that stakeholders involved in a therapeutic garden have an inherent wisdom about the elements that are critical to a successful project. Therefore, ensure that stakeholders are involved in a meaningful way during the design process and creation of their therapeutic garden.

Get the potential stakeholders involved early on to ensure "buy in" as the project is evolving rather than having to "sell" the project after the design is finished. Traditional design and planning results are rarely implemented when they have been "sold." As many members of the community who are interested should join in the process of creating the garden design or master plan. Developing a plan with as many members of the community as are interested will capture the wisdom of the many who love the community and who have an investment in the outcomes of a therapeutic garden.

## SMALL GROUP PROCESS FACILITATION

Small group process facilitation is a method of meeting and information gathering that engages the whole person and taps into their deep inner wisdom (Dalar and Associates 1999). This process can be used to guide the holistic approach to the design process.

A meeting that uses process facilitation has several distinct features. In this method, chairs are organized in a circle. The premise of using a circle is to create a nonhierarchical climate for encouraging communication. Circles are a natural form of encouraging communication and enhancing learning, and are seen in many native traditions. Tables are kept outside the healing circle, as they are barriers to communication.

As part of the circle, the realization by the participants of the interconnectedness of all things helps to bring harmony and balance to the group. The relationships encouraged by this process include the relationship to self and relationship of self and nature; the relationship to another person, and the relationship of the two people to nature; the relationship of a collective, and the relationship of the collective to nature.

In native traditions, work on relationship to self is often called Rainbow Hoop Work. Relationship with one other is often called Gratitude Hoop Work. Moreover, relationship to the collective, to the earth, to the Universe, to God is called Medicine Wheel Work (Dalar and Associates 1999).

The facilitation process was also designed to attend to all five senses through memory visualization while stimulating memory and imagination. Again, this emphasizes people's relationship to nature and their environment. This process is directed toward facilitating learning and uncovering the inner wisdom of the participants. As part of this understanding, the physical abilities of the participants are critical. To accommodate the needs of all attending, the process was designed so all could be comfortable and fully engaged. For example, those in wheelchairs were assisted in retrieving objects from the center of the circle and accommodated during the design of the process.

The facilitation process was designed to stimulate right-brain activity (the creative side of the brain) and intuition as well as uncover the participants' inherent knowledge of their dreams, needs, and desires of their environment. The remainder of this paper will explain the process and the learning theory behind this process and present an example.

## STIMULATION OF PARTICIPANTS' IMAGINATION

The first step to stimulating the imagination of the participants is a right-brain exercise or "transfer in." The use of this type of exercise stimulates right-brain activity (the creative side of the brain). When right-brain activity is stimulated, people are called into the present moment. Generally in western culture, left-brain activity dominates. For example, the decision-making process to brush our teeth in the morning is a left-brain activity. By stimulating our imagination with this exercise, the whole brain (and whole person) is called into the room. It also calls one's heart into the room—which is essential to the success of this process. This exercise encourages faster comprehension, allows for grasping concepts more quickly, and allows for full attention. The right-brain exercise is specifically chosen and is directly related to the purpose of coming together.

The following excerpt is from the *Small Group Process Facilitation Manual* (Dalar and Associates 1999):

Comments from the (learning) group about what they learned by taking the time to do a transfer in to the meeting.

- Throws you to pick up from the unconscious—you pick up that which you need to learn for yourself, without even asking that as a question. It heightens personal awareness about learning needs and learning intention.
- Gets away from "my position, my title."
- Caused me to reveal more about myself than I normally would and to bring my real self to the group very early in our time together.
- Helped me to reflect why I am here.
- Challenged me to take that which I recognize intuitively and try to express it. I found this hard and recognize that it is something that I need to practice so I can do it.
- Helped me be in touch with my intuition.
- Helped me recognize that as a facilitator one must recognize that everything one does is open to interpretation.
- Fascinating in the way this breaks through group tension in a more real way by bringing acknowledgement to ourselves as people.
- Introduction by another is a respectful way of acknowledging another or having myself acknowledged.

- When we introduce someone else we attend to it more carefully.
- As a listener we learn to listen.
- I was quickly jolted out of my pre-meeting anxiety.
- A way of understanding that we are unique.
- Gets us focused on who we are now and who we want to become.

These comments reflect the opportunity given to the group members to acknowledge their personal desires, needs, and fears related to the meeting.

## THEMES GATHERED THROUGH STIMULATION OF IMAGINATION OF PARTICIPANTS

Unfortunately, too often designs and plans are "put on the shelf" because of budget cuts or difficulty getting "buy-in" from decision makers. Small group process facilitation uncovers the passion of those who are in the room. This process and another methodology called Open Space Technology (Owen 1992) (from the area of expertise known as Organizational Development and Large Group Intervention) engages the whole person with the result of uncovering the individual and collective passion of the group while getting directly relevant information and input on important issues. This holds true for large projects such as the design and creation of master plans for retreats and sacred spaces involving hundreds of acres or for small projects like the creation of a therapeutic garden in a space one-half the size of a residential city lot.

Therapeutic gardens and the process of designing them must fully engage the people who are affected by the garden and are part of the program of administering the garden. This is especially true when the mission and values of those in the garden are directly related to the garden and the land. Developing a plan with the usual planning committee rarely captures the wisdom of the many who will use the garden, love the place/community, and have an investment in the outcome of the garden (mentally, emotionally, spiritually, or physically).

In one large-scale project of Visionscapes, it was learned prior to beginning the process that there was an individual who always sabotaged every initiative to renovate and update the landscape based on historical issues. During the storytelling evening that was arranged to launch this particular process, this person ended up in tears saying that for the first time, they had been asked to express their dreams, ideas, and thoughts before any initiative was taken. From that time forward, this person became the biggest supporter of the overall project.

In the first step, the group answered the question posed after reflection by themselves, then in groups of two people, and then returned to the group. In this second step, themes of both personal reflection and those related to the question, in this case, healing gardens, were separated. The data were then analyzed to determine the themes and assumptions about a successful healing garden. For example, an assumption was made that the current operating budget would address the maintenance of a new garden. Themes included attention to memories with plants and elements and creating physically comforting spaces.

This part of the process begins to reveal what truly makes a successful therapeutic healing garden rather than an assumption that everyone may know what a therapeutic healing garden is.

## STORYTELLING

For dreams and imagination for the future to occur, both past and present must be acknowledged. Understanding the client and community and its connection to the land is uncovered through the telling of stories. Learning and telling the story of the land and its community of people attend to our connectedness with nature and natural evolution.

At this point, participants are asked to reflect on a healing garden that they knew of and what that garden's story is. This taps into our own knowledge and causes reflection upon the lessons and depth of learning we have gained during our human experience. Again, this is done with one's self in silence and through reflection then discussed in a group of two then in a group of four, drawn on paper, and discussed. Finally, several of the groups would present the story to the overall group. In this instance of moving people through their story(ies), the method of small group process facilitation is illustrated as a dance of energy where the facilitator attends to each of the four Kolb's adult learning styles (Litzinger and Osif 1992).

## KOLB'S THEORY OF LEARNING STYLES

Learning styles are defined as "the different ways in which children and adults think and learn" (Litzinger and Osif 1992). It has been proven that each of us develops a preferred and consistent set of behaviors or approaches to learning. Kolb (as stated in Litzinger and Osif 1992) showed that learning styles could be seen on a continuum running from:

1. Concrete experience: being involved in a new experience (doer).
2. Reflective observation: watching others or developing observations about one's own experience (watcher).
3. Abstract conceptualization: creating theories to explain observations (thinker).
4. Active experimentation: using theories to solve problems or make decisions (feeler).

When we attend to each of these learning styles during a process to obtain input or to teach/involve people about/in any topic, the majority of participants thoroughly grasp the subject matter. For more information on learning styles and how they are incorporated into speaking and teaching, see http://www.cyg.net/~jblackmo/diglib/.

## SMALL GROUP PROCESS FACILITATION EXAMPLE

In 1999, Visionscapes was hired by a long-term care facility that was undergoing a revision of its outdoor garden spaces to address aspects that were deemed inconsistent with their quality objectives since its completion 5 years earlier. As a part of the ongoing commitment of the facility to provide quality long-term care, it was their wish to have the input of residents, their families, staff, and volunteers into the creation of the new garden at the front entry.

The current front entry was dedicated more to cars than people. It needed to be reconsidered as a space for the people who live in, work at, and visit the long-term care facility. Management was not interested in repeating the same autocratic process of implementation previously done. A 1-hour small group process facilitation meeting was conducted with residents, family members, staff, volunteers, and visitors to the facility to uncover the inner wisdom of those who wished to offer their input for future design work.

A circle of chairs was created with spaces for those in wheelchairs and with walkers. All of the user groups were present, as were those assisting the residents, as well as those with an interest in the garden area.

The process began with an exercise designed to "bring the whole brain" into the room. It transferred attendees into the work they had come together to do. In order to do a visioning process in the short time frame of 1 hour, with a group that had never met before in this workshop style, it was imperative that people focus on the garden as quickly as possible.

With the stories of both past and present told, the attendees were then ready to imagine their wishes and dreams of the kind of place they wanted the garden to be. The following are elements stated by participants:

- Nooks to sit and read
- Flower beds
- Old English garden with plants popping up everywhere
- Pond (fish, plants, water lilies)
- Masses of color
- Informal
- Areas of grass
- Display beds
- Dome of different trees and flowers/species from different countries
- Flowering shrubs and trees
- Vines (flowering)—morning glories
- Roses/violets and primroses
- Cool places
- Birdbath—fountain
- Water
- Butterflies and hummingbirds

- Honey (bees)
- Hooting owls at night
- Love of God is wonderful
- Cardinals and other birds
- Quiet and peaceful
- Changes of seasons
- Hollyhocks along rail fence
- Gentle music
- A garden to watch from her room
- Plant own seeds
- Vegetables/asparagus/tomatoes
- Perennial gardens—good idea (with blooms each month)
- Barn—with vines up side
- Plume poppies 6 feet tall
- Rockery—hens and chickens/sedum with rocks
- Dwarf iris
- Duck nests
- Fruit trees/shade trees
- Cedar hedge
- Boxwood hedge
- Euonymus hedge—green all winter
- Walnut trees, white and red pine trees
- Cherry
- May need to purchase new property to fit all the garden wishes
- Flagstone—quiet sitting spot with bench and red mother of thyme
- Lavender
- Magnolia
- Grasses
- Lemon balm/thyme
- Rose of Sharon (pink)—sheltered place
- Petunias—mauve with mixture of pink and white
- No weeping willows
- Herbs
- Flowering crab apple

The input was rich and highly applicable to the facility's commitment to a therapeutic healing garden by giving specific, tangible, and relevant input to the design. It has also contributed to several garden areas for the facility. The entry still accommodates cars, with future construction works planned, including a complete redesign and replacement of all paved surfaces to improve access and create accessible places under a porch roof for greeting others. Garden areas have already been renovated using the species and colors contributed during the process.

## CONCLUSION

In conclusion, consider the following when embarking on the journey of creating a therapeutic healing garden: attend to the story; attend to the design from the mental, emotional, physical, and spiritual aspects; attend to the process; and treat all people as human beings by accessing their whole selves—body, mind, and spirit.

## LITERATURE CITED

Dalar and Associates. 1999. *Small group process facilitation manual.* Dalar and Associates, Ancaster, Ontario.

Litzinger, M., and B. Osif. 1992. Accommodating diverse learning styles. In *What is good instruction now? Library Instruction for the '90s* (Library Orientation Series, No. 23), ed. Linda Shirato. Pierian Press, Ann Arbor, MI.

Owen, Harrison. 1992. *Open space technology, a user's guide.* Abbott Publishing, Potomac, MD.

# 8

# The Medical Center Gardens Project

**Royce K. Ragland**

## INTRODUCTION

One autumn day in 1997, a small group of enthusiasts gathered to explore possibilities for a new dimension of health care at Georgetown University. There were no professional planners, medical researchers, or market strategists present. Instead, the group, composed of teachers, a journalist, an artist, a fundraiser, horticulturists, and a seminarian, was bound simply by the compelling idea that this preeminent medical center in the nation's capital needed gardens to complement their healthcare spectrum. Their personal experiences united them in believing that gardens would contribute to the health and well-being of everyone at the medical center. There was no budget, no strategic plan, and no staff. What they did have was a clear vision of a new environment for wellness and the conviction that they could make this vision a reality. This paper presents the evolution of this effort to date.

## THE PROJECT

The Medical Center Gardens Project at Georgetown University, launched in 1998, was intended to be an integral part of the healthcare environment, with gardens of beauty and harmony. The gardens are intended to impact social and emotional well-being as part of a holistic approach to modern health care. They are also intended to enhance the quality of daily life for all who serve and are served by the medical center—patients and their families, employees, students, staff, and administrators. The gardens will educate, inspire, and restore those who use them. To advance these objectives, descriptive brochures and garden tours will promote awareness of the gardens, their purpose, and their features. Plant labels will list botanical names, common names, and medical significance. Plaques will list garden names and major donors.

In providing spaces for peacefulness, beauty, company, solitude, contemplation, and prayer, the gardens will afford a respite for anyone who seeks it. Benches and chairs thoughtfully placed in garden areas will offer a change of atmosphere for patients, families, and staff. Tables and chairs available for lunch, study, or coffee break will be located in areas frequented by medical students to provide a refreshing break from the stresses of the classroom. Classes in the medical school curriculum will explore the role of medicinal plants in complementary medicine, as well as the role of emotions in health and well-being.

Patient registration packets will include garden maps detailing information such as usage and access directions. Garden maps will also be available to campus visitors, prospective students and staff, and potential donors. Both maps and garden brochures will be included in university recruitment and promotional materials.

Through these efforts, it is hoped that the gardens will contribute to the health and well-being of everyone who uses the medical center. It is also intended that these efforts will highlight the contributions of nature to the broad realm of health care. All of these goals are in keeping with Georgetown University's Jesuit tradition of caring for the whole person.

The project is reflective of a broad range of literature documenting the positive therapeutic impact of gardens, nature, and landscapes. Ulrich and Parsons (1992) noted that since ancient times, nature and elements of nature have been believed to reduce stress. Ulrich (1979) found that views of nature and visual landscapes have positive psychological effects. Kaplan and Kaplan (1989) reported that the experience of nature has positive psychological effects. Work by Cooper Marcus and Barnes (1995) indicates that stress reduction may be experienced by people using gardens in healthcare facilities.

## THE PROCESS

The goal of creating gardens on the medical center campus was undertaken by a committee of 10 volunteers, all of whom have relationships to the medical center such as alumnus, employee, medical student, and volunteer worker. They and their families have also been treated at the medical center. Committee members share a conviction that gardens are good for people in a myriad of ways, and each has experienced the impact of such complements to modern health care. These individuals see themselves not only as catalysts for a new idea, but also as stakeholders in an important aspect of health care at Georgetown, which began as a series of informal discussions and emerged as a formal committee called the Medical Center Gardens Project. After receiving endorsement of their idea from top university and medical center executives, the group began exploring the possibilities for gardens, designs, and overall strategies.

They read and discussed a wide range of books, articles, and research items on gardens, landscapes, and healing and restorative environments and their impact on health and wellness. Of particular interest was *Healing Gardens: A Natural Haven for Emotional and Physical Well-Being* (Mintner 1993) based on the London Physic Gardens in London. Several committee members visited the site and interviewed the curator. Work by Cooper Marcus and Barnes (1999) served as a major reference. *The Meaning of Gardens* (Francis and Hester 1990) and *Contemplative Gardens* (Messervy 1990) were also helpful. For a medical perspective, tapes of the television production "Healing and the Mind" (Moyers 1993) were viewed, and an interview with a featured participant from the program, Candice Pert, MD, of Georgetown University Medical School was instructive. *Timeless Healing* (Benson 1997), a book on emotions and wellness, was also crucial in providing medical theories to support the committee's ideas, as was Ulrich's (1984) research on the restorative value of a view of nature from a hospital window.

The committee reviewed numerous items on similar projects existing at healthcare facilities in the United States and abroad. Cooper Marcus and Barnes' (1995) work on several healthcare facilities in California was particularly helpful. Also important were visits and interviews regarding gardens at Stanford University Medical Center and Lucille Packard Children's Hospital, both in Stanford, California.

Informal interviews of patients, students, employees, and faculty to solicit ideas, suggestions, opinions, and reactions were conducted, followed by informal observations throughout the campus. Concerns, reactions, and needs of those interviewed lent credibility to decisions made regarding design issues. This portion of the process facilitated the gathering of ideas, opinions, and support throughout the medical center.

Committee members visited all prospective garden spots at different times of day and night, taking notes and photographing views from various buildings, windows, doors, and walkways. They observed traffic patterns, temperatures, sounds, wind, sunlight, shade, long views, and airplane noise. This firsthand research bolstered the hopes and ambitions of everyone involved. The group became even more convinced of the value of the Medical Center Gardens Project.

Early in the project, the committee received advice and assistance from the medical center development office. However, other larger projects soon took precedence, and the committee members found themselves functioning independently, without guidance, assistance, funding, or supervision. Feeling a strong commitment to the project, they continued their work independently of the medical center administration.

At the onset of the project, it was agreed that all funds would be raised through gifts and donations because the medical center was experiencing financial difficulties.

To educate, inform, and build support throughout the medical center community, committee members held informational meetings with numerous staff and administrators. Administrators, faculty members, and students were invited to committee meetings, and presentations were made at board and trustee meetings and at employee workshops. In addition, several campus publications and medical center newsletters carried reports and updates on the project and its progress.

The project encountered a number of obstacles during this period. There was considerable turmoil and distraction in the medical center administration. Severe cost cutting in all parts of the university and the medical center were taking place. The medical center grounds were not considered a priority and had been neglected by the university for some time. The garden projects were not included in the university's *Master Plan.*

Although a major university-wide fund-raising campaign was under way, the Gardens Project was not one of the priorities. Because of several rapid staff changes, the committee sometimes had less than ideal working relationships with medical center liaisons. Several high-profile medical center faculty members were waging lawsuits against the university. Changes in top leadership occurred, and a medical center merger was in process. These problems were reflective of many institutions in the healthcare industry at this time.

In contrast, the project enjoyed some very positive elements. The University Landscape Manager, a national leader in campus landscaping and a horticulturalist by training, became an early supporter of the Gardens Project. He became an active member of the committee. He has been an advisor and advocate for each garden undertaken. The project could not have attained its high level of quality without his experience, expertise, and popularity within the university community.

Throughout the process, the Gardens Project has consistently received the support of the medical center executive staff. Several deans and faculty members are active advocates. Some of the most gratifying support comes from the medical school students and student council members, who have been strong participants since the project began. The parent council of the medical school has also been supportive.

Another area of support comes from the Jesuit leadership on the campus, which maintains close communication and participation in project activities and is a source of encouragement to committee members and donors alike.

The source of this support is the nature of the work itself—a noble purpose, spreading goodwill and creating a beautiful, healthful, rejuvenating place—a task mirroring what people like to do in their private lives.

Lastly, the composition of the committee itself is rich in talent, resourcefulness, generosity, enthusiasm, perseverance, and adventurousness. Without these attributes, the project would not have been successful.

## THE SIX GARDENS

After informal research and considerable reading and discussion, the committee concluded that there should be six separate garden areas, each serving a different purpose and a different constituency. The committee decided to create a landscape master plan for the entire medical center area that could be used as a long-term guide, to ensure that long-term efforts, as well as short-term or small projects, would be undertaken in a consistent fashion, even though the facilitators might change over time. A landscape architect was then retained for the collaborative effort of interpreting their ideas into a master plan for the entire medical center campus.

Of the six gardens in the master plan, three have been implemented. All six gardens are within the medical center campus, enclosed by the Georgetown University Hospital, the medical school, the Lombardi Cancer Center, the Life Sciences Building, the Basic Research Building, and the administration building, with the medical school's Dahlgren Library located in the center of the complex. Most of the garden's surface area is actually a rooftop and has classrooms, laboratories, or libraries beneath it. All gardens are based on the principles of universal design. Campus maps will be available to patients, visitors, and employees, indicating locations and access routes to all the gardens (Figure 8.1).

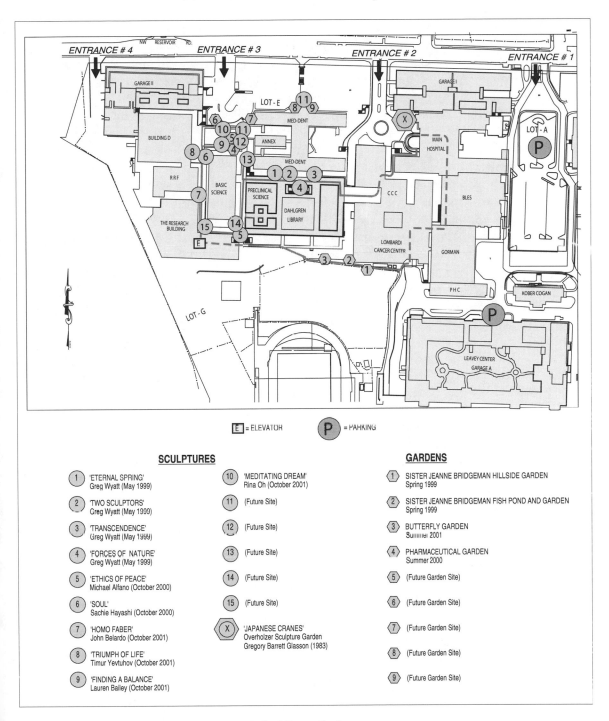

Figure 8.1.    Georgetown University—Medical Center Gardens.

The *Plaza Garden*, still in the planning stage, will be in front of the hospital's Concentrated Care Center (CCC) Building, currently a 1,800-square-foot empty and open brick rooftop plaza. The area is enclosed on three sides by five-story hospital buildings, and open on the fourth side to a long view westward across campus. This area can be very hot during the summer so provisions for shade must be included. It is a natural wind tunnel during most of the year, so there will be some protection from wind, as well as a possible wind/kinetic feature to help mask the constant airplane noise. Seating for socializing and private conversation will be included to serve patients, families, employees, and caregivers. Other uses will include space for lunch or a place to take a break from the day's activities. All plantings will be in containers. Views will be considered from inside the waiting room of the CCC Building, as well as down from the hospital rooms on three sides of the space.

The *Meditation Garden*, also in the planning stage, includes a structural bridge and a large raised bed, approximately 3,600 square feet, with a 15-inch depth of soil. The bridge is currently in need of structural repair. This area will contain four quadrants planted with perennials, including those with medical, cultural, and historical significance. Seating areas will include both social and private spaces, and a water/wind feature will be a central element. Patients, families, students, and faculty are expected to use this space.

The *Frisbee Park*, in the planning stage, is approximately 7,200 square feet. It is a raised bed with 15-inch depth of soil and is currently in need of structural repair. This will be a large, open, grassy space surrounded by low shrubs and canopy trees. The Frisbee Park will be a place where students can play a game, faculty can take a break, and employees can enjoy their lunch in pleasant environs.

The *Sister Jeanne Bridgeman Memorial Garden* borders one side of the medical campus. It consists of nearly an acre of space, unusual for an urban garden. This was the first completed garden, installed in the spring of 1999, and is in honor of Sister Jeanne, a beloved chaplain who served the Georgetown Medical Community for many years. The garden was created by a special fund-raising effort, directed by a committee member who had been formerly counseled by Sister Jeanne. The garden was funded by gifts from former patients, physicians and staff, and friends and family of Sister Jeanne.

The garden is on a large and steep hillside that had been a muddy and barren section of a fenced-in sports field. Many of the campus utilities are buried here, which called for special consideration of the plantings. Plants with shallow root systems were used so that repairs to utilities could be made easily and quickly. The plants were selected for year-round interest and adaptability to the hillside conditions. Two bird-houses are located here to capitalize on the natural feel of this location. The focal point of the garden is a fishpond and waterfall at the bottom of the hill. The garden is bisected by a paved walkway that has experienced heavy foot traffic. It is a popular spot for parents and children to visit, employees to take group photos, and physicians to conclude their grand rounds, coffee cup in hand. This area has been expanded to include a butterfly garden with shrubs, perennials, and a butterfly house. This will be developed as a seventh garden as more plantings and seating are added.

The *Sculpture Garden*, installed in 1999, is a collaborative effort with an education program for talented young sculptors, sponsored and funded by a private foundation. An important goal of this garden is to unite the concept of art and science. To date, the garden has 10 bronze sculptures, and will receive an additional five pieces over the next 2 years. The garden begins at the south entrance to the medical school in an area of approximately 1,200 square feet and has sculptures located throughout the medical center campus. Teak benches and potted trees are arranged near the sculptures, providing a restful and provocative place to relax, meditate, or study. The artists who create the bronze sculptures meet periodically with committee members, medical students, and faculty to discuss ideas and themes appropriate for the designs. The final pieces are presented to the institution during a formal dedication ceremony.

The *Pharmaceutical Garden* is the newest garden, installed in the summer of 2000 and located between the medical school and the administration building. This 900-square-foot garden contains 12 plants, currently used in modern medicines, selected by medical school students and faculty. All plants will be labeled with their botanical name, common name, and current medical use. This garden is intended to be educational as well as decorative. It was made possible by a gift from a local garden club. The garden is a source of pride to the medical school and a delight to all who pass by it. A second garden club has recently donated funds to expand this area.

## CONCLUSION

Much has been accomplished during the first 3 years of the Gardens Project. Three gardens have been implemented, and three more are in the planning stage. Several additional small gardens and indoor gardens are being considered; however, the project will continue for several years before all plans are completed.

The Medical Center Gardens Project has become highly visible and a source of pride to students, faculty, and administration. Despite the many changes in the medical center and the university, the project now enjoys popular support, acknowledgment, and encouragement from all parts of those organizations, and the merger partner now collaborates with the project committee on upcoming projects.

Over time, the committee of volunteers has remained consistent in membership and mission. All members view their work as a psychological boost for the medical center, an aesthetic contribution to the campus, and an appropriate addition to the concept of modern health care at Georgetown Medical Center. Most importantly, they know the satisfaction of seeing their volunteer efforts come to fruition, and their visions become a reality.

## LITERATURE CITED

Benson, Herbert. 1997. *Timeless healing.* New York: Simon & Schuster.
Cooper Marcus, C., and M. Barnes. 1995. *Gardens in healthcare facilities: Uses, therapeutic benefits, and design recommendations.* Martinez, CA: The Center for Health Design.

Cooper Marcus, C., and M. Barnes, eds. 1999. *Healing gardens: Therapeutic benefits and design recommendations.* New York: John Wiley & Sons.

Francis, M., and R. Hester, Jr. 1990. *The meaning of gardens.* Cambridge: MIT Press.

Kaplan, R., and S. Kaplan. 1989. *The experience of nature: A psychological perspective.* Cambridge: Cambridge University Press.

Messervy, J. 1990. *Contemplative gardens.* Charlottsville: Holwell Press.

Mintner, S. 1993. *Healing gardens: A natural haven for emotional and physical well-being.* London: Headline Book Publishing.

Moyers, B. 1993. Healing and the mind. Public Affairs Television, Inc., and David Grubin Productions, Inc.

Ulrich, R. S. 1979. "Visual landscapes and psychological well-being." *Landscape Research* 4(1):17-23.

Ulrich, R. S. 1984. "View through a window may influence recovery from surgery." *Science* 224:420-421.

Ulrich, R. S., and R. Parsons. 1992. Influence of passive experiences with plants on individual well-being and health. In *The role of horticulture in human well-being and social development*, ed. D. Relf. Portland, OR: Timber Press, pp. 93-105.

# 9

# Participatory Design of a Terrace Garden in an Acute In-Patient Unit

Shelagh Rae Smith

## INTRODUCTION

The British Columbia Professional Fire Fighters' Burns and Plastic Surgery Unit at Vancouver General Hospital is a 15-bed trauma unit that serves all of British Columbia and the Yukon Territories. Patients are 7 years of age and older and must be able to breathe on their own. Admission to the unit is a result of various burns (fire, oil, water, chemicals, etc.), vehicle and other accidents, animal mauling, pressure sores requiring treatment for paraplegics and quadriplegics, and other medically required plastic surgery. Patients are admitted for a couple of days to several months. An average stay on the unit is 2 to 3 weeks.

The Burns and Plastics Unit is on the second floor of a large tower and includes an outdoor terrace of just under 1,000 square feet. The terrace faces south onto a residential area separated from the hospital by a major thoroughfare, and west to a view of trees, the sunset, and residential housing. The British Columbia Professional Fire Fighters' organization provided funds to change what was a completely inaccessible rooftop covered with stone ballast into an accessible terrace with a door, safety railing, and concrete base. The improvements stopped at this stage until the summer of 2000 when the Universal Garden Society (UGS) donated a bench, table, and several containers of plants. UGS is a nonprofit, charitable organization assisting in the creation of healing gardens at the Vancouver General Hospital.

This paper reports on a project to develop and implement a participatory garden design process. The purpose of the project was twofold: to discover how the end users (staff, patients, visitors, and volunteers) would like the garden designed, and to provide therapeutic benefits to those participating in the design process through the process itself.

## PARTICIPATORY DESIGN

The benefits of using participatory design (Carpman and Grant 1993) include clarifying design objectives, making better design decisions, lowering construction costs through avoidance of errors, stimulating positive behavior and attitudes, and helping to create a sense of community. Participatory design is also known as co-design (King

1989) and community design (Hester 1990). In Hester (1990), reference to the thera-peutic nature of the process of participatory design is found: "Community design includes a participatory process; the resulting design decisions are informed by the nuances of the cultural landscape of the community, producing places that fit the unique needs of the users. The community design process is intended to be therapeu-tic for individuals and the group as a whole." Hester (1990) also describes several exercises or "contemplations" to assist designers in understanding their own needs and spatial values: "In order to do empathic design, we must understand how our own inner psychological landscapes influence the outer landscapes we design. Otherwise we remain captives of our environmental pasts and can't design very well for other people." This idea seems fitting since one of the basic tenets of counseling is that the better counselors understand themselves, the more they are able to put their own needs and judgments aside in helping others.

One of Hester's contemplations, titled "Favorite Spaces," describes a simple method of guided visualization to access peoples' conscious and unconscious design preferences. The theory behind it is that it is relatively easy for people to say what they want in terms of function (e.g., a bench for sitting), but many important design values or preferences can be hidden in the unconscious. An adaptation of this method proved very useful in drawing out of people what they might appreciate most for the Burn Unit terrace.

This guided visualization exercise of exploring favorite outdoor spaces was includ-ed in the participatory design process with the hope that it would provide stress-reliev-ing benefits of "temporary escape" to the participants. In *Healing Gardens* (Cooper Marcus and Barnes 1999), Roger Ulrich discusses the "control-related benefit called *temporary escape* . . . as being of high importance in restoration [from stress]. Temporary escape might be passive, such as gazing out of a window at a pleasant gar-den view, or in the mind only, as when daydreaming about a favorite nature area that one could escape to."

## METHODOLOGY

Forty-three people were interviewed, including 24 individual interviews (seven staff members, four volunteers, nine patients, four family members of patients) and five group interviews that included a total of 19 staff members. The participatory design process also included a written questionnaire asking people to indicate which of many activities they would like to do on the terrace, and which of many features they would like to see on the terrace. Thirty-three people filled out questionnaires (20 staff, one volunteer, 11 patients, and one visitor).

### Individual Interview Process

A series of questions was asked of the interviewees:

1. If you could design the outdoor terrace just the way you would like, what might you do with it? Think about activities you would enjoy doing out there, and whom you would like to spend time with. What physical features would you include? Would you choose a theme? How would you like to feel when you are out on the terrace?

2. Tell me about some favorite outdoor places you have enjoyed, either from your childhood, places you spend time in now, or anything in between. For each place: What did you particularly enjoy about this outdoor place? What activities did you do? What were the smells, sounds, colors, textures, feelings, etc? About one-half of the interviews included a guided visualization about a favorite outdoor place, with similar questions. (Not all interviews included this because of time constraints or a feeling by the interviewer that it wasn't appropriate with the interviewee.)

3. The first question was revisited: If you could design the outdoor terrace just the way you would like, what would you do with it? Have any of your answers changed? Physical features, activities, companions, feelings?

## Group Interview Process

As a result of the staff's limited time, they were interviewed in small groups for a shorter period, using a similar procedure to the individual interviews. Instead of recording their verbal details after the guided visualization, they were asked to write down their ideas, one per Post-it note. The ideas were assembled on a wall and placed in categories as determined by the staff.

## RESULTS

### Interview Results

When people were initially asked for ideas about developing the terrace, the responses were few and focused for the most part on elements of cultivated gardens. However, when people were asked about their favorite outdoor places, the responses revealed a strong preference for our natural, local surroundings—forests, mountains, and natural water features. The guided visualizations provided similar nature content, while the responses (given while they still had their eyes closed) tended to be about different outdoor places and included more vivid and detailed descriptions. Unsolicited comments about the process included

- "Powerful."
- "I was having a bad day and now I feel so much better; it was fun."
- "You see now, you bring me back memories—that was fourteen years ago."

Many of the interviewees made unsolicited comments about the potential benefits of using the terrace garden. Fresh air was most often mentioned (windows in the unit do not open), closely followed by relief from pain and stress, escape from the hospital environment, and peace and restoration. Other comments included:

- "The garden might make [the patients] remember pleasant memories [and] provide inspiration to get back home."
- "A patient plants something, leaves a living thing behind, becomes part of the garden."
- ". . . a sense of ownership from planning and building it, [and] meaning from plants grown and given by family and friends."

## Combined Interview and Questionnaire Results

Certain design objectives were identified as essential to the physical and psychological comfort of the users and to satisfy safety considerations:

- Shade, to mitigate the southwestern exposure and because the burn patients' wounds cannot be exposed to sun
- Protection from wind and rain
- Space for access by people using wheelchairs
- Cozy areas for one or two people and a larger area for a group to gather
- A balance between privacy and a view for the patients whose windows face onto the terrace
- A water feature, by far the most requested item, for its refreshing, cooling, and sometimes life-saving qualities
- A barbecue, suggested by a number of people but vetoed because the sound and smell of sizzling flesh was deemed inappropriate
- Smoking not to be permitted
- Evening lighting to provide safe passage around the garden without disturbing patients in adjacent rooms
- Safe plant material: No thorns, sharply edged leaves, skin-irritating sap, highly toxic or allergenic plants
- No lightweight furniture or shallow-rooted trees for the wind to knock over
- An automatic watering system for ease of maintenance

The following summary list and Table 9.1 show the most requested items compiled from both the interviews and the questionnaire results, listed in order of preference:

1. *Theme:* West Coast temperate forest, including water, mountains or rocks, and some elements of a cultivated garden.
2. *Atmosphere:* Shady, quiet, peaceful, relaxing, with rays of sunlight.
3. *Plants:* Trees, vines, British Columbia native plants, moss, flowers, shrubs, lawn.
4. *Colors:* Shades of green, earth tones, natural West Coast seasonal colors.

Table 9.1.   The most requested items from interviews and questionnaires in a participatory  design process in planning a terrace garden in an acute inpatient unit

**Features (in order of preference)**

| From interviews: | From questionnaires: |
|---|---|
| Water feature (pool, stream, waterfall) | Colorful flowers |
| Trees | Shelter from rain |
| Birds | Sunny spots |
| Vines | Shady spots |
| Green | Water feature (fountain, pool) |
| Native plants | Movable chairs |
| Benches | Wheelchair accessibility |
| Moss | Tables |
| Flowers | Benches |
| Shade | Protection from wind |
| Private spaces | Quiet |
| Lattice/trellises | Evening lighting |
| Rocks | Views of greenery |
| Sound of moving water | Arbors with vines |
| Shrubs | Privacy |
| Quiet | Theme garden |
| Chairs | Fragrance |
| Covered area | Large trees |

**Activities**

| From interviews: | From questionnaires: |
|---|---|
| Sit | Sit and relax |
| Children playing | Enjoy the plants |
| Watch the sun go down | Spend time alone |
| Barbecue | Visit with friends and family (patients) |
| Watch the plants grow | Be surrounded by nature |
| Take a tour around | Look at the view |
| Look out the window | Spend time with one or two others |
| Nature crafts | Take a break from work (staff) |
| Eat, drink, socialize, watch night sky, etc. | Eat and/or drink |

## NEXT STEPS

The results of the participatory design process were shared with a class of landscape architecture students at the University of British Columbia. Under the direction of landscape architect Elizabeth Watts, each student created a design using the principles of universal design (Simson and Straus 1998) and guidelines for designing healing gardens (Cooper Marcus and Barnes 1999). The design created by student Avril Woodend was chosen (Figures 9.1 and 9.2) as most appropriate. Avril placed her design on a 45° angle to the building, providing several benches that allow for some semblance of privacy, with a central area for larger groups. Overhead structures, trees, and shrubs provide shade and wind protection. Plants in front of windows provide privacy and a view for the patients inside. A small water feature is included, and the plant material includes several Pacific Northwest native plants.

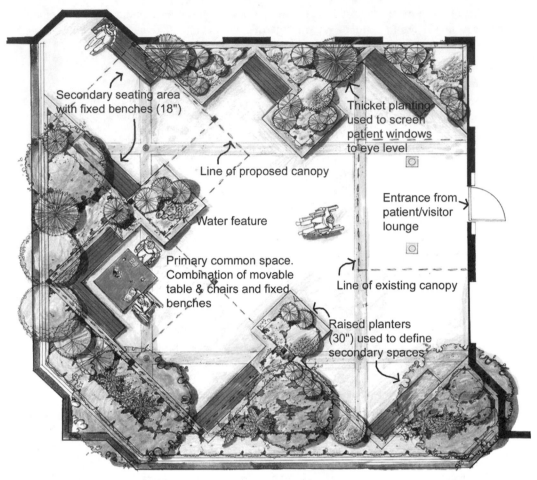

Figure 9.1.   Plan view of the outdoor terrace for the British Columbia Professional Fire Fighters' Burns and Plastic Surgery Unit of Vancouver General Hospital.

Figure 9.2.   Elevations of the outdoor terrace for the British Columbia Professional Fire Fighters' Burns and Plastic Surgery Unit of Vancouver General Hospital.

Cornelia Oberlander, one of Canada's foremost landscape architects, donated her time to fine-tune the design and create working drawings. The funds to build the garden have been raised through donations from the British Columbia Fire Fighters and private donors. The garden was constructed in the spring of 2002, followed by a community planting party and gala opening. Postoccupancy evaluation is planned.

## CONCLUSION

The participatory design process used at the burn unit provided both anticipated and unexpected results:

- Several design criteria were identified as essential to the physical and psychological comfort of the users and to satisfy safety considerations.
- The eventual users of the terrace garden had the opportunity to express their design preferences.
- Most of the interview participants were observed to relax, enjoy the process, and experience the benefits of "temporary escape"; those who participated in the guided visualization were perceived to have the most beneficial and restorative experience.
- The visualization process allowed people to think in an unconventional way about the possibilities for the terrace.
- This raised the question—though it is beyond the scope of this project to answer it—of how particular design processes might influence results.
- It became obvious that many of the people interviewed see the development of the terrace garden as a profoundly beneficial resource for patients, visitors, and staff.

## LITERATURE CITED

Carpman, Janet, and Myron Grant. 1993. *Design that cares: Planning health facilities for patients and visitors*, 2nd ed. American Hospital Publishing, Chicago.

Cooper Marcus, Clare, and Marni Barnes, eds. 1999. *Healing gardens: Therapeutic benefits and design recommendations*. John Wiley & Sons, New York.

Hester, Randolph T. 1990. *Community design primer.* Ridge Times Press, Mendocino, CA.

King, Stanley. 1989. *Co-design: A process of design participation*. Van Nostrand Reinhold, New York.

Simson, Sharon P., and Martha C. Straus, eds. 1998. *Horticulture as therapy: Principles and practice*. The Haworth Press, Binghamton, NY.

Ulrich, Roger S. 1999. Effects of gardens on health outcomes: Theory and research. In *Healing gardens: Therapeutic benefits and design recommendations*, eds. Clare Cooper Marcus and Marni Barnes. John Wiley & Sons, New York, pp. 27–86.

# 10

# Building an Alzheimer's Garden in a Public Park

**Mark Epstein**

The American Society of Landscape Architects (ASLA) in 1999 initiated a program to build 100 parks through volunteer efforts to commemorate their 100th anniversary. The "100 Years, 100 Parks" program included nine therapeutic garden projects in different states in collaboration with the Alzheimer's Association. Two of the nine Alzheimer's gardens were planned for public spaces. This paper will describe the unique public process undertaken for one of those public gardens, the Portland Memory Garden, in Portland, Oregon. The Portland Memory Garden illustrates an innovative process of site selection and site design of an Alzheimer's garden within an existing city of Portland park.

The multidisciplinary partnership for the Memory Garden is comprised of the following individuals:

- Several members of the Oregon ASLA chapter, including Brian Bainnson, Richard Zita, Mark Hadley, and Mark Epstein
- John Sewell and Gregg Everhart from Portland Parks and Recreation
- Eunice Noell of the Center for the Design of an Aging Society
- Teresia Hazen, a horticultural therapist for Legacy Health System
- Landscape designer Bob Adams
- Nancy Chapman of Portland State University's School of Urban Studies
- Andi Miller from the Alzheimer's Association

These individuals formed the steering committee that has guided the all-volunteer effort. The following four objectives were agreed upon at the first meeting and have directed the project to the present:

- Create a restorative environment meeting the needs of those with dementia, their families, and their caregivers.
- Promote the project as a model or demonstration garden to teach how to design restorative gardens.

99

- Use the garden as a tool for public education about Alzheimer's disease.
- Provide increased recreational opportunities to the citizens of the Portland metropolitan area.

Approximately 40,000 people are living with Alzheimer's disease in the Portland metropolitan area, and about 75% of them are living or being cared for at home. Caring for a loved one with Alzheimer's is a demanding, stressful task, and this garden is envisioned as a safe place to supervise a loved one in a relaxing outdoor environment. The garden would also provide an opportunity to educate family members about the disease, through signage or exercise suggestions, in a passive, nonthreatening manner.

Construction projects typically proceed in a linear sequence of funding, site selection, design, and then construction. The Memory Garden diverged from the typical path, resulting in a unique public education process that engaged and informed a wider body of design professionals and lay public than the usual construction project.

Early in the process, it became evident that the Portland Parks Department would require a lengthy public involvement process to ensure community acceptance of the project. They were wary from stiff public resistance to two recent proposals for memorials in existing park land, and did not wish to repeat the process. The steering committee therefore decided to pursue three tracks independently but simultaneously: site selection, garden design, and fundraising (Figure 10.1).

## SITE SELECTION PROCESS

The site selection process was innovative because it involved a citywide nomination process and subsequently generated citizen lobbying for the project to be located in their neighborhood. After deciding on a name for the garden, the most important step was formulating a specific list of criteria for nomination and selection of a site (Figure 10.2).

The criteria formed the basis for the educational materials and allowed the logical and defensible final selection of a site. The criteria form was incorporated into a nomination form used by neighborhood groups and individuals to nominate their neighborhood park to host the garden (Figure 10.3).

The criteria and nomination forms were included within an attractive brochure that was repeatedly used for public awareness and outreach efforts (Figure 10.4).

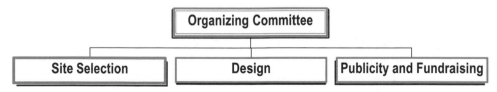

Figure 10.1. Tracks pursued simultaneously by the steering committee for the Portland Memory Garden.

**SITE CRITERIA/INFORMATION**

*Check if site meets criteria. If site partially meets criteria or does not meet criteria, please explain under "comment."*

☐ There are no nearby hazards such as ponds or steep slopes.
Comment ______

☐ The site is flat. *(If it is not flat, how steep is the slope?)*
Comment ______

☐ There is handicap accessible parking in a parking lot within 100 feet of the garden site.
Comment ______

☐ Curb cuts exist.
Comment ______

☐ There is access to the site on paved streets.
Comment ______

☐ There is an existing restroom on the site.
Comment ______
*(Do you know if the restroom is handicap accessible?)*

☐ There are utility services to the site. *(to the best of your knowledge)*
☐ Water service ☐ Electrical service ☐ Sanitary sewer
Comment ______

☐ There are existing shade trees on the site. *(describe)*
Comment ______

☐ The site is removed from traffic or other noise impacts.
Comment ______

☐ The site is protected from strong winds.
Comment ______

☐ Other nearby park uses are compatible with the Memory Garden.
Comment ______

☐ Ballfields, basketball courts, playgrounds are not located adjacent to the site.
Comment ______

☐ Other beneficial site features: *(describe)*
______
______
______

NARRATIVE: *(please enclose a separate sheet)*
Please tell us briefly why you think a Memory Garden would benefit your community. What garden features sounded particularly attractive? Is there support for this idea from the neighorhood association and/or immediate neighbors? *(describe)* What potential concerns have been raised or might be anticipated? If the site is now undeveloped, is there support in the community for completing a park master plan in conjunction with this project?

NOMINATOR

______
NAME
______
ORGANIZATION *(if relevant)*
______
MAILING ADDRESS:
______
______
DAYTIME TELEPHONE: FAX

NOMINATIONS ARE DUE JUNE 1, 1999.
*Mail or deliver to:* Portland Parks and Recreation
1120 SW Fifth, Room 1302 Portland, Oregon 97204 823-5588

Figure 10.2. Criteria used for selection of site for the Portland Memory Garden.

The nomination form served not only to inform community members about the need for the garden, but also to engage their curiosity and allow the individual, rather than an outside expert, to determine if the garden would be appropriate for their park. It subtly educated the reader as to some of the needs, limitations, and special requirements of those with Alzheimer's disease, and perhaps of those with other disabilities.

Before the nomination forms were sent out, the steering committee did some reconnaissance of its own. Fearing receiving no nominations, committee members previewed many of the parks in the targeted area to determine likely candidates and to alert the appropriate neighborhood associations that they might have a good park for this type of garden. At the deadline, the Parks Department received nominations for eight park sites. The process switched from trying to find one suitable site to justifiably eliminating seven park sites.

Committee members made presentations to each of the eight neighborhood groups, explaining the garden's goals and attributes in more detail and answering questions. We were also listening to the audience, attempting to gauge the level of interest and enthusiasm within each local community group, as we understand that the long-term

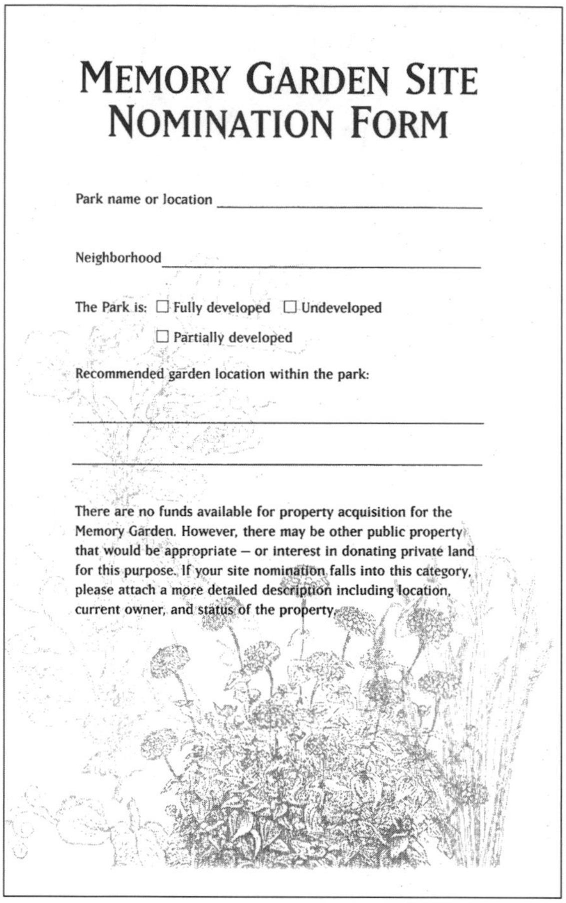

# MEMORY GARDEN SITE NOMINATION FORM

Park name or location ______________________________

Neighborhood ______________________________

The Park is: ☐ Fully developed ☐ Undeveloped
☐ Partially developed

Recommended garden location within the park:

______________________________

______________________________

There are no funds available for property acquisition for the Memory Garden. However, there may be other public property that would be appropriate – or interest in donating private land for this purpose. If your site nomination falls into this category, please attach a more detailed description including location, current owner, and status of the property.

Figure 10.3. Nomination form for neighborhood groups and individuals to nominate their neighbohood park.

success of the garden depends on one or more committed individuals to the project.

The committee eliminated three sites because of site constraints in those parks, leaving five sites on the shortlist. Another newsletter announcing the finalists was published and distributed to the residents living within a certain radius of the nominated site (Figure 10.5), and another round of presentations was made to each group.

The committee made it clear that the decision on the garden location would be based on community support as well as site constraints and design opportunities. At this point, a rivalry developed among some of the neighborhood associations, with the Parks Department receiving letters explaining why the garden should be located in their neighborhood, and why the other sites were lacking. Some better-connected citizens attempted to use their political ties to influence the decision. We believe the open dialogue between the decision makers and the potential users of the project, and the active participation of the public in the site search process, eliminated any "not in

Figure 10.4.   Nomination brochure for the Portland Memory Garden.

my backyard" sentiment experienced in recent park proposals.

The winning park site was selected in October 1999. The Parks Department Director and the Parks Commisioner then quickly approved the project without dissent.

## DESIGN PROCESS

The site selection process took 10 months. To keep the project progressing, we decided to start the design process before deciding on a site. Because the site selection process was used as a public education tool, the design process was used as a designer education tool.

Given the specific criteria stated for the Alzheimer's garden, many elements of the

# Imagine a Garden

that takes you back to your child-hood. Perhaps it was your grandmother's garden. That's the kind of garden that might become a part of your neighborhood . . .

## A GARDEN IN SEARCH OF A PLACE

Last spring, Portland Parks & Recreation announced that it was seeking site nominations for a Memory Garden. The garden is to be small, about the size of two tennis courts. Its serene atmosphere and medley of flowers, trees and shrubs will be designed to enchant users of every age and ability. Its use of old fashioned plant materials and wide looping pathways will be especially appealing to visitors with Alzheimer's disease and other memory disorders.

## Nominations have been narrowed to five sites.

Several nominations were received and evaluated using specific design criteria. Five potential sites were selected for further consideration. The final decision will be based on degree of community support, site constraints and design opportunities.

## Presentations on the project

are scheduled during September and October - one for each of the five sites. Please plan to join us to learn more about the project, ask questions, and let us know what you think about a Memory Garden in your neighborhood park!

*Locations within the parks being considered for the Memory Garden are illustrated below, along with information on each of the five meetings. Those interested are welcome to attend any or all of the meetings. The sites are listed in the order the meetings are scheduled. For the larger sites, potential garden locations are marked with a star.*

### 1. Ventura Park
SE 115th & Ventura

Hazelwood Neighborhood Association Meeting
Monday, September 20th,
7:00 pm
East Portland Police Precinct - community room
737 SE 106th Avenue

### 2. Ed Benedict Park
SE 99th & Powell

Lents Neighborhood Association Meeting - with Powellhurst-Gilbert Neighborhood Assn.
Tuesday, September 28th
7:00 pm
Pilgrim Lutheran Church
4244 SE 92st Avenue
(parking lot at SE 92nd & Cora St.)

### 4. Ivon Street Park
SE 47th & Ivon

Richmond Neighborhood Association Board Meeting
Monday, October 11th
7:00 pm
Central Christian Church
1844 SE 29th Avenue

### 3. Essex Park
SE 79th & Center

Foster-Powell Neighborhood Association Meeting
Tuesday, October 5th,
7:00 pm
St. Anthony's Village – Parish Hall
3618 SE 79th Avenue

### 5. Eastmoreland Golf Course Overflow Parking Lot
SE 24th & Bybee

Eastmoreland Neighborhood Association Mtg.
Thursday, October 21th
7:00 pm
Duniway Grade School – library
7700 Reed College Place

Figure 10.5.    Short-list brochure of possible sites for the Portland Memory Garden.

garden could be designed without a site. Brian Bainnson and Richard Zita headed the design effort of a "generic" Alzheimer's garden that began as a separate process from site selection, but eventually merged into that process as the selected community became more involved in adapting the design to the selected site. Teresia Hazen developed the program elements that led to features designed by the landscape architects and would be easily adapted to the opportunities and constraints of a selected site.

The generic design was created through ASLA workshops and open meetings that involved about a dozen local landscape architects. These sessions informed the designers of the specific needs of Alzheimer's patients and generated creative ideas and thoughtful discussions on the merits of proposed elements. The generic design included a circular main path with secondary paths leading to a central lawn area, and small intimate seating areas off the main path (Figure 10.6).

A trellis marks the garden entrance and allows a vista over the entire garden. Plantings are lush, yet ordered, including some in raised planters of different heights. A "rain-catcher" water feature, a touching garden, and a symbolic orchard were pro-

Figure 10.6. Generic Alzheimer's garden design.

posed as memorable features within the enclosed space of the garden.

The generic plan became useful for subsequent publicity of the project, for early review by the Parks Department, and was especially helpful opening a design dialogue with the selected neighborhood association. Landscape architects and community members refined the generic design a couple of times through open community design meetings after the site was selected, culminating in a final site design (Figures 10.7 and 10.8).

## PUBLICITY AND FUND-RAISING

The project has generated supportive enthusiasm by all who have been involved. The open process and clear objectives for the garden contributed to the broad acceptance and continued donation of time and resources from this all-volunteer effort. The various skills and abilities of the steering committee members have allowed the project to proceed more smoothly and with more success than originally anticipated.

The widespread publicity of the project can be credited to the Portland Parks Department, the Alzheimer's Association, and the Center for Design of an Aging

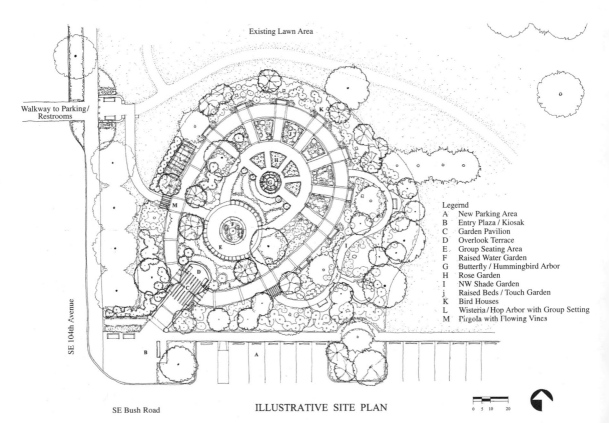

Figure 10.7.   Plan view of the final garden design plan for the Portland Memory Garden.

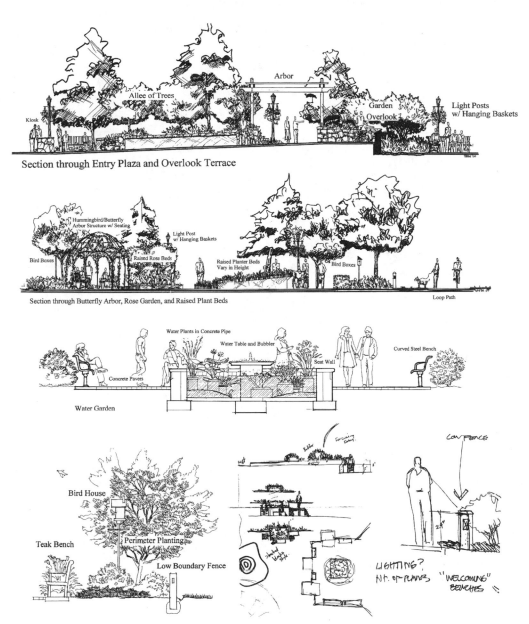

Figure 10.8. Final garden design details for the Portland Memory Garden.

Society. The Parks Department facilitated the creation and mailing of the initial public brochures and press releases (Figure 10.9). The Alzheimer's Association publicized the project in their newsletter (Figure 10.10) and at various fund-raising events. The Center for Design of an Aging Society donated display material for open houses and exhibitions, including a booth at the Portland Home and Garden Show. That exhibit led to a nice article in *The Oregonian*, the region's major newspaper (Figure 10.11).

In addition to the professional in-kind donations already received, local contractors have promised to donate labor and materials for the construction of the garden.

## PORTLAND PARKS & RECREATION
www.portlandparks.org

**For Immediate Release**
April 12, _____

# A Garden In Search of a Place

The American Society of Landscape Architects and the Alzheimer's Association have proposed designing and building a Memory Garden as part of ASLA's "100 Parks, 100 Years" Centennial Celebration. The garden will become a public park and could be sited on land already designated for park purposes or on a donated lot. Nominations are now being sought for the perfect Memory Garden site.

## What is a Memory Garden?
Memory gardens are generally small - about the size of two tennis courts. Their lush, fragrant plantings of "old-fashioned" flowers, trees and shrubs (lilacs, climbing roses, honeysuckle, and daphne) provide color and seasonal interest and attract a variety of birds and butterflies. Wide gentle pathways circle through the garden and there are plenty of benches and garden furnishings to add to the serene atmosphere. While Memory Gardens are beautiful, special places for everyone to enjoy, they are particularly important for people with Alzheimer's disease and other memory problems.

## To nominate a site . . .
Nomination forms are available from Portland Parks and Recreation. Nominated sites will be evaluated based on the degree to which the site meets most of the garden criteria and the extent of support for the nomination from the neighborhood association and/or immediate neighbors. For example, sites should be relatively flat and accessible, protected from strong winds and located away from heavy traffic or other noise impacts. *Nominations are due June 1,_____.* They may be submitted by individuals or organizations.

## To learn more . . .
ASLA, Portland Parks & Recreation, and the Alzheimer's Association are hosting an informational meeting to share more specifics about this project.

*Thursday, April 22,_____*
*7:00 - 9:00 pm*
*East Portland Community Center*
*740 SE 106th Ave.*
*(Just south of Stark)*

This is a good time to learn more about memory gardens, talk informally about possible sites, make suggestions, and ask questions. Those attending will be treated to an inspiring slide show!

Figure 10.9. Press release from the Portland Parks Department.

Figure 10.10. Alzheimer's Association newsletter.

Figure 10.11. Part of a newspaper article in *The Oregonian*.

Portland State University is applying for neighborhood grants to support the project, and university students are looking to perform postconstruction evaluations of the garden. Fund-raising has surpassed the requirements to build the garden, which was completed by spring 2002.

## CONCLUSION

The Portland Memory Garden has been a successful volunteer effort to this point because of the clear direction and goals of the project, the multidisciplinary talents and commitment of the steering committee, and the unusual process of site selection and project design (Figure 10.12). The community effort put into this project has already yielded educational benefits before any construction has begun and promises to contribute a valuable new resource to the city and region upon completion.

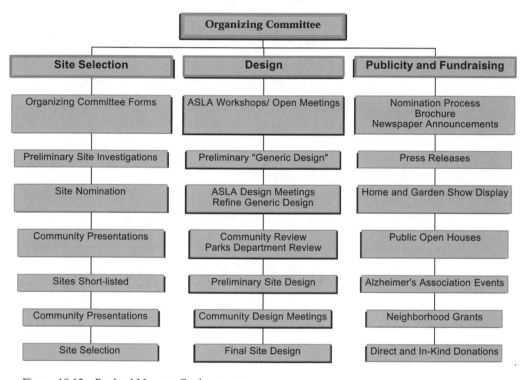

Figure 10.12   Portland Memory Garden process.

# 11

# Alzheimer's Garden Project of the American Society of Landscape Architects and the National Alzheimer's Association

Jack Carman

People benefit from interaction with the natural environment throughout their life. Simply stated, our involvement with nature is good for our physical, mental, and spiritual well-being. The fulfillment of this concept is the basis for developing a project to create positive outdoor environments for individuals with Alzheimer's disease.

The American Society of Landscape Architects (ASLA) commemorated 100 years as a professional association by establishing a program to design 100 parks in 100 cities. One component of the 100 parks in 100 cities program was to create Alzheimer's gardens throughout the United States in collaboration with the National Alzheimer's Association. Eight sites were selected that featured a variety of settings in which to create an Alzheimer's garden. The goal of the Alzheimer's Garden Project has been to highlight the need to create pleasant, safe, and secure outdoor areas for individuals with Alzheimer's disease for their benefit and enjoyment.

Alzheimer's disease continues to be one of the great scourges of abnormal aging. Although most older adults move into the eighth decade cognitively intact, many older adults (approximately 4 million in the United States today according to the National Alzheimer's Association) are struggling with this progressive decline of mental function. This challenging journey into the loss of selfhood is taken by both the victim and their family caregivers. In the past few years, healthcare providers have come to understand the effects this disease has upon the older individual's memory, learning, thinking, and behavior. In response, supportive care facilities such as assisted living residences, adult day care programs, nursing homes, etc., have incorporated design changes indoors to enable the older adult with Alzheimer's disease to function as independently and pleasantly as possible. Although these innovations have been progressive, little or nothing has been designed and implemented to enable the cognitively impaired individual to access and enjoy the out-of-doors. It is almost as if the collective unspoken feeling is to keep the older adult with Alzheimer's disease secured within the four walls of the senior residence "for their own good."

It is projected that within the next 30 years, thousands of aging "Baby Boomers"

111

well into or approaching their eighties will develop Alzheimer's disease...14 million to be exact (National Alzheimer's Association). We can no longer afford to ignore the outdoor environment and its potential for positive support for the demented older adult, as well as their family members and healthcare caregivers.

Studies have shown that good environmental design creates (1) a sense of control, (2) social support, and (3) positive distraction (Hoover 1995; Fandall et al. 1990). In Alzheimer's disease, the short-term memory is the first casualty. For many, the ability to walk and be physically active remains intact, far into the disease process. This physical activity, coupled with the remaining long-term memory, can actually be a resource to enable the older adult to form a connectedness with the familiarity of the outdoor environment, resulting in a sense of comfort and thereby offering a feeling of "being in control." Various elements such as common plants, a clothesline, a garden chair, or an American flag are familiar to many.

As the disease progresses and more and more language is lost, family members and facility caregivers may be at a loss as to how to spend time with their loved one. The garden provides a safe, comforting environment where visits with family and friends can be enjoyed. A properly designed garden offers the individual with Alzheimer's disease the opportunity to socialize, as well as take some time to remain alone safely, in perceived solitude.

In addition, many individuals with Alzheimer's disease experience what are called "catastrophic events." These sporadic episodes result in agitated, emotional outbursts in response to negative triggers in the environment. The outdoor garden can provide positive distractions. These are everyday elements that produce positive feelings. These also effortlessly hold the individual's attention and interest and block worrisome thoughts. Traditionally, the most effective positive distractions have been elements that have been important to humans over time. Nature elements, such as trees, plants, water, and sunshine, have proven to be soothing to older adults with Alzheimer's disease (Mooney and Nicell 1992; Beckwith and Gilster 1997; Mooney and Hoover 1996).

Edward O. Wilson, an environmental psychologist, postulated the "Biophilia Hypothesis" that says that all human beings are innately attracted to nature (1984). In essence, nature is good for us. This natural attraction stems from our evolution, producing a "Savannah Gestalt." The best way to describe a savannah landscape is to ask the reader to think of the movie *Out of Africa* This landscape encompasses a wide grassy plain with an oasis of trees and water. In times gone by, man could hide in the bank of trees, gathering food and water, while surveying the plain for signs of oncoming predators. In modern times, people have been asked in studies to pick pleasing landscapes out of groups of paintings. Inevitably, the choices are of rolling grassy plains with a section of lush trees in one part of the painting. The least liked landscape paintings are the ones with dark, ominous forests, such as the ones seen in the movie *The Wizard of Oz.*

Patrick Mooney, MLA, CSLA, an assistant professor of Landscape Architecture, University of British Columbia, Vancouver, British Columbia, compared access to the outdoor areas of five Alzheimer's residences (1992). Residents in two of the facilities had access to the outdoor environment, and residents in three facilities did not. Not surprisingly, staff in the first two residences saw a 19% (cumulative) decline in catastrophic reactions by the residents who spent time outdoors. In the three residences with no access to the outdoor environment, the catastrophic incidents increased by 681% (cumulative).

Today the need to create positive outdoor living environments for individuals with Alzheimer's disease is greater than ever. The ASLA/Alzheimer's Disease Garden Project has begun to reach their goal of creating eight Alzheimer's gardens. One of the first garden projects completed was the *Alzheimer's Memory Garden* in Macon, Georgia. This garden was a collaboration among ASLA, the National Alzheimer's Association, the Macon Park and Recreation Department, Habitat for Humanity, and Nickelodeon cable television. Nickelodeon sponsored the construction of the garden and provided hundreds of child volunteers to help install the plants for the garden, as part of a nationwide community outreach program. The *Portland Memory Garden* is also a public garden space and involves the participation of ASLA, the National Alzheimer's Association, Legacy Hospital, the Portland Parks Department, and the Center of Design for an Aging Society. Both gardens are significant, not only because of their community involvement and cooperation but because they are also public parks. Approximately 75% of people with Alzheimer's disease live at home with a family caregiver. These gardens enable caregivers the opportunity to visit a pleasant, safe, and secure outdoor area with their loved ones.

A third garden is located in Rochester, New York, at Monroe Community Hospital. This is a garden for the residents of the Alzheimer's Unit at the hospital and is accessible to the public. It offers, like the other gardens, a safe, level walking path that is on grade to reduce the risk of falls. The plant materials are nontoxic, so someone does not ingest something that could be harmful to them. The plants have been selected to attract birds and butterflies into the garden. The garden is also being built with community input and collaboration, with help from groups such as United Way. Another garden is being created in partnership with a healthcare institution, Regina Medical Center in Hastings, Minnesota. This garden incorporates visual features for men and women, particularly farmers and homemakers, including crops, clotheslines, an area for animals, and flower gardens.

The 80th Street Residence in New York City; DaySpring Assisted Living Residence in Muskegon County, Michigan; Alterra Clare Bridge in Oklahoma City, Oklahoma; and Solheim Lutheran Home in Los Angeles, California, are all special care facilities specifically designed to offer special care for individuals with dementia. The gardens located in each of these four residences offer design elements similar to the other gardens mentioned. They are enclosed with a fence, or in the case of the New York City courtyard garden, a wall, to prevent elopement. There are raised planters for horticultural therapy activities. The gardens have bird feeders and birdbaths to add interest and entertainment, naturally. Wind chimes and hanging baskets adorn the walkways. In addition, benches are placed at regular intervals to offer a person a place to stop and rest as well as sit and enjoy the garden. Some of the gardens offer a water feature. The sound of water can be very soothing as well as provide a "white noise" to block out other unwanted sounds.

In addition to calling attention to the need to develop safe outdoor areas for victims of Alzheimer's disease, the ASLA/National Alzheimer's Association Garden Project will also study the effects that gardens can have on individuals with dementia. These gardens will be part of a research project to determine the benefits outdoor spaces can have on such things as maintaining circadian rhythms, reducing agitation, improving appetite, normalizing sleep patterns, and related behaviors.

We have known for years that nature can have a positive effect upon our behavior and how we relate to the world around us. This reasoning should also be applied to people with special needs and conditions. It seems reasonable to conclude that if we are truly part of a larger, living environment, i.e., nature, it is essential that all persons (whether demented or not) experience the outdoor environment as part of a therapeutic "tonic."

## LITERATURE CITED

Beckwith, Margarette E., and Susan D. Gilster. 1997. Annual meeting proceedings of the American Society of Landscape Architects, pp 198-201.

Fandall, Priscilla, Sandra S. J. Burkhardt, and Joan Kutcher. 1990. Exterior space for patients with Alzheimer's disease and related disorders. *American Journal of Alzheimer's Care and Related Disorders and Research*, July/August, pp 31-37.

Hoover, Robert C. 1995. Healing gardens and Alzheimer's disease. *American Journal of Alzheimer's Disease,* March/April, pp 1-9.

Mooney, Patrick, and Robert C. Hoover. 1996. The design of restorative landscapes for Alzheimer's patients. Annual meeting proceedings of the American Society of Landscape Architects, pp 50-54.

Mooney, Patrick, and Leonore Nicell, CHE. 1992. The importance of exterior environment for Alzheimer, residents: Effective care and risk management. *Healthcare Management FORUM* 5 (2):23-29.

Wilson, Edward O. 1984. *Biophilia*. Cambridge: Harvard University Press.

# 12

# Healing Landscapes: Design Guidelines for Mental Health Facilities

**Myra Kovary**

In the latter half of the 19th century, the concept that the natural environment had a positive effect on psychological well-being was promoted by the designers of mental health facilities. Landscape Architect Frederick Law Olmsted and Dr. Thomas Kirkbride (a medical doctor) were pioneers in using extensively landscaped grounds as an adjunct to therapy in their designs for American asylums. Twentieth century designers continue to recognize the value of the natural environment in stress reduction for people who live in urban environments. Hospital design, however, has been primarily influenced by technological advances in the fields of medicine, engineering, and climate control. As a result, patients have limited access to the natural environment. In the last decade, designers and researchers are again recognizing the significant therapeutic value of access to the outdoors for patients in medical hospitals. Very little research, however, relates specifically to psychiatric patients.

Designers should be aware of several issues that are important to people who are confined in mental health facilities that differ from issues for patients in medical hospitals. Medical patients are generally presumed to be competent to make decisions that will be in their best interests. People experiencing mental health crises (whose judgment is considered impaired) may find themselves confined involuntarily. The mental health facility then functions in a custodial role. Patients still need to retain their self-respect and retain control over as much of their lives as possible. Even in the midst of a serious mental health crisis, people need surroundings that allow and encourage them to make choices that are part of everyday life. Access to the outdoors should be an option that is available.

In the field of mental health, the major 20th century advances are in the area of psychopharmacology. New medicines, the process of deinstitutionalization that began in the 1960s, and changes in funding are rapidly changing how mental health facilities are being used. In-patient units are providing more services to people in acute crisis and less long-term care. There is, therefore, a more intense requirement for security and direct supervision of patients. As a result, patients have even less access to the natural environment. Staff members also may not be able to get outside during their working hours.

115

In-patient mental health units are often housed within general hospitals that were designed primarily to accommodate patients with physical problems. The hospitals are frequently designed as multistory buildings surrounded by attractively landscaped grounds. The mental health patients have very limited access to the grounds due to the lack of staff available for supervision. Independent psychiatric facilities generally provide somewhat better access to the outdoors.

## REVIEW OF THE LITERATURE AND DESIGN RECOMMENDATIONS

Ulrich (1999) suggests that the capability of gardens to promote healing is based on their effectiveness in facilitating a person's ability to cope with stress. He theorizes that gardens in healthcare settings provide patients with a sense of control and access to privacy, spaces for social support, opportunities for physical exercise, and positive distractions. Ulrich (1984) also reported that simply viewing nature from a hospital room window can promote a patient's recovery.

Kaplan and Kaplan (1989) and Kaplan et al. (1998) have conducted research that also supports the idea that access to nature can provide a significant contribution to the restoration of mental health. Their theory concludes that mental fatigue is a result of the effort required to focus one's attention by blocking out distracting stimuli. The process for recovering from such fatigue involves allowing the mind to rest by experiencing something that is pleasurable and interesting. Kaplan and Kaplan cite four components of restorative settings:

1. Being away (escaping into another world)
2. Extent (being in a large enough space that the boundaries are not immediately evident)
3. Fascination (capturing one's attention with little or no mental effort)
4. Compatibility (being in an environment that is supportive of one's efforts)

Since the natural environment has been shown to be a preferred setting and includes all four components, they conclude that being in nature will be restorative.

### The Exterior of the Facility

The exterior building and grounds provide information about the facility: whether it is a welcoming place, a place for healing and respite, or a place for confinement and a place where people do not want to be. The chain link fences and razor wire of a prison, for example, convey a message to the community members that they are safe from the criminals inside. A mental health facility, on the other hand, should convey the message that it is a place of caring and healing. Patients may desire refuge, but they also need to stay connected to their lives in the community. A visual relationship between the hospital and the community fosters a psychological link for patients who may be feeling disconnected from an environment that is familiar to them (Gruffydd 1967).

Design recommendations follow:

- Include elements of neighboring buildings and landscapes into the exterior design.
- Use landscape elements and plants to create a welcoming entrance.
- Include colorful plant material in the exterior design. Color brings a sense of life and energy.
- When surrounding fences are necessary, cover the fences with dense climbing plant material.

Architect Frank Lloyd Wright's concept of organic design could be a valuable concept in mental health facility design. Organic design involves "a true liberation of life and light within walls; a new structural integrity; outside coming in; and the space within, to be lived in, going out. Space outside becomes a natural part of space within the building. . . . Walls are now apparent more as humanized screens. They do differentiate, but never confine or obliterate" (Pfeiffer and Nordland 1988). Create transparency by designing exterior walls that contain large windows. Include materials and sculptural elements that are part of the exterior landscape in the interior design.

## *Security*

Goshen (1959) collected several articles relating to the architecture of psychiatric facilities. He indicated that security was the dominant component of most mental hospital designs. The assumption underlying the design of the hospital was that psychiatric patients are violent. He found that only approximately 5% of psychiatric patients behaved violently and those 5% required "special measures of protection" only part of the time. Reducing the emphasis on control reduces the risk of activating frustration in the patients. Goshen suggested, "hospital design in itself can offer many opportunities to obviate the institutional atmosphere which fosters destructive patient expression" (1959).

The history of the design of the internal quadrangle is also valuable in dealing with security issues. The history of the quadrangle can be traced back to the monastery cloister and to Persian paradise gardens (Jellicoe and Jellicoe 1987). Islamic gardens and cloister gardens include open-air patios, atriums, and courtyards. These spaces focus inward and provide a sense of protection from the outside world.

- If the site and program allow for the construction of a one-, two-, or three-story building, design the buildings to completely enclose an open-air courtyard.
- If direct visual supervision in all the outdoor spaces is not possible, install electronic surveillance devices. Electronic surveillance is preferable to restricting outdoor access to only those times when a staff member can accompany patients.
- If the climate is such that open-air courtyards are not feasible, consider designing an indoor street. Include skylights and interior trees.
- If the site is in an urban setting that requires a multistory building, design the building around an interior landscaped atrium.

## Privacy versus Social Interaction

Seating for privacy within a psychiatric hospital is not easily found. The placement of outdoor furniture can provide people with the choice of being alone or interacting with others.

- Offer a variety of seating arrangements for use by patients, staff, and visitors. Include some fixed solitary benches as well as some fixed L-shaped benches. To reduce the feeling of being in a fishbowl, locate benches with their backs toward the outer edges of the outdoor rooms.
- Include moveable furniture to allow for more options and choice in seating arrangements. To prevent the possibility of freestanding furniture being used as a weapon, the furniture may need to be secured. Flexibility can still be provided by tethering one leg of the furniture to the ground with a short chain.

## Active Interaction with Nature

Patients need rest, but they also need relief from boredom. Cooper Marcus and Francis (1990) found that well-designed play spaces have a profound influence on the style and spirit of children's play. Well-organized spaces can promote cooperative interactions and reduce disruptive behavior and discipline problems. Dattner (1969) hypothesized that the qualities that create successful play environments for children could also help adults with mental health problems gain knowledge about themselves and their physical environment.

- Include lawn areas for picnics and various sports and a paved area for basketball and dancing.
- Include "playground" areas for adults and visiting children.
- Locate toilets and water fountains near exterior spaces.

Horticultural therapy provides an excellent option for outdoor activity. Horticultural therapy brings people closer to the beauty of nature and to the mystery of growth and development. Therapists can use metaphors related to gardening as a way of connecting with patients. Patients can experience a sense of accomplishment, gain self-esteem, and gain a sense of control over their surroundings. Gardening also provides sensory pleasure through colors, smells, and tactile experiences. Even very small gardens can provide these benefits.

Involving patients in a horticultural therapy program can encourage direct participation by patients in the design of the landscape. Such a program could also be expanded to include the design and maintenance of the grounds.

- Locate a water source and storage space for tools near the garden site.
- Locate moveable chairs near the garden to facilitate working in the garden and to provide seating for others who would like to enjoy viewing the garden and the activity that goes on there.

- In urban settings where space may not be available for gardening at ground level, install rooftop gardens. Safety features such as fences and/or walls must be installed to prevent access to the edge of the roof.

## Passive Interaction with Nature

To achieve the goal of recovering from mental fatigue, people need to be able to choose the level of sensory stimulation that is appropriate for them. Patients need sensory stimulation but not so much that they become overloaded. Patients should also be provided with opportunities to choose between active or passive interaction with nature. Access to outdoor spaces increases the available options. Photosensitivity of the skin and eyes is a serious side effect of many psychotropic medications. Shaded areas must be provided.

- Choose plant materials that will rustle in the breeze. Ornamental grasses and trees that retain their dried leaves into the fall and winter make rustling sounds in the breeze.
- Install fountains or other forms of water sculptures that will provide water sounds. For safety reasons, the water features and fountains must be very shallow. The budget must include provision for continued maintenance.
- Choose native plant materials that will attract birds and butterflies. Avoid plants, however, with toxic berries that may be eaten by patients.
- Provide arcades and porches to create shade.
- Choose small-leafed canopy trees to provide dappled shade.
- Include arbors that are covered with climbing plant material.
- Offer a variety of seating arrangements in both sunny and shaded areas.

## Color

Goshen (1959) indicated that color could be an effective tool in preventing a sense of dreariness. He also noted that the insensitive use of color could cause undesirable effects.

- Use earth tones when selecting colors for exterior walls, seating walls, and pathways.
- Select flowering plant material to create interest, but not to be the overwhelming focus of the experience of the landscape.
- Use warm colors in cool, shaded areas, and use cool colors in warm, sunny areas.

## Acoustics

In psychiatric units, great importance is placed upon verbal interaction among patients as well as among patients and staff members.

- Avoid highly reflective surfaces that can create confusing acoustic conditions with respect to echoes and footsteps (Spivack 1984).
- Avoid cobbled pathways that can create poor acoustic conditions for wheelchair users.
- Install dense plantings where needed to minimize the reflection off courtyard walls and windows.

## Time Flow

An awareness of the passage of time contradicts the sense of stasis that typically permeates the inside of an artificially lit mental health facility. Natural light in the interior environment fosters awareness of the weather outside and of the passing of the days from morning to afternoon, to evening, and into night.

- Install skylights and large windows where feasible.
- Select deciduous as well as evergreen plants to emphasize the changes in the seasons.
- Include flowering plants that will accent the changes in the seasons.
- Select plants that will attract birds and butterflies to enhance awareness of the cycles of life in the natural environment.

## Solitude, Serenity, and Reflection

Institutional settings rarely allow for opportunities for solitude. The need for solitude as an important component of mental health is not often addressed in the designs of facilities where security and confinement play major roles. Prisoners have deliberately broken rules in order to be put into solitary confinement because they wanted an opportunity to get away from the harshness of prison life (McIntosh 1996). Access to outdoor spaces in mental health facilities can provide patients, staff, and visitors with much-needed opportunities for solitude, serenity, and reflection.

The Japanese garden concept of *yugen* may be of value in mental health facility design. *Yugen* is the profound mystery of things being revealed indirectly and experienced through a process of discovery. Olmsted's design for Prospect Park, in Brooklyn, New York, serves as a good example of how spaces unfold to create a sense of freedom through discovery. The high level of abstraction in Zen meditation gardens inspires the use of one's imagination. In contrast, creating opportunities to view literal reflections may inspire figurative reflection.

- Plan the outdoor pathways with curves so that the destination is not visible from the point of entry.
- Include special places as destinations by installing fountains, bird feeders, and other artwork. Be cognizant, however, of possible negative connotations and patients fears when using abstract forms (Cooper Marcus and Barnes 1999).

- Create special places where there is a particularly interesting or unusual tree, thereby providing a natural sculpture.
- Include shallow water features that can function like mirrors, with the opportunity for reflecting something of interest.
- Provide a comfortable place to sit and reflect on the special places.

## RECOMMENDATIONS FOR FURTHER RESEARCH

Current theories of landscape preference, perception, and assessment are inadequate for providing designers with necessary information for creating therapeutic landscapes (Uzzell 1991). Further research is needed to determine several issues: why individuals prefer certain landscapes; how different landscapes influence emotional states; and how different landscapes can have a therapeutic effect on people with specific psychiatric disorders. Collaborative work by psychologists, sociologists, medical professionals, recreation specialists, as well as designers will be required to find answers to these questions (Parry-Jones 1990).

The designer of a mental health facility must be sensitive to multiple users' needs. Users include the patients, the staff, and visitors of all ages and abilities. Designers will also need to

- Conduct preoccupancy and postoccupancy evaluations with patients, visitors, staff, and the operators of the facility and incorporate the resulting feedback into their designs.
- Work with the staff to develop and maintain effective use of designed spaces after the completion of a project and as the use of the facility changes over time.

As trained facilitators and problem solvers, landscape architects are in a position to take the lead in initiating collaboration between various professionals working in the field of mental health. Landscape architects should be involved in all aspects of mental health facility design to facilitate access to the outdoors. The creation of healing environments can meet the needs of people who are experiencing serious mental health crises, as well as the needs of the staff and visitors, and presents an exciting challenge for researchers and designers in the 21st century.

## LITERATURE CITED

Cooper Marcus, Clare, and Marni Barnes, eds. 1999. *Healing gardens: Therapeutic benefits and design recommendations.* John Wiley & Sons, Inc., New York.

Cooper Marcus, Clare, and Carolyn Francis, eds. 1990. *People places: Design guidelines for urban open spaces.* Van Nostrand Reinhold, New York.

Dattner, Richard. 1969. *Design for play.* Van Nostrand Reinhold, New York.

Goshen, Charles E., MD, ed. 1959. *Psychiatric architecture.* American Psychiatric Association.

Gruffydd, Bodfan. 1967. *Landscape architecture for new hospitals: A King's Fund Report.* London.

Jellicoe, Geoffrey, and Susan Jellicoe. 1987. *The landscape of man: Shaping the environment from prehistory to the present day.* Thames and Hudson, London.

Kaplan, Rachel, and Stephen Kaplan. 1989. *The experience of nature: A psychological perspective.* Cambridge University Press, Cambridge, England.

Kaplan, Rachel, Stephen Kaplan, and Robert L. Ryan. 1998. *With people in mind: Design and management of everyday nature.* Island Press, Washington DC.

McIntosh, Hugh. 1996. Solitude provides an emotional tune-up. *APA Monitor 27* (March): 1-10.

Parry-Jones, William. 1990. Natural landscape, psychological well-being and mental health. *Landscape Research 15* (2):7-11.

Pfeiffer, Bruce Brooks, and Gerald Nordland, eds. 1988. *Frank Lloyd Wright: In the realm of ideas.* Southern Illinois University Press, Carbondale, IL.

Spivack, Mayer. 1984. *Institutional settings: An environmental design approach.* Human Sciences Press, New York.

Ulrich, Roger S. 1984. A view through a window may influence recovery from surgery. *Science 224*:420-421.

Ulrich, Roger S. 1999. Effects of gardens on health outcomes: Theory and research. In *Healing gardens: Therapeutic benefits and design recommendations*, eds. Cooper Marcus, Clare, and Marni Barnes. John Wiley and Sons, Inc., New York, pp. 27–86.

Uzzell, David L. 1991. Environmental psychological perspectives on landscape. *Landscape Research 16* (1):3-10.

# 13

# A Children's PlayGarden at a Rehabilitation Hospital—A Successful Collaboration Produces a Successful Outcome

**Sonja Johansson**
**Nancy Chambers**

PlayGardens at rehabilitation hospitals, when creatively designed, are the ideal places to support and encourage a wide range of therapeutic benefits for children with disabilities.

## BACKGROUND

Children begin their lives by using their senses and motor abilities to gather and interpret information about their environment. When motivated, they engage in activities, repeating things again and again, to test and absorb the experiences. They respond to stimulation in their surroundings by using their senses to give them feedback and cues, so that they can navigate and orient themselves—by voices, landmarks, boundaries, shapes, and mass. Through this movement, repetition, and accident, they grow and develop.

Professionals who work with children with disabilities recognize that active, creative play is essential for development. They understand the need for creating environments that stimulate the senses and offer physical, social, emotional, and cognitive experiences.

According to the March of Dimes, each year in the United States approximately 150,000 babies are born with birth defects and another 230,000 become disabled at some point during their childhood. Some babies have restricted motor abilities and reduced endurance, which limits their ability to interact, experience, and play, especially outdoors in nature. Others have sensory or perceptual deficits that make them unable to integrate or accurately organize colors, lights, shapes, and sounds that their senses receive, or orient themselves to their environment.

Traditional playgrounds and gardens do not sufficiently meet the needs of children with disabilities. Spaces must be specifically designed to increase the capacity of children to interact with the environment. The environment itself must provide motivation for them to spontaneously move about, exercise fully, and interact with all the elements of the space. It must be consciously designed to develop and integrate their perceptual, motor, and cognitive skills with enough diversity to engage all children regardless of their abilities or disabilities.

## DEVELOPMENT OF THE THERAPEUTIC PLAYGARDEN

Rusk Institute of Rehabilitation Medicine, at New York University Medical Center, is one of the world's leading centers for rehabilitation medicine, providing comprehensive treatment for adults and children who have a wide range of physical disabilities. One of the unique features of Rusk is the Glass Garden built in 1959 and subsequently expanded with outside gardens. The Glass Garden has been a pioneer in the field of horticultural therapy as well as setting standards for the design and use of gardens in a medical setting.

In 1994, a project team was created at Rusk Institute to develop a new PlayGarden with the landscape architecture firm Johansson & Walcavage, now called Johansson Design Collaborative, Inc. The project team included recreational, occupational, horticultural, and physical therapists; teachers; and physicians (every specialist working with children) as well as a project manager representing the Plant Maintenance and Construction Department at the Medical Center.

The scope of work for the landscape architect was clearly defined from the outset to include the following:

- Work with a designated client team and project supervisors.
- Work within a designated budget.
- Integrate nature into all play elements.
- Create topography and three-dimensional space.
- Provide play equipment, gardens, and water and sand play.
- Create multiple-use space.
- Extend accessibility and risk-taking activities.
- Ensure all surfaces are accessible, safe, low maintenance, usable in all weather, aesthetic, durable.
- Provide an additional accessible entry from the building.
- Provide a car barrier and privacy fence along the sidewalk.
- Provide storage space.

The users were clearly defined. They included children up to 12 years of age (toddlers, preschoolers), in- and outpatients, children's families (including siblings), therapists working with children (occupational, physical, horticultural, recreational, and music therapists), teachers, and families in the local community. The commitment to maintenance, staffing, and usage was in place. The PlayGarden would be open to the public as well, if adults were there to supervise all children.

The landscape architect visited the children during treatment and in class, and held meetings and design discussions with the staff to understand the desired outcomes relating to rehabilitation, development, and education. The landscape architect suggested ways of achieving these goals through design, and offered sketches, photographs, and other materials to illustrate ideas or feelings for the creation of the space. (*The sketches that follow are some of those that were shown to the team to get their reaction and input.*)

From this process, the rehabilitation team and landscape architects decided that they needed something different from the standard play experience. They developed the concept that this little corner of New York City should be transformed into a naturalistic interactive PlayGarden, where children would be encouraged to explore and enjoy activities and materials at their own pace and in their individual ways.

## DESIGN IMPLEMENTATION

### *Design for Safety, Motor-Planning, and Physical Movement*

To create a safety car barrier along the sidewalk that was the northern perimeter of the PlayGarden, the existing flat site was altered to create a 3-foot rise against a new exterior retaining wall. This rise was topped by a 6-foot privacy wood fence that enclosed the space on three sides. This altered topography became the key design solution to creating the multilevel space that added visual interest and physical challenge to the users of the PlayGarden. See Figure 13.1.

A lush grassy hill became the focal point of the PlayGarden, onto which and around which the children could flow. The hill invites tumbling, rolling, and stretching in the sun. This sloping lawn also gives the small PlayGarden a sense of openness. See Figure 13.2.

Custom-designed slides—one for toddlers and one for older children—are tucked into the hillside to add interactive play experiences while giving the children sloping, vertical, and horizontal levels on which to play. There are several ways to get up or down the slides, some more difficult than others. The children decide what route to take on their own, balancing the need for challenge against their need for security. See Figure 13.3.

To enhance motor-planning, body positioning, balancing, and upper body strength, a playhouse with overhead rungs and side hanging bars was designed. This would enable children to exercise by swinging, climbing, and turning while being monitored by a therapist or parent (Figure 13.4). A sandbox edged with large rocks was created. Children learn coordination and motor skills as they climb over large rock steps to get into the sandbox (Figure 13.5). Swings were installed to challenge children's balance and to cradle a child with spinal injuries.

A colorful safety surface pathway system was designed to aid orientation as it curves around the grassy slope, under arbors, and rises over an arched bridge to cross a babbling brook (Figure 13.6). Another span crosses a lush, plant-filled bog. The path is the unifying element in organizing the PlayGarden space. The safety surface pre-

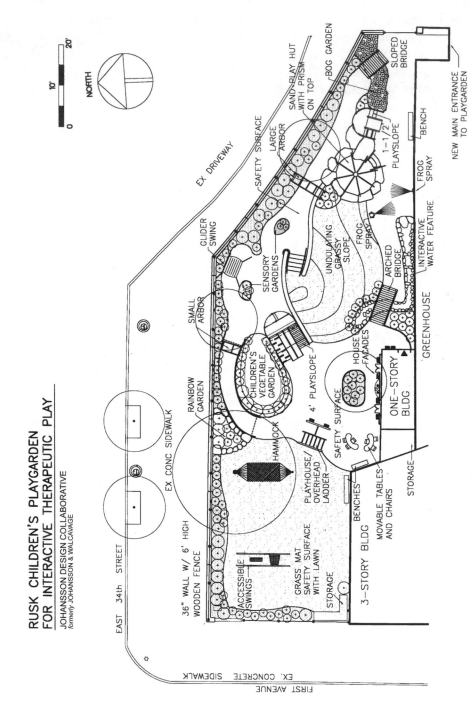

Figure 13.1.　Plan View of Rusk Children's PlayGarden.

Figure  13.2.    Sloping lawn.

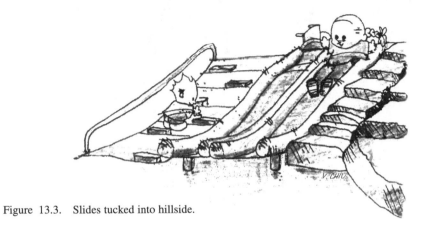

Figure  13.3.    Slides tucked into hillside.

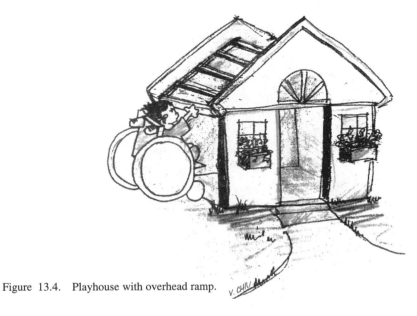

Figure  13.4.    Playhouse with overhead ramp.

Figure 13.5.    Sandbox.

Figure 13.6.    Pathway under arbor.

vents scrapes and bruises from falls. A grass mat provides a more natural-looking accessible safety surface under the swings and hammocks.

Thus, the range of topography, surfaces, and play equipment motivates the children to exercise all their muscles by running, crawling, sitting, bending, turning, swinging, and jumping. All the natural and manufactured play elements foster challenges to motor-planning, eye-hand-foot coordination, balancing, spatial awareness, body positioning, and a multitude of challenging opportunities for a full range of gross motor and coordination skills.

## Design for Sensory Stimulation

The PlayGarden was deliberately designed to enhance stimulation of the entire range of senses for children with disabilities. The entire space was designed to be touched, explored, and experienced by young children in or out of wheelchairs. The grassy hill is the largest sensory-tactile element for children to smell, roll on, and feel on their skin. A multilevel natural stone water channel was created to add to the sensory mix as the children receive the tactile pleasure of water play and hear the water song (Figure 13.7). The children can touch and manipulate materials to take in the sensory information that can provide a base for cognitive learning.

The shrubs, trees, vines, annuals, and perennials used in the space were specifical-

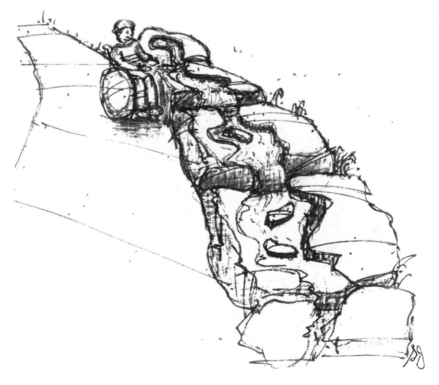

Figure 13.7.   Multilevel stone water channel.

ly selected for their textures, scents, tastes, colors, shapes, sizes, and even sounds (Table 13.1). The plants were also selected to provide a habitat for butterflies, birds, and beneficial garden creatures to augment and enhance visual and auditory stimulation. Two small plant beds, cut into the pathway, overflow with various mints and lavender that children can brush against as they wheel by.

The sandbox juxtaposes the textures of play sand and large boulders, and the swings and colorful playhouses have natural elements integrated into their manufactured appearance—trellises covered with colorful and scented vines shade the swings and sandbox, and the window boxes on the house facades overflow with small gardens that the children planted.

A unique feature of the PlayGarden is a prism sculpture atop the sandbox trellis. This prism gently rotates in the wind, capturing the sun's rays and casting rainbows throughout the PlayGarden. The scented herbs, bright flowers, dazzling grass, dancing rainbows, running brook, flitting butterflies, and ringing chimes engage the entire range of senses.

## *Design for Discovery and Learning*

The PlayGarden was designed for children to cause things to happen. When they open the doors of the fanciful playhouse facades, the children discover specially designed manipulative objects that stimulate their fine motor and decision-making skills (Figure 13.8). Elsewhere, they turn the frog spray and the waterfall on or off; they move

Figure 13.8.  Fanciful playhouse facades.

Table 13.1.    Plant list for the Rusk Institute of Rehabilitation Medicine PlayGarden

**Trees**
*Chionanthus retusus* (Chinese fringe tree)
*Salix alba* (Weeping willow)
*Koelreuteria paniculata* (Golden-rain tree)
*Oxydendrum arboreum* (Sourwood)
*Prunus* × *subhirtella* (Weeping Higan cherry)

**Shrubs**
*Buddleia davidii* (Butterfly bush)
*Calycanthus floridus* (Carolina allspice)
*Cedrus atlantica* 'Glauca Pendula' (Weeping
    Blue Atlas cedar)
*Cotinus coggygria* 'Royal Purple' (Royal Purple
    Smoke bush)
*Hamamelis* × *intermedia* (Witch hazel)
*Hibiscus syriacus* 'Diana' (Rose of Sharon)
*Hydrangea quercifolia* (Oakleaf hydrangea)
*Hydrangea macrophylla* 'Blue Wave' (Lacecap
    hydrangea)
*Philadelphus coronarius* (Mock orange)
*Prunus laurocerasus* 'Otto Luyken' (Cherry lau-
    rel)
*Syringa pubescens* subsp. *patula* 'Miss Kim'
    (Lilac)
*Syringa villosa* 'Donald Wyman' (Late lilac)
*Vaccinium corymbosum*, mixed, early-late
    (Highbush blueberry)
*Viburnum carlesii* (Koreanspice viburnum)

**Vines**
*Actinidia kolmikta* (Variegated kiwi vine)
*Clematis* spp. Assorted varieties
*Lonicera* spp. Assorted varieties (Honeysuckle)
*Hydrangea anomala* subsp. *petiolaris* (Climb-
    ing hydrangea)
*Polygonum aubertii* (Silver lace vine)

**Bog Plants**
*Acorus* sp.
*Typha latifolia* (Cattail)
*Equisetum hyemale* (Horsetail)
*Canna* spp. Assorted varieties (Hardy canna)
*Iris* 'Caesars Brother'
*Liatris microcephala*
*Lobelia cardinalis* (Cardinal flower)
*Lobelia siphilitica* (Blue cardinal flower)
*Saggitaria* (Arrowhead)
*Marsilea quadrifolia* (Water clover)
*Petroselinum* (Water-parsley)

**Perennials**
*Achillea* 'Pink Deb' (Yarrow)
*Aquilegia flabellata* 'MiniStar' (Columbine)
*Aruncus dioicus* (Goatsbeard)
*Aster* × *frikartii* 'Mönch'
*Astilbe* spp. Assorted varieties
*Aurinia saxatilis* 'Compacta' (Basket of gold)
*Cimicifuga racemosa* 'Autopurpurea' (Purple
    cohosh)
*Coreopsis* 'Moonbeam' (Tickseed)
*Corydalis lutea*
*Echinacea purpurea* 'Magnus' (Purple cone-
    flower)
*Gentiana asclepiadea* (Gentian)
*Hemerocallis* 'Hyperion' (Daylily)
*Hibiscus moscheutos* (Common rose mallow)
*Iris chrysographes* (Siberian iris)
*Iris kaempferi* (Japanese iris)
*Lavandula angustifolia* 'Hidcote' (Lavender)
*Lychnis coronaria*
*Lysimachia clethroides* (Gooseneck loosestrife)
*Mertensia sibirica*
*Monarda didyma* 'Marshals pink' (Bee balm)
*Nepeta* × *faassenii* 'Blue Wonder' (Cat mint)
*Paeonia* 'Bowl of Beauty' and 'Sarah Bern-
    hardt' (Peony)
*Perovskia atriplicifolia* (Russian sage)
*Phlox paniculata* 'Bright Eyes' (Garden phlox)
*Phlox subulata* (Moss phlox)
*Physostegia virginiana* (Obedient plant)
*Rudbeckia hirta* (Black-eyed Susan)
*Salvia* 'Wild Watermelon' (Sage)
*Santolina chamaecyparissus* (Lavender cotton)
*Santolina virens*
*Saponaria officinalis* (Soapwort)
*Solidago rugosa* 'Fireworks' (Goldenrod)
*Stachys* 'Helen von Stein' (Lamb's ears)
*Stokesia laevis* 'Blue Danube' (Stokes' aster)
*Thymus* × *citriodorus* 'Gold Edge' (Golden
    thyme)
*Thymus vulgaris* (Thyme)
*Tricyrtis formosana* 'Amethystina' (Toad lily)

**Grasses**
*Festuca glaca* (Blue fescue)
*Miscanthus* 'Yaka Jima'
*Panicum virgatum* 'Heavy Metal' (Switch
    grass)
*Pennisetum alopecuroides* (Fountain grass)

rocks to change the water's flow; they unfasten locks and clasps to discover shells, feathers, and stones behind secret doors.

The children learn about nature by planting seeds in their garden and watching them grow. They explore the materials of the earth in the soil, sand, small and giant rocks, and cascading water.

Every area of the PlayGarden offers opportunities for different learning experiences. Children read or listen to stories while swinging quietly in the hammock. They propagate plants and make nature craft projects at the stackable tables and chairs. They learn about light refraction and wind when watching the dancing rainbows from the prism sculpture. They follow the path of the insects flitting through the bog. They learn how to play creatively with others inside the playhouse. They move their bodies and objects about and through space as they explore the limits of the space.

These experiences enhance organizational and planning skills, sequencing, understanding of cause and effect, motor planning, and initiation by providing a safe environment in which to explore, experiment, make decisions, and learn independently, without the need for help unless warranted.

## Design for Social Activities and Quiet Restoration

Creating diverse opportunities was a major design concept of the PlayGarden. All of the elements in the PlayGarden foster social interactions or independent contemplation. The children can work cooperatively in groups around the tables. They play roles in the little house. They go down the slide holding hands with each other and swing in the hammock in groups. The glider swing, designed for wheelchair accessibility, is a special place where parents can actively play with their children. There are colorful benches for additional caregivers.

The PlayGarden also offers restorative, soothing elements to reduce the stress of the hospital environment. This is enhanced by the harmony of design and scale, the calming pastel colors, fresh-smelling green grass to lie on, gentle curving pathways, swings to dream on, and a house to hide in.

Every place in the PlayGarden offers easy visibility for adult supervisors; there are no totally hidden places, just symbolic ones in which kids would feel comfortable. Benches are available so that parents can easily watch the children while also relaxing and visiting with other adults.

## SUMMARY

The success of the PlayGarden's design was a result of the collaboration among the landscape architects and the treatment team at the rehabilitation hospital. The focus was constantly on desired outcomes for children with disabilities. The space and all the elements were designed as interactive features to which children can respond and from which they receive information.

The space is clearly a unified, child-friendly environment, child-scaled, fully comprehensible so that they can plan and carry out activities independently. It is a safe environment in which to explore, experiment, make decisions, and play. The Play-Garden opens opportunities for building competencies that lead to self-esteem and independence.

Equally important, the PlayGarden introduces the children to natural elements that they might never experience otherwise (particularly those growing up in the city). It allows them to experience these elements in a safe and therapeutic way. These learning experiences and the knowledge and skills gained from the interaction between play and the natural environment not only help in the healing process, but also enhance these children's future lives by instilling in them an appreciation for nature.

The impact of the PlayGarden on hospital policy has been significant. The donors are extremely happy with the results and plan to continue supporting future projects. The PlayGarden has become one of the key venues for promoting the outreach efforts of the medical center. Neighborhood families have discovered the PlayGarden, so now community toddlers play there along with the Rusk children.

Because the therapists were involved with the design from the beginning, they regard it as truly their place. They use it eagerly and inventively. The children look forward to and enjoy their therapy sessions in the PlayGarden. The PlayGarden has become a favorite spot as well for parents and siblings. In addition, therapists working with adults use the gentle slopes of the garden paths to practice wheelchair mobility and ambulation. The PlayGarden adds diversity to the treatment mix and engenders a sense of pride in this special place.

It is a place for nature in the city. The energy of the wind and sun is evident in the rainbows and the rustling grasses. It is a place to feel the cool shade while lying in a hammock. It is a place to hear the chirping of insects in a bog. A place to see the color changes from season to season, a place to feel the materials of the earth in the water, the soil, sand, and rocks.

Most of all, the new PlayGarden has become the focus of attention for the Rusk children and their families and all the others working with them.

## ACKNOWLEDGMENT

The author wishes to acknowledge Vincent Chiu and Sonja Johansson for having drawn the sketches.

# 14

# The Role of Perception in the Designing of Outdoor Environments

**Marni Barnes**

Design guides perception through space. Psychotherapy guides emotions over time. Therapeutic design guides the movement of emotions over time and through space.

This chapter examines the links between emotional restoration and the physical environment. The interplay between the outdoor environment, perception, and emotional responses will be explored. "Bridging the gap between what is internal and what is external is the function of the senses. They convey messages from the three-dimensional world into the complex corridors of our mind, that subconscious matrix through which all we see must pass" (Lewis 1996). The conveyance from the three-dimensional world—that which designers mold and shape—to the internal perceived world is not as simple or straightforward as merely taking in the lines, forms, colors, and shapes that comprise our environment. What the individual experiences is enhanced or added to at two stages of incorporation into the human psyche. The first level of enhancement is contained by what the individual *observes* and the second by how it is *interpreted*.

Environmental psychologists Ittelson, Proshansky, and Winkel (1974) address this when they discuss human perception. In summary, they report that there is a layer of "symbolic value" inherent in the environment. Part of the information picked up by our sensory organs—the sounds, sights, and smells in the environment—are external stimuli that are not directly identified by the conscious mind. They are cues that enter the psyche at a subliminal level. The meaning of these stimuli is "felt" more than deduced by thought. This symbolic communication is implicit in most environments and is an inherent part of observing the surroundings. (An open gate is inviting, while a closed one says, "Keep out.") This symbolism, which is primarily culturally determined, becomes part of how we define ourselves in relation to our environment. (We are either included or excluded in the example above.) Thereby we define our role in that environment (participant or outsider/observer). These symbolic meanings can be imbued by very subtle cues. Nevertheless, subtlety does not diminish the strength of the message.

Beyond this *observed* environment is a level of filtering that takes place within the individual—automatically, often instantaneously, and without awareness. The translation from the observed objects to the internal perception of those objects is *interpreted* in light of past experiences and present expectations, and therefore inherently incorporates our individual distortions.

## EMOTION AND PERCEPTION

People tend to perceive things in a way that is congruent with their emotional state. Author Laurie Fox (1998) captures this phenomenon in the words of the high school-aged protagonist when she has just lost her first love in *My Sister from the Black Lagoon*:

> Wearing headphones protects you from the radioactive fallout of family life but it does a lousy job of screening out the miseries of your love life. On the contrary, while you think you are cloaking yourself in songs that keep you warm and safe and distracted, *you are actually homing in on every syllable from every song that has everything to do with your sorrow. You become a beacon for sadness, a magnet for woe* (italics added).

This level of filtering, that of interpreting one's surroundings based on personal experiences and expectations, is particularly relevant to design in medical facilities. Patients are vulnerable, partly because they are unwell and partly because they are in a stressful environment. Clinics and hospitals, even the best ones, serve to compound a sense of dependency, thereby increasing feelings of helplessness and rendering people less confident than would normally be the case. Upon being admitted to an in-patient facility, patients are deprived of their ability to control their environment, particularly in relation to access to privacy and the ability to make choices; they are removed from their social support network of family and friends; and their physical mobility and opportunities to exercise are often severely curtailed.

As discussed by Ulrich (1999) and others, these deprivations and the infliction of the medical procedures themselves produce a need for psychological escape and desire for a positive distraction. Any depressive or anxious components in a person's emotional state will likely be projected onto the environment, because it is used to mirror the internal state. In this vulnerable condition, people are more apt to perceive threatening or depressive messages in the objects around them.

> A poignant example of this aspect of perception is that of a 79-year-old woman whose husband had just died from cancer. This woman attended a bereavement group for family members sponsored by the hospital. At the end of the session, a flower from the bouquet that had been on the table was given to each person as he or she left. This woman put her carnation in water when she got home, but in the morning found that it had wilted. She was extremely distressed by this and became very angry. She repeatedly expressed how uncaring and hurtful it had been to be given what she could only see as "a *dying* flower." Her experience as a recent widow was one of death and abandonment, and this flower was seen by her to embody both of these messages—the death of the flower and the "abandonment" of the group by perpetrating such a hurtful act. She was so upset that she talked about it for many days. Despite acknowledging

that she had benefited from attending the gathering, she refused to return to any subsequent meetings, thus cutting herself off from a source of caring and support (Barnes 1999).

People seek to validate their experiences as a way to find comfort from others who have felt the same. However, without realizing it, a self-reinforcing cycle is being established. This is good if it is a positive experience; however, it is not so good if one is caught in a negative situation.

These two examples illustrate part of the complexity of the "subconscious matrix through which all we see must pass" (Lewis 1996). This complexity can be diagrammed as a cycle-of-awareness (Figure 14.1). This cycle encompasses the three contributing components of Perception, Environment, and Reaction, each in turn influencing the other.

## PHYSIOLOGICAL REACTIONS

In addition to the emotional component of awareness, humans respond to their physical surroundings with bodily chemical changes as well. These chemical or physiological changes are quite evident when the cycle-of-awareness is on a downward spin. In a stressful situation, whether chronic or acute, our instinctive physiological reaction kicks in automatically. This defense reaction is commonly referred to as the fight/flight reaction. The body prepares itself to take action through increased heart rate, stimulation of the adrenal glands, and other manifestations of a hyperalert state. Historically, this made sense. In earlier times, sources of stress typically required immediate or sustained activity. In the case of the sudden appearance of a predator (a mountain lion, perhaps a crocodile) escape was called for: swift decisive exertion of physical energy. Other life-threatening situations required sustained physical activity, such as in the struggle to survive required by a drought-induced famine, or in an aggressive encounter with others where physical combat was required. Our genetically encoded defense reaction is an adaptation that prepares us for *action*.

Today the situation is very different. A review of the stressors in our contemporary culture indicates that most are those that *inhibit action*. Oftentimes our level of stress increases because we cannot take charge, cannot work, or even cannot move. Sources of stress in the workplace, for instance, often involve conflict in a social setting (perhaps a staff meeting) where the "resolution" is through negotiation or even subjugation. Many people also experience the stresses and strains of attempting to get com-

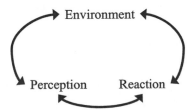

Figure 14.1.  Cycle-of-awareness.

puters to perform needed tasks. All too frequently, computers illogically and indeter-
minately shut down, erasing, delaying, or preventing completion of tasks and inter-
rupting communication. Sitting in traffic has become a stressor in our time-pressured
culture. Road rage is one manifestation of the encoded physiological changes that are
prompting action, regardless of its inappropriate application to the situation.

In these instances, the fight/flight response is maladaptive, and we are inhibited
from attaining resolutions that provide positive outcomes. The more we are blocked
by stressful situations, the less we experience (1) task completion, (2) success, (3)
value/self-worth/ability to make a difference, and (4) reward and recognition. When
these experiences are lacking in our life, they become their own stressors and feed
again into a negative spin in the cycle-of-awareness of our environment.

## ENVIRONMENTAL INTERVENTION

It becomes evident that there is a cycle-of-awareness that is self-reinforcing. It is pro-
pelled by the human tendency to seek validation of our experiences of the moment
and the outdated physiological fight/flight reaction, which more often than not
prompts maladaptive responses. The need for outside intervention to break this cycle
becomes critical when a negative cycle dominates an individual's experience of the
world (Figure 14.2).

There are two logical points of intervention in this cycle that can be supported by
appropriate therapeutic design. The first is to design to help compensate for the "lacking" experiences. Planning and management of spaces to help compensate for these
insufficiencies can be thought of as compensatory design.

### *Compensatory Design Alternatives*

The lacking experiences can be supplied through the provision of activities, opportunities, and appropriate challenges. Within the categories the specifics will vary
according to the needs of the people being served, but the goals are clear. *Activities*
can provide a sense of self-worth, ability to care for self and others, and task completion, when success is set within easy reach. An example of an activity would be the
provision of a washing line in an Alzheimer's facility for the use of the residents.

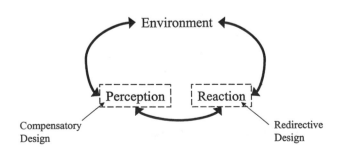

Figure 14.2. Cycle-of-awareness points of intervention.

*Opportunities* such as access to a cutting garden or a chance to work in a greenhouse can supply a chance to nurture and care for a living entity that responds and grows, or to provide gifts for friends. It provides an opportunity to direct and control an aspect of life, a chance to make a difference and to obtain recognition of one's achievements. The offering of *challenges* such as a maze in a children's garden or a rising path to a vista for older adults can provide success, completion, and a sense of self-worth, while perhaps also offering recognition from others and social contact. As these experiences, which are otherwise lacking, are supplanted, the individual begins to seek validation of these more rewarding experiences in their larger environment. This can reestablish a positive spin in the cycle, which in turn also can become self-reinforcing.

## Redirective Design Intervention

The second point of intervention is at the point of reaction, the moment the defensive response begins to surge. Appropriate therapeutic design can stem these physiological changes. By offering spaces that evoke positive responses, relaxing distractions, and opportunities for meditation or self-reflection, the environment can trigger countervailing influences, which help to shift a person into a positive perspective.

This is a well-practiced form of intervention. In Western culture, it has roots back as far as the Middle Ages, while Eastern religions are grounded in the power of nature and gardens that have historically provided space for spiritual and physical healing. Recent laboratory and empirical studies are beginning to examine how to redirect emotional patterns through environmental design. Helpful to the conceptual understanding of this type of design is the work of Herbert Benson. Dr. Benson and his colleagues (Klutz et al. 1985) have studied the fight/flight reaction and its opposite (and antidote), the "relaxation response." This response is characterized by a drop in heart rate and blood pressure, an increase in alpha waves, and the physiological indicators of a wakeful hypermetabolic state "to which nonverbal intuitive spatial-holistic, now-sequential qualities are attributed. This plasticity of cognition. . . loosens the defenses . . . These emotional alterations (provide) a sense of self or centeredness experienced as feelings of inner trust and serenity" (Klutz et al. 1985).

In exploring various methods of evoking the relaxation response including biofeedback, psychotherapy, medication, hypnosis, and others, Benson, Beary, and Carol (1994) determined that mindfulness meditation was the most effective procedure. This is described as a process involving four elements:

1. A shift in thought patterns to a more primitive receptively oriented mode of cognition (Klutz et al. 1985).
2. A passive attitude that allows one to stay with the experience of the moment.
3. A relaxed position leading to decreased muscle tonus.
4. An environment with minimal intrusive stimulus (Benson et al. 1994).

Research by Barnes (1994) indicated that the process of mindful meditation could be stimulated by a series of environmental cues. By designing physical surroundings to support this physiological shift, outdoor spaces can short-circuit the arousal of the defensive fight/flight response and instead guide one's emotional and physiological state toward an adaptive, healthy position of internal relaxation and external readiness. Drawing from design principles established to stimulate the relaxation response involves moving the individual garden user through a series of emotional phases: the Journey, Sensory Awakening, Self Awareness, and Spiritual Attunement (Barnes 1994). Closely allied with the steps of mindful meditation, these phases can evoke a "positive sense of self" described by Klutz et al. (1985). The positive shift can reverse the spin of a negative cycle-of-awareness, improving an individual's outlook. This emotional benefit in turn has ramifications for physical health, productivity, and an overall sense of well-being.

## CONCLUSION

Perception is as varied as each individual, his or her situation, and his or her mood fluctuations. Understanding the forces that drive human perception provides a conceptual framework that enables landscape designers to sensitively design spaces that provide therapeutic benefit. The cycle-of-awareness model suggests two points of intervention that can be influenced by environmental factors. Compensatory design and redirective design emerge as two distinct elements of therapeutic design. Each is compatible with the other. Establishing the primary design modality for each project will clarify design goals and provide direction throughout the implementation, thus enabling the creation of potent healing spaces.

## LITERATURE CITED

Barnes, Margaret A. 1994. A study of the process of emotional healing in outdoor spaces and the concomitant landscape design implications. Masters thesis, Dept. of Landscape Architecture, Univ. of California, Berkeley, CA.

Barnes, Marni. 1999. Design philosophy. In *Healing gardens: Therapeutic benefits and design recommendations*, eds. Cooper Marcus, Clare, and M. Barnes. J. Wiley & Sons, New York, pp. 87–114.

Benson, Herbert, John Beary, and Mark Carol. 1994. The relaxation response. *Psychiatry* 37:38.

Fox, Laurie. 1998. *My sister from the black lagoon*. Simon and Schuster, Inc., New York, p. 211.

Ittelson, William, L. Proshansky, and G. Winkel. 1974. *An introduction to environmental psychology*. Holt Rinehart and Winston, New York.

Klutz, Ilan, Joan Borysenko, and Herbert Benson. 1985. Meditation and psychotherapy: A rationale for the integration of dynamic psychotherapy, the relaxation response and mindfulness meditation. *American Journal of Psychiatry* 142(1):1-8.

Lewis, Charles A. 1996. *Green nature/human nature: The meaning of plants in our lives*. University of Illinois Press, Champaign, IL., p. 7.

Ulrich, Roger. 1999. Effects of gardens on health outcomes: Theory and research. In *Healing gardens: Therapeutic benefits and design recommendations*, eds. Cooper Marcus, Clare, and M. Barnes. J. Wiley & Sons, New York, pp. 27–86.

# 15

# Design Strategies for Integrating Natural Elements with Building Design

Phillip G. Mead

According to the American Lung Association, roughly 90% of our time is spent indoors (http://www.alaw.org/air_quality/). Compare this to the turn of the 20th century when working outside predominately in agriculture was the norm. Socioeconomic advances such as the industrial revolution and inventions like electric lighting and air conditioning moved many jobs to a more comfortable and productive indoors. But the move was abrupt, and many environmental factors changed dramatically. What kinds of side effects can this sudden move inside do to our overall health? What happens to a brain that evolved for millenniums in outside light conditions that are 10 to 100 times brighter than the average inside luminance today? What kind of respiratory problems can develop in lungs that are used to taking in ever changing outdoor air on a daily basis? What kinds of stress-related diseases result from a brain that evolved with natural views, which now takes in simplistic and monochromatic interiors? The following environmental research may provide some clues.

In the United States and most industrial nations, the two leading causes of death are heart disease and cancer. In contrast, these diseases were very low on the mortality list at the turn of the 20th century and appear to be absent from forensic investigations of archeological excavations. Many environmental factors changed dramatically during the 20th century that may provide insight. These factors include diet, work conditions, and built environmental conditions. Although strong evidence suggests that diet can influence heart disease and cancer while work conditions can lead to greater stress, there is also strong medical evidence that today's built environment could make significant contributions to heart disease, cancer, and other maladies. Despite the increased life span of today's population (which may add to one's susceptibility to both heart disease and cancer), there is strong medical evidence pointing to our built environments as the cause of these illnesses—as well as other serious health problems. These findings must be visited more widely and seriously considered.

One culprit may be indoor lighting. Indoor fluorescent light compared to daylight lacks full spectrum ultraviolet B (UV B) light, which is essential for the production of vitamin D3, a sunshine-, not dietary-induced hormone that helps bones and the rest of the body to absorb calcium[1]. Although dietary vitamin D2 (which can be toxic at high levels) can be metabolized into vitamin D3, the process is only 60% as effective as outdoor light exposure (Holick 1996). This combination of calcium and vitamin D3 not only helps form strong bones and teeth, but research also shows the combination helps regulate the body's immune system, for which bones play an integral part (Holwich and Dieckhues 1980). This contribution to the immune system has resulted in a multitude of promising studies since the 1980s that show strong links to UV B light and its role in combating heart disease and cancer (breast, prostate, and bowel).[2] Reinforcing this research is a number of surprising studies that show lower incidences of both cancer and heart disease in global areas and times of the year where the sun shines strongest (higher altitudes, lower latitudes, and in the summer) (Scragg 1981; Bartsch et al. 1993). Follow-up studies tracked immigrants who moved from lower to higher latitudes and vice versa and found the subjects appeared to assume the risks associated with the new location.

The sunshine vitamin D3 links to cancer and heart disease overshadow the UV B link with melanoma. Although excessive exposure to UV B light (as demonstrated in the 1970s suntan fad) has been linked with melanoma, there is contrary evidence that significant lack of exposure to UV B can also contribute to skin cancer.[3] Melanoma researcher Dupont Guerry, MD, states in *Melanoma Prevention, Detection and Treatment* (1998) that "mild exposure to the sun is not harmful to most people. Indeed, it may have beneficial effects. . . . It has been reported that the sort of mild continual sun exposure that produces a bit of a tan, but no burn may even protect you from melanoma." With this evidence, physicians recommend a balanced dosage of sunlight of 15 minutes a day either before 10 am or after 2 pm (Liberman 1991).

An additional concern for medical researchers is extremely low indoor light levels. According to sleep researcher Dr. Daniel Kripke of the University of California, San Diego (personal communication), humans function normally in the wake/sleep cycle when exposed regularly to light conditions of 1,500 to 2,500 lux (a unit of illumination). Typical outside light is between 5,000 to 10,000 lux. Indoor light measurements in office buildings are far lower, 300 to 500 lux (10 to 100 times lower), whereas home levels are 100 lux. According to Norman Rosenthal, a pioneer in winter depression, these low levels do not fully activate the production of the daytime neurotransmitter serotonin (Rosenthal 1998).

Serotonin is a crucial neurotransmitter that is central for giving the brain a sense of well-being. Those who suffer depression as well as anxiety lack crucial amounts of serotonin (Norden 1996). Studies show that alcoholics, violent offenders, sex offenders, and those who commit suicide show marked deficiencies of serotonin. With this evidence, the psychiatric community in the 1980s spawned a new breed of serotonin-inducing drugs known as Prozac, Paxil, and Zoloft, which are all significantly more effective than their predecessors Valium and Xanax (Norden 1996). However, serotonin levels are difficult to measure. Because light therapy has similar effects as antidepressant drugs and since winter depressives crave carbohydrates, which also raises serotonin levels, Rosenthal has hypothesized that bright light raises serotonin levels.

Serotonin converts to its chief metabolite, melatonin, at night, which is crucial for inducing sleep. Both neurotransmitters are essential for lowering stress. However, when melatonin levels are present in our brain during the day, lethargy and depression result. Bright light acts to suppress melatonin (which is easily measured) while perhaps raising serotonin in order for the brain to function normally (Rosenthal 1998).

More recent environmental research has strong precedence dating back to the ancient medical/environmental writings of Hippocrates, and the ancient Roman architect Vitruvius and the Renaissance architect Alberti, who all wrote on the health impacts of light and air.[4] The link to healthy environments was still strong at the turn of the 20th century when Neils Finson was given the Nobel Prize for scientifically proving that sunlight cured tuberculosis (Holick 1998).[5] From this era emerged the health conscious work of Modern architects Frank Lloyd Wright, Irving Gill, Alvar Aalto, Rudolf Schindler, and Richard Neutra—all who expressed how their designs promoted fresh air, light, and views.[6] Sanitariums served as model building types because doctors used sunlight and fresh air for patient recovery of tuberculosis and lupus. Outstanding examples include Aalto's Pamio Sanitarium in Pamio, Finland, and Neutra's Lovell Health House in Hollywood, California, United States. However, in the late 1930s, light and fresh air lost medical potency when sulfanilamide drugs proved a quicker cure for tuberculosis. Additionally, Alexander Flemming in 1928 discovered penicillin, which in the 1940s firmly established biochemical drugs as the first mode of treatment over environmental cures (Becker 1985).

Hospital design also changed in the early part of the 20th century from air, light, and garden-friendly winged pavilions and courtyard designs to more compact box configurations. These compact designs provided space-saving steps for nurses and doctors (Barnes and Cooper Marcus 1996), which has led to the disorienting collection of windowless rooms that make up most of today's hospitals. Additionally, this planning ran counter to earlier courtyard and pavilion designs that focused on patient recovery through the restorative qualities of light, air, and gardens (Gerlach-Spriggs et al. 1998).

In 1976, a new environmental health revolution began to emerge. The first shot came from Philadelphia when Legionnaire's disease broke out in a convention hotel and rudely awoke the profession to the realization that built environments could once again dramatically affect our health. In the 1980s, building occupants began to report allergy and cold-like symptoms that came mostly from energy efficient, hermetically sealed work places without operable windows. The early 1980s was also the time the Environmental Protection Agency announced that radon gas emitted from uranium-contaminated soils beneath buildings could cause lung and possibly stomach cancer in residents.

In 1984, the impact of environmental forces picked up momentum when the anti-depressant quality of light was scientifically recognized by Dr. Norman Rosenthal's diagnosis of Seasonal Affective Disorder (SAD) (1984). However, in 1860, Florence Nightingale foreshadowed Rosenthal in her book *Notes on Nursing*, where she wrote that patients felt better and recovered sooner in sunny rooms over dull ones. In 1984,

Ulrich's landmark study on how window views of nature can accelerate the healing process while lowering pain medication was published. From this rich assortment of research and medical practice, it appears that the more we can expose ourselves to outside natural views, light, and air, the more healthy we may be.

Building practices today are influenced by strong short-term economic forces that separate people from the outdoors. This kind of planning has resulted in oversimplified economic design strategies. One of the strongest reasons for the existence of these short-term strategies is "energy conservation," which takes place on three levels: (1) heating and air conditioning energy conservation, (2) ease and speed of building construction, and (3) ease of planning and drawing production.

In cold climates like Chicago, the ideal plan configuration that minimizes energy consumption, building construction time, and planning time is the compact thick building that is nearly square in plan, with many windowless interior rooms. From a short-term energy/production viewpoint, the most burdensome building configuration would be the opposite kind of plan with many courtyards and pavilion wings accompanied by multiple windows and doors to access the outside.

The historian Reyner Banham in his book *The Architecture of the Well-Tempered Environment* (1969) hails the 1900 Royal Victorian Hospital in Belfast, Ireland, as exemplary mechanical design because the structure maximized heating distribution fully by compacting its plan into more of a warehouse configuration. This energy-efficient design limited heat-leaking windows and turned outside-facing walls that formerly defined courtyards into easily constructed energy efficient interior dividers. By compacting the plan, heating duct runs could be shorter resulting in less heat delivery loss. From a planning production perspective, this kind of building requires fewer drawings because there are not as many elevations and exterior walls to weatherproof and detail. From a construction perspective, a reduction in exterior wall space means less assembly because outside walls usually require more structural stability, insulation, and waterproofing. Compare this to interior walls that require minimal structural stability and no waterproofing and insulation. Logistics are greatly simplified.

From a lighting perspective, thick boxy buildings are very problematic. According to sleep researcher Dr. Kripke (1996), our brain, which is accustomed to outdoor light levels of 1,500 to 2,500 lux or higher (compared to indoor levels of 500 lux), performs best when regularly exposed to these levels. However, to bring artificial light sources to that level of brightness in a windowless room would be cost prohibitive in both energy use and maintenance. Although full-spectrum lights contain UV A and in some cases UV B light, they still lack significant amounts of the color blue and cost substantially more than regular fluorescent lights. The most reliable way to achieve a full dose of natural light is to be outside. The next best solution is to be near a window or skylight.

Another significant health concern of compacted thick windowless box buildings is that of indoor air quality. Rooms located in the deep interior are at the mercy of the air conditioner for fresh air. If there are moisture and mold problems or the air conditioner is infected with bacteria or fungus, no windows can be opened to temporarily relieve the situation.

Since a high percentage of our time is spent inside, it appears imperative that our buildings provide regular opportunities for us to occupy the outside. If planned right, a building can encourage people to linger in outdoor spaces or at least take shortcuts across courtyard gardens. If proper comfort and convenience items are planned between interior and exterior spaces, then people will naturally want to be outside or next to a window. Fortunately, there are a number of precedents and design strategies that exist for casual outside interaction. Buildings built before the turn of the 20th century were designed to easily take in natural views, light, and air. Interior spaces had to be lit and ventilated by the outside, so spaces had to be shallow to capture ample light and fresh air. This resulted in a proliferation of winged pavilion and courtyard designs.

From a building plan strategy perspective, Christopher Alexander's book *Pattern Language* (1977) lists many "patterns" or plan configurations that can accommodate the outside. His patterns of Positive Outdoor Space, Wings of Light, Outdoor Rooms, Six Foot Balcony, and Courtyards that Live are but a few that encourage people to step outside.

Frank Lloyd Wright designed a number of buildings that gracefully integrated the interior with the outside. Contrast this with many modern architects who use large scale-less panes of glass to bring the outdoors in abruptly. Wright would generally make his outdoor spaces a refuge with an overhang and possibly a bench with plantings. The author calls these areas "fuzzy spaces" (commonly known as porches, decks, arcades, loggias, and verandas) because they are not totally inside or outside but occupy the nebulous area between the two. Indoor places next to a window can also be inviting. Window seats entice us to sit comfortably while taking in natural views, high levels of daylight, and fresh air—if a window is opened. Historically, Victorian and bungalow styles often integrated seating next to windows in the form of bay windows. Frank Lloyd Wright's work, which evolved from Victorian architecture, used the window seat masterfully in many of his projects.

The following projects illustrate how recent medical research and 19th and early 20th century design precedents can be applied to a beginning design studio. Neophyte architecture college students tend to think of design in terms of elevation and style. This kind of thinking can easily lead to "decorated sheds" with little outside interaction. To break this instinct, students are encouraged to see their designs beyond the isolated building plan and visualize the entire site and floor plan as a well-composed graphic image. This is not so unusual, for if one looks at the majority of master architects' site and floor plans, they usually hold together as a fine graphic composition. Wright was a master at this where his site/building plans incorporated not only the building itself but the whole site, including plants and water. The Martin and Barstall houses (Buffalo, New York, and Hollywood, California, respectively) are excellent examples. After students realize this concept, they are given an assignment, which allows them to piggyback off a master's plan to create black and white compositions. Students are then told to look at their new composition as an assemblage of closely related indoor and outdoor spaces. Some shapes can be plants, others water, a roof or an interior space. This contrasts with most beginning design problems that concentrate on the building itself with little or no emphasis for the design of exterior spaces.

The second part of this project takes a part of the design and makes it habitable. It is here where students are highly encouraged to make comfortable "fuzzy" outdoor spaces. These "rooms" can be both partially roofed and open to the sky to allow users a choice of how exposed they want to be. Planted areas are encouraged for not only relaxation and stimulation, but to help define places that offer pleasing views. The incorporation of water is encouraged to help users relax or cool off. Other amenities may include an outdoor fireplace, ceiling fans, and concrete-encased thermal water pipes. Conversely, on the inside, window seats are encouraged to ensure high levels of daylight and views of nature.

In the third part of the assignment, students are told to take a part of their graphic image and design a tower. Towers are healthy for a number of reasons. Because of their height, occupants have commanding views of neighboring buildings, gardens, and streets that may contribute to greater relaxation and a sense of control. The landscape theorist Jay Appleton (1975) calls these views a prospect, which is analogous to when our prehistoric ancestors would sit high in trees and survey the African Savannah. From this vantage point, they could feel safe from harmful animals while spotting potential food. Because we stand upright, our own bodies are like a tower, which gives us superior views over animals, who are close to the ground and rely on a sense of smell for orientation. From the height of a tower, one can be at the same level as the elegant tracery of tree branches while surveying the two-dimensional layout of a garden below. Part of the appeal of Frank Lloyd Wright's buildings is that he often raised living spaces up one floor to allow for superior views and prospect while creating a sense of privacy.

The second reason towers are healthy is that sunlight can enter from all four sides thus raising brightness levels to near-outside conditions. It is most conceivable that a person who suffers from regular depression and winter depression could sit more comfortably in the natural light levels of this space rather than staring into the glaring light of a light box.

Stairs are another way to invite interaction with the outdoors as well as promoting exercise. According to new government exercise guidelines, workouts no longer have to last for 30 continuous minutes, but can be spread out during the day. According to Dr. Michael J. Norden, author of *Beyond Prozac* (1996), more than a dozen studies support the idea that exercise can be spread out at small intervals during the day. Additionally, Norden writes that exercise also raises serotonin levels. To this end, the National Institutes of Health (NIH) in Bethesda, Maryland, has placed signs near elevators urging its office workers to take the stairs instead. Since most commercial and institutional buildings over one story must have two fire exit stairs that are separated by a 2-hour firewall, why not plan these stairs to encourage people to be near the outdoors? Most fire exit stairs are dreary concrete towers with no windows and closed entry doors that essentially say "keep out." However, these doors need not be closed when there is not a fire, so heat sensitive magnetic door closers can be installed to invite participation. (A magnetic door closer is electrical, so when a fire is sensed, the electricity is shut off, thus de-magnetizes the magnet, shutting the door). Additionally, because there is no need to separate the stairs completely from the outside, the stairs can be open to a garden. Intermediate rest areas can encourage people to linger, inviting further outside use.

## CONCLUSION

The painful image of SAD patients blindly staring into a 10,000-lux light box prompts designers to think of more humane medical design responses to those suffering from depression. The sight of hundreds of workers and students stuck inside vast interior spaces with no windows prompts one to think of healthier design responses to improve worker satisfaction and productivity. The profession needs to rethink its design strategies beyond the seduction of simplistic economies and strive for more responsible, healthful designs.

## ENDNOTES

1. While some glass will admit UV A light, all glass blocks out UV B light.

2. Findings reported in medical journals such as *Cancer Causes and Control, Lancet, International Journal of Epidemiology, American Journal of Physiology, Circulation,* and *Hypertension* in the 1980s and 90s have been reported. Dr. Michael Holick of Boston College of Medicine has sponsored a symposium on these and other light-related findings that are compiled in his two conference proceedings entitled *The Biologic Effects of Light 95,* M. Holick and E. Jung, Eds. (New York, Walter de Gruyter and Co. 1996) and *The Biologic Effects of Light 98,* M. Holick and E. Jung, Eds. (Norwell, MA, Kluwer Academic Publishers 1999).

3. Melanoma and other skin cancers have been linked with UV B light, but research has also shown that those jobs that take place mostly inside can have a higher risk of skin cancers than those jobs that are outside. F.C. Garland, M. C. White, C. F. Garland, E. Shaw, and E.D. Gorman, *Arch Environmental Health* 45: 1990, 261-267. Additionally, reports show that melanoma mostly takes place on the torso and legs, places that are not regularly exposed to sunlight. The face and hands, which are nearly always exposed, rarely contract melanoma. M. Braun and M. Tucker (1995) "Do Photoproducts of Vitamin D Play a Role in the Etiology of Cutaneous Melanoma," in M. Holick and E. Jung (Eds.), *The Biologic Effects of Light 95* (New York, Walter de Gruyter 1996). Although excessive exposure to UV B light is not advisable, many health-related articles and books recommend daily skin exposure of 15 minutes of sunlight.

4. Vitruvius and Alberti's writings are in many instances derived from the observations of Hippocrates. In *Airs, Waters and Places* and *Aphorisms,* Hippocrates writes on "place" specific diseases as they relate to wind direction, humidity, and water conditions, which are strongly echoed in the publications of Vitruvius and Alberti.

5. Although exposure to outdoor light, in particular sunshine, was generally recognized as a sanitizing agent, prevention for rickets, an antidepressant, and as a curative agent for tuberculosis, it wasn't until 1903 that Dr. Niels Finson was awarded the Nobel Prize for his scientific research on sunshine as a cure for tuberculosis. Michael Holick, "The Biologic Effects of Light, Historical and New Perspectives" in Michael Holick and Ernest Jung (Eds.), *The Bio-*

*logic Effects of Light 98* (Norwell, MA Boston, Kluwer Academic Publishers 1999).

6. F. L. Wright wrote about the unhealthful qualities of basements because of the lack of fresh air and light in his 1954 book *The Natural House* (New York, Horizon Press) while Irving Gill wrote about the sanitizing qualities of concrete floors in his article entitled "New Ideas about Concrete Floors" in *Sunset*, December 1915. R. M Schindler wrote in place of Dr. Lovell in the weekly *Los Angeles Times* column "Care of the Body" on March 14 and April 11, 1926, and Richard Neutra wrote on various health issues in his book *Survival through Design* (New York, Oxford Press, 1954).

## LITERATURE CITED

Alexander, Christopher, Sara Ishikawa, and Murray Silverstein. 1977. *Pattern language.* Oxford University Press, NY.

Appleton, Jay. 1975. *The experience of landscape.* John Wiley, London.

Banham, Reyner. 1969. *The architecture of the well-tempered environment.* The Architectural Press, London.

Barnes, Marni, and Clare Cooper Marcus. 1996. Research report: Applying the therapeutic benefits of gardens. *Journal of healthcare design*, Volume 8.

Bartsch, H., C. Bartsch, D. Mecke, and T. H. Lippert. 1993. The relationship between the pineal gland and cancer: seasonal aspects. In *Light and biological rhythms in man*, ed. J. Wetterberg. Pergamon Press, New York, NY.

Becker, Robert. 1985. *The body electric.* Morrow, New York.

Gerlach-Spriggs, Nancy, Richard Kaufman, and Sam Warner. 1998. *Restorative gardens.* Yale University Press, New Haven, CT.

Guerry, Dupont. 1998. *Melanoma prevention, detection and treatment.* Yale University Press, New Haven, CT.

Holick, M. 1996. Introduction. In *The biologic effects of light 95*, eds. Holick, M., and E. Jung. Walter de Gruyter and Co., New York.

Holick, Michael. 1998. The biologic effects of light, historical and new perspectives. In *The biologic effects of light 98*, eds. Holick, M., and E. Jung. Kluwer Academic Publishers, Norwell, MA.

Holwich, F., and B. Dieckhues. 1980. The effect of natural and artificial light via the eye on the hormonal metabolic balance of animal and man. *Ophthalmologica* 180(4):188-197.

Kripke, Daniel, and S. D. Youngstedt. 1996. Illumination levels in wake and sleep. In *Biologic effects of light 95,* Eds. Holick, M., and E. Jung. Walter de Gruyter and Co., New York.

Liberman, Jacob. 1991. *Light, medicine of the future.* Bear and Co., Santa Fe, NM.

Nightingale, Florence. 1860. *Notes on nursing.* Appleton, NY.

Norden, Michael. 1996. *Beyond Prozac: Brain-toxic lifestyles, natural antidotes and new generation antidepressants.* Regan Books, New York.

Rosenthal, Norman. 1984. *Seasons of the mind.* Bantam, New York.

Rosenthal, Norman. 1998. *Winter blues.* Guileford Press, New York.

Scragg, R. 1981. Seasonality of cardiovascular disease mortality and the possible protective effect of ultra-violet radiation. *International Journal of Epidemiology* 10: 337-341.

Ulrich, Roger. 1984. View through a window may influence recovery from surgery. *Science* 24:420-421.

# 16

# Designing Natural Therapeutic Environments

**Katie Johnson**

Instinctively, we feel a connection with nature. When we become sick, we tend to lose a sense of control over our lives and lose a sense of who we are. Nature helps to restore this connection, bringing us back to who we are and our purpose in society. The purpose of this essay is to promote the creation of healing spaces in hospital facilities. My goal as a designer is to improve the quality of life of patients through active and passive interactions with nature.

For purposes here, "nature" refers to an environment containing features found in the untouched landscape. "Healing" is defined as a restorative mental and physical process.

For centuries, the healing power of nature has been eminent. "In the middle ages hospitals incorporated courtyards where residents could find shelter, sun, or shade in a human-scale enclosed setting" (Barnes and Cooper Marcus 1999). Cypress trees were used as a symbol of healing in the 15th century (Marberry 1995). During the 14th century, the outbreak of the plague gave rise to several hospitals in Europe (Marberry 1995). Typical hospitals had courtyards filled with living plants and floor-to-ceiling windows, providing a view of nature. In the United States, a Philadelphia, Pennsylvania, hospital, built in 1856, had a linear plan that allowed views of nature from patient rooms. The modern architecture and modern technology movement in the 1920s yielded clean sterile lines, lack of color, texture, and ornamentation. This movement portrayed the human body as a machine to be repaired through technology. Death was viewed as a failure of science (Marberry 1995). This technological approach to health care sparked an interest by some designers to create healing spaces that work with the environmental and functional needs of the patient. Today, the American Horticultural Therapy Association (AHTA), the People-Plant Council (PPC), and the Therapeutic Design Committee of the American Society of Landscape Architects (ASLA) provide students and professionals of landscape architecture, horticultural therapy, urban planning, healthcare administration, social science, and psychology a forum to discuss research studies and design projects. Currently, there are nearly 100 hospitals in the United States with therapeutic gardens.

This essay will discuss the process of creating a natural therapeutic space, including the benefits of having a natural therapeutic space, the physical elements that compose an ideal space, and the experiential elements essential to a natural therapeutic space (Figure 16.1). Two case studies with good examples of how a healing space is designed will then be presented.

## BENEFITS

Studies have shown that interaction with nature improves our quality of life by lowering stress levels, muscle tension, and blood pressure, and by raising our tolerance to pain (Ulrich 1984). Ulrich (1984) also reported that patients who had a view of trees had shorter hospital stays, took less pain killers, had a minor number of complications with the surgery, and had fewer negative comments on their nurse reports compared to the patients with a view of a brick wall. Ulrich's "restoration hypothesis" was one of the first studies that scientifically demonstrated the healing qualities of nature, and became a catalyst for future horticultural therapy studies.

## CRITERIA

The physical elements of a space must be understood to successfully design a natural therapeutic space. The first consideration should be the patient group. Understanding the physical needs of the patient group is important to design a space that is inclusive and will allow for active and passive rehabilitation within the garden. Many patients may use wheelchairs or may have other mobility problems. Understanding the general range of motion for patients in wheelchairs and the range of lifting and bending that can typically be achieved with various mobility problems is necessary.

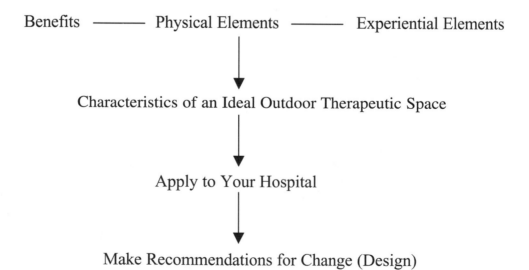

Figure 16.1.   Design process diagram for a natural therapeutic space.

Once the patient group's abilities are understood, the next step is to analyze the needs of the patients and the hospital. Most hospitals have group therapy programs as well as individual therapy programs. A therapeutic garden can provide spaces for both functions. Group sessions need a location central to the patients' rooms to easily be identified and reached, a large space (approximately 40 square feet per patient) to accommodate large group sessions, and a dynamic space that fosters conversation. Individual spaces often need to have a subcentral location for privacy and a smaller space for a more one-on-one experience.

The following characteristics should be used for both individual and group natural therapeutic environments:

- Wide spaces with gradual variations in slope are needed for patients using wheelchairs or walkers.
- Transitions between spaces and pathways should be well defined to avoid confusion. Spaces can be defined with various elements including vertical walls, benches, flower beds, arbors, paving, edges, etc.
- Paths should lead the patient throughout the space without obstructing the view to the destination. Encircling pathways are always successful as they lead the patient around the garden and back to where the journey began.
- Paving on the pathway should be distinctive from other areas that also have paving material.
- A focal point is often used to help patients stay focused on their destination and to strike their memory, orienting them to where they are going.
- A comfortable environment is important. Provide spaces for sitting in the sun as well as shady areas for comfort. Keep in mind that some patients may be on medications that cause sensitivity to sunlight.
- Incorporate flower beds of varying heights and sizes to best accommodate patients with varying abilities. Beds 24 inches long by 48 inches wide provide a comfortable gardening space for wheelchair users (Rothert 1994).
- A ledge around the bed provides a place to sit while patients are gardening.
- Vertical beds can define space and provide access for patients who cannot bend.
- Containers in different sizes can provide varying interest and access.

Active and passive rehabilitation are important concepts to consider when designing a natural therapeutic environment. As Ulrich showed, providing a visual connection from the hospital room to a natural space will allow patients to participate in passive rehabilitation. Stimulating the senses in the garden is another passive approach to get patients interested and involved in the healing process. Therefore, in designing a therapeutic garden, use plants that contrast in texture, color, smell, and sound to create interest and appeal throughout the natural therapeutic environment. Plants used should be suitable to the region. Avoid using plants that are thorny or poisonous. Creating a dynamic living environment allows the focus to be taken off the patient's illness.

## EXPERIENTIAL ELEMENTS

A healing space should be restorative. Natural therapeutic environments encompass one or more of the following qualities: quiet fascination; wandering in small spaces; separation from distraction, using natural materials; and the view from the window (Kaplan, Kaplan, and Ryan 1998).

Quiet fascination comes from an environment that permits reflection, but does not dominate one's thoughts. Activities such as walking, fishing, or gardening can bring one into quiet fascination. Using sound patterns, motion, and intensity in forms and color enhance the feeling of reflection.

Wandering in small spaces can instill the feeling of extent. A space does not have to be vast, literally, to feel a sense of escape, but it does have to constitute a "world" of its own. Extent is evoked by the feeling or suggestion that more lies behind the currently viewed space. One strategy used to feel extent encompasses the positioning of spaces so the user cannot see the entire space from any one point.

Separation from distraction is a freedom from interference with objects in the design that do not fit in its context. Enclosures or the sense of enclosure achieves separation and shifts the focus to the user. Using overhead, vertical, and understory planes brings the feeling of separation (Kaplan, Kaplan, and Ryan 1998). Materials influence the healing process. Using wood, stone, and old materials adds to natural settings rather than detracts from them.

A patient does not have to be in a setting to obtain its healing benefits. Viewing the setting from a window can evoke the feelings of escape. "A view of a little grove of trees, a natural area, a garden, a pond . . . all these provide a context for the mind to wander" (Kaplan, Kaplan, and Ryan 1998).

## CASE STUDIES

Hospitals and other institutions throughout the United States are using these concepts to create healing environments. Two examples that display the criteria mentioned earlier are the Enid A. Haupt Glass Garden at the Rusk Institute in New York City, and the Beuhler Enabling Garden at the Chicago Botanic Garden in Glencoe, Illinois.

### *The Rusk Institute*

The Rusk Institute, located in New York City, was established in 1950. The hospital is in partnership with the New York City Medical Center. The four gardens currently at Rusk were created after the building was erected. They include the Enid A. Haupt Glass Garden, the Enid A. Haupt Perennial Garden, the Alva and Bernard F. Gimbel Garden, and the children's play yard. The Rusk Institute gardens are an example of what a successful design can achieve in a harsh urban environment.

The Enid A. Haupt Glass Garden provides a place of solitude in a very busy environment and serves as a transition from the hospital room to the outside world. Nearly 25,000 horticultural therapy sessions take place here each year. Upon entering the glass garden, the cascading greenery surrounding the garden from floor to ceiling is astounding. The glass-paned roof allows the natural light to move and dance throughout the space. Birds, ranging from cockatoos to zebra finches, fly freely throughout the garden. A wide variety of plants are used. These plants can be grown in many New York City home environments. Patients can learn how to care for these plants prior to leaving the hospital.

## Beuhler Enabling Garden

Another example of an excellent creation in healing gardens is the Beuhler Enabling Garden at the Chicago Botanic Garden in Glencoe, Illinois. This garden was designed as a place for people of all abilities. This space contains many of the necessary elements that were mentioned earlier. A cascade of water is noticed when first entering the garden. Fountains and reflection ponds are centrally located and at varying heights, allowing all people to experience them. The planting beds mirror the fountains, also at varying heights. Paving directs the visitor through the space and identifies pathways from group spaces by incorporating different colors and patterns. Braille signage is used throughout the garden and a raised-relief site plan at the entrance is used to orient the visually impaired. Plants are used that vary in color, texture, and smell to evoke the senses.

## CONCLUSION

Nature can aid in the healing process of patients. Interaction with plants provides many benefits such as a reduction in stress, blood pressure, and anxiety. Natural therapeutic spaces require specific physical and experiential elements to be successful. With the research and design work that has been accomplished recently, we have several good examples of successful natural therapeutic environments. This information can be used as a starting point in designing natural therapeutic environments for hospitals or even in your own home.

## LITERATURE CITED

Barnes, Marni, and Clare Cooper Marcus, eds. 1999. *Healing gardens: Therapeutic benefits and design recommendations.* John Wiley and Sons, Inc., NY.

Kaplan, R., Kaplan, S., and Ryan, R. 1998. *With people in mind: Design and management of everyday nature.* Island Press, Washington DC.

Marberry, Sara. 1995. *Innovations in healthcare design.* Van Nostrand Reinhold, NY.

Rothert, Gene. 1994. *The enabling garden: Creating barrier-free gardens.* Taylor Publishing, Dallas, TX.

Ulrich, R. S. 1984. View through a window may influence recovery from surgery. *Science* 224:420-421.

# III

# Therapeutic Design Applications

# 17

# Implementation of Therapeutic Design Applications

**Jane Stoneham**

## INTRODUCTION

The therapeutic benefits that can come from contact with plants and the natural world have been recognized informally for centuries, but it is only in recent years that the subject has received formal research interest (Relf 1992). Much of this work has focused on people's responses to the natural landscape, while some has given attention to gardens and public open space (e.g., Francis and Hester 1993; Kaplan and Kaplan 1998; Lewis 1996). There is a limited amount of research exploring the responses and preferences of disabled or elderly people (e.g., Grahn 1994; Stoneham and Jones 1997).

This paper explores how increased interest, anecdotal evidence, and research attention over recent times have helped to highlight design and management approaches that are particularly successful in creating landscapes that aim to provide benefits to people's well-being. There is a particular focus on people who are often excluded from contact with the natural world because of disability, age, or background, and for whom therefore the benefits can be especially important. Exploring key issues such as accessibility, participation, and interpretation, it draws upon examples of good practice in design from a wide range of situations such as housing, public landscape, and institutional environments such as hospitals and schools. It also draws on the recent Making Connections study that has been exploring the benefits of contact with the natural world, and the types of barriers that prevent or dissuade use by disabled and older people (Price and Stoneham 2001).

## THE NEED FOR MORE RESEARCH

Horticultural therapy research has tended to focus on intervention and activities rather than the setting within which those things occur. There has been an increasing focus on the role of landscape as a restorative setting and the concomitant creation of "healing gardens" (e.g., Gerlach-Spriggs et al. 1998). However, there is a need for more research to look at how landscape settings can influence such issues as behavior, personal development, and social interaction and how this applies to different settings such as schools, residential homes, and public green space. For example, there is interest in the influence of landscape on the behavior and development of children, particularly as behavioral difficulties have been increasingly common (Titman 1994; Moore and Wong 1997).

There is considerable scope for applying simple small-scale survey and evaluation techniques to a wide range of projects. This work can provide useful feedback about successful design elements, user responses, spectrum of people provided for, and how to make improvements. Such lessons can be valuable in helping new projects avoid design mistakes and meet user needs more closely.

## ACCESSIBILITY IN DESIGN

To design and manage environments in ways that optimize their benefits, two key goals must be met—they must be accessible and worth visiting. Good accessibility is fundamental if people are to be able to enjoy the benefits of a therapeutic environment. There has been a tendency in the United Kingdom (UK) for an exclusive focus on one or two forms of disability (usually wheelchair users) and the physical aspects of site design (such as ramp and path design). However, accessibility is a complex issue and relies on both physical factors (such as distance from home) and sociocultural factors (such as people wanting to go somewhere and feeling comfortable there). These social factors are generally less obvious but often very significant in making disabled and older people feel excluded from the landscape.

Understandably, the first and overriding concern has been to ensure that there are no physical barriers to people using a site and to design paths and access routes to accommodate disabled people. The UK has been slow to follow the United States in introducing legislation placing a responsibility on service providers to make their facilities accessible, and the Disability Discrimination Act will not fully come into force until 2004. However, it is already having an impact, particularly in raising the whole issue of accessibility as a basic human right.

The main barriers to the use of landscape (Carr 1996; Price and Stoneham 2001) tend to include the following:

- *Physical barriers* such as steps, slopes, lack of toilet facilities, inaccessible public or private transport, and lack of accessible car parking.
- *Cultural and psychological barriers* such as lack of confidence, fear over personal safety, sticking with what is familiar, and lack of motivation.
- *Organizational barriers* such as lack of information, lack of guide dog facilities, and absence of on-site helpers or guides.

There is an increasing range of guidance material available on the topic of outdoor access. However, different guidelines are not always fully in accord, and there is rarely a set of nationally agreed upon standards. The main reason is the difficulty in trying to identify a rigid set of prescriptions that can accommodate such a diverse range of landscape types and people's needs. It is obvious that the setting of standards for a mountain trail cannot be viewed in the same light as a hospital garden. In the same way, the design requirements of someone who is blind will not always correspond with those of a wheelchair user. With technology improvements, there will also be a need to revise access standards. A current example in the UK is the use of battery-powered pavement vehicles by people with limited mobility. These vehicles have different dimensions to wheelchairs but are becoming increasingly popular and will therefore require adjustments be made to site access design. Use guidelines with some degree of caution and consider their appropriateness carefully with regard to a particular site or visitor profile. Involvement of the target audience in the planning and design stages is widely advocated as an effective way of identifying best approaches to access while also helping to develop feelings of involvement and ownership.

Other access-related issues, often overlooked, are equally important to consider. For example, access to information is important. People need to know in advance whether a landscape is accessible to them, what experience they can expect, and how they can get to and around it. There is increasing awareness of the importance of making off-site and on-site interpretation materials accessible for everyone by using a range of different techniques such as audio, pictorial illustration, large print, tactile materials, and guides. Distance is also an important factor for the significant number of people with limited mobility or stamina. It is widely acknowledged that visits to the countryside involve significant travel distances for many urban dwellers, and this reduces the number of visits made by older and disabled people. Significant distances can also be involved in visits to urban green space or within large hospital grounds. To overcome these problems, some projects have developed accessible transport initiatives both within the site and to people's homes. Another approach is to bring the experience to the people, i.e., to develop opportunities for people to have contact with nature near their homes, for example by the development of community gardens and local walks.

When discussing access, there is a tendency to regard people's travel as a problem to be minimized or eliminated. However, journey itself may offer the valuable experience, e.g., a walk or a scenic car journey. The requirement in those instances is to make sure that good access ensures that the journey is as comfortable, easy, and safe as possible.

## ASPECTS OF THERAPEUTIC DESIGN

Ensuring easy and safe access to a landscape is an important aspect of therapeutic design. Equally important is that people feel the site is worth visiting, i.e., that it offers an experience that satisfies the effort involved in going there. The following key aspects are associated with the implementation of therapeutic design.

### Designing for Contact with the Outside World

The landscape can provide a valuable means of contact with the outside world, particularly for people who spend much of their time indoors. There has been important work showing the health benefits of a view of a landscape from hospital windows

(Ulrich 1984). In a survey conducted with elderly residents of retirement housing, 96% said they enjoyed looking out of the window of their apartments and most of these views were of the landscape (Stoneham 1998). It would be useful to explore the impact on residents' mental well-being and attitudes in such environments.

This area requires collaboration among the designers of both the landscape and building. Careful window design is critical in ensuring that people can sit and look out. The landscape design is important in ensuring that there are attractive views from inside, for example, by using raised beds to present plants at a height where they can be seen.

It is important to consider that a person's decision to go outside is made indoors. The view can be important in motivating someone to venture outside. A view of an accessible path, attractive plantings, or perhaps a seat or feature within a short distance can all help provide the incentives for someone to make the effort to go out.

People who are frail or very elderly are often more sensitive to temperature extremes. Conservatories are popular features in residential and hospital design as they provide a comfortable space with light levels similar to the outdoors but with a more controlled temperature.

## Designing for Relaxation and Restoration

One of the most frequently cited benefits of contact with nature relates to restoration through lack of stress and the opportunity to unwind. Some landscapes, such as healing gardens, are created primarily for this purpose; others are designed to provide it as one of a range of options.

People who are in hospitals or residential care may find it reassuring to be within easy reach of toilets, telephones, and help and therefore close to the building. This requires the design of attractive spaces where people can sit just outside the doors of wards or rooms.

In public green space, it can be a challenge to design areas that feel quiet and private but without making the areas feel isolated and unsafe. The incorporation of water features, attractive plantings, seating, shade, and some open views can all help to create a peaceful ambience where people can relax and feel secure.

## Designing for All the Senses

The idea of multisensory design has become associated mainly with people with visual impairments. However, it is reasonable to expect that the experience of enjoying a landscape can be enhanced for people in general by considering more than just visual aspects of design.

Artwork, water, patterns of light and shade, different ambience, sounds, and an imaginative mix of plants and artifacts can all help to create a varied, rich, and stimulating experience. Scent, tactile interest, and sound are especially beneficial to people with sensory impairments as long as they are within easy reach of access paths. In school grounds, the development of multisensory landscapes has been found to be highly beneficial to the development of children with special needs (Stoneham 1996) and has the potential for combining a wide range of educational and communication techniques.

## Designing for Social Contact

Various workers (e.g., Gehl 1987) have reported the role of landscape in initiating and sustaining social contact. This can be particularly important for people who are isolated from their local community, perhaps because of limited mobility, or for people who have moved into a new residential or healthcare environment where they do not know anyone. The landscape can be a subtle vehicle for social interaction; it can provide an easy and changing subject of conversation—a place where people arc likely to mcct others without overtly seeking to do so and the focus of a social or working group.

## Designing for Involvement

Some of the most powerful personal experiences that people recount about the environment relate to feelings of involvement or sense of belonging. Unfortunately, these are often the most difficult aspects to design for, not least because their success often relies on processes beyond the design phase and in particular on the style of the management and operational systems in place. For example, in a retirement home, staff members often heavily influence the use and perception of the grounds by elderly residents. At worst, staff can deny access to gardens or create a feeling of "hands off" control. At best, they can encourage residents to feel that the garden is theirs and motivate them to enjoy it in whatever ways they wish.

Sometimes activities are introduced to stimulate involvement. There is increasing use of art as a medium to encourage participation in the landscape, for example, by running events where people create sculptures from natural materials. Some sites provide the opportunity for private activities, such as allotments where people can garden.

Contact with wildlife is another motivation for involvement. One of the most popular hobbies associated with wildlife in the UK is bird watching. The design of accessible bird hides (blinds), ensuring easy access from parking lots, and providing on-site information and guides all help to make the provision as universally accessible as possible.

One of the best examples of projects designed to encourage involvement in the UK are the City Farms. These projects were initiated in response to a concern that many urban dwellers had lost contact with the natural world, particularly the food cycle. The farms are typically small-scale, located in urban areas and containing farm animals, horticultural areas, and community gardens. They have become very successful and popular projects, involving local people of all ages and backgrounds in animal husbandry and horticultural work, and running education and volunteer programs. They have been particularly successful in attracting people who often do not participate in community programs, for example teenagers.

## Designing a Setting for Therapy

There has been a tendency for horticultural therapy programs to be designed with an exclusive focus on their operational criteria rather than considering the whole environment in which they are operating. However, the landscape can provide a supportive environment within which to conduct horticultural therapy programs, and at the same time offer social space, enclosure, shelter, and resource material for program activities.

## Designing for Connecting with the Past

Powerful associations with the past can be triggered by landscape, for example, the scent of a flower or landscape scene associated with a particular time in a person's life. There is considerable scope for incorporating this aspect into the design of landscapes for older people, for example, by selecting plants that would have been popular when those people had their first gardens or incorporating artifacts associated with that time. Such strategies have potential applications in reminiscence therapy and reality orientation programs for people with dementia. In Denmark, the interior design of nursing homes has incorporated traditional styles (e.g., from the 1950s) with the aim of creating a more familiar environment for residents with dementia (Regnies 1994). It would be interesting to apply a similar approach to residential landscapes, together with evaluation to ascertain the effect on residents' moods and behaviors.

### TOWARD THE FUTURE

This conference has shown the significant progress over the last 10 years in the development of accessible and therapeutic environments, assisted by the increasing research interest in this area internationally. However, there is a need to disseminate more widely the research that exists and the excellent therapeutic designs that have been created. We also need to identify where there are gaps in the research and encourage more evaluation of design and management approaches.

In particular, we need to communicate with people working in associated areas, such as health care, education, and social services, and to provide them with evidence and real examples of the benefits of landscape. This relies on communicating in appropriate ways. For example, healthcare practitioners will be convinced primarily by benefits to people's well-being, while educators will want to see evidence of learning and developmental outcomes. Such collaboration is essential if we are to ensure that the full range of benefits of a well-designed environment are as widely and easily available as we would all like them to be.

## LITERATURE CITED

Carr, M. 1996. *Countryside recreation for disabled people*. Cheltenham, UK: Countryside Agency.

Francis, M., and R. T. Hester. 1993. *The meaning of gardens*. Cambridge, MA: MIT Press.

Gehl, J. 1987. *Life between buildings*. New York: Van Nostrand Reinhold.

Gerlach-Spriggs, N., R. E. Kaufmann, and S. B. Warner. 1998. *Restorative gardens: The healing landscape*. New Haven, London: Yale University Press.

Grahn, P. 1994. *Green structures—The importance for health of nature areas and parks*. Planning and development in urban areas related to various patterns of life—Questions of infrastructure. The challenges facing European society. Council of Europe.

Kaplan, R., and S. Kaplan. 1998. *With people in mind: Design and management of everyday nature*. Washington DC: Island Press.

Lewis, C. 1996. *Green nature human nature: The meaning of plants in our lives*. Chicago: University of Illinois Press.

Moore, R. C., and H. H. Wong. 1997. *Natural learning: Creating environments for rediscovering nature's way of teaching*. Berkeley, CA: MIG Communications.

Price, R. C., and J. A. Stoneham. 2001. *Making connections: A guide to accessible greenspace*. Bath, UK: The Sensory Trust.

Regnies, V. A. 1994. *Assisted living housing for the elderly. Design innovations from the United States and Europe*. New York: Van Nostrand Reinhold.

Relf, D., ed. 1992. *The role of horticulture in human well-being and social development*. Portland, OR: Timber Press.

Stoneham, J. A. 1996. *Grounds for sharing: A guide to developing special school sites*. Winchester, UK: Learning through Landscapes.

Stoneham, J. A. 1998. The design and management of sheltered housing landscapes. M. Phil. Thesis, University of Reading, UK.

Stoneham, J. A., and R. Jones. 1997. *Residential landscapes: Their contribution to the quality of older people's lives*. New York: The Haworth Press.

Titman, W. 1994. *Special places special people. The hidden curriculum of school grounds*. Surrey, UK: World Wide Fund for Nature.

Ulrich, R. S. 1984. View through a window may influence recovery from surgery. *Science* 224:420-421.

# 18

# Long-Term Memory Response of Severely Cognitive-Impaired Elderly to Horticultural and Reminiscence Therapies

**Emi Kiyota**

As the number of people with cognitive impairment continues to increase, their caregivers must continue to seek meaningful and enjoyable activities to meet their needs (Namazi and Haynes 1994). Considering their age cohort and geographical background, many older adults have past experiences with horticultural activities.

Many elderly residents with cognitive impairment "live in the past" by recalling their earlier lives, former accomplishments, and significant relationships (Butler 1963). Memories can be comforting to older adults who are insecure in their present environment (Kovach and Henschel 1996). Activities involving plants can evoke self-appraisal reminiscence linking past and present (Richard Mattson, personal communication). By working with plants, older adults have an opportunity to feel proud of accomplishments at a time when they are likely to feel some degree of uselessness and powerlessness.

Research concerning therapeutic intervention during cognitive deterioration provides information on integral components of healing garden designs (Tyson 1987). Researchers have examined and described benefits of facilitating gardens in long-term care facilities for elderly residents with cognitive impairments (Mooney and Mulstein 1994; Namazi and Johnson 1992). Landscape architects have designed special gardens for people with Alzheimer's disease (Hoover 1995; Beckwith and Gilster 1997). In this research, horticultural activities involving garden design were used to measure individual long- and short-term memories. A case study approach was used to record conversation during the activity to complete content analysis. Case studies are the most effective way to examine individual differences in response.

## METHODOLOGY

### Subjects

This case study was conducted at a long-term care facility with a census of 80 elderly residents. Four elderly female residents with severe cognitive impairments partici-

165

pated in this study. Three were European-American, and one was African-American. Age range was 87 to 92 (age mean = 88.8). The Mini Mental State Exam (MMSE) scores provided by the facility were used to evaluate the cognitive abilities for three elderly residents (Folstein et al. 1975). The MMSE score was not available for subject "B." Cockrell and Folstein (1988) indicated that an MMSE score of 17 or lower frequently is an indication of severe cognitive impairment.

## *Procedure*

From a larger sample, case studies were conducted of four elderly people with severe cognitive impairment (had MMSE scores of 17 or lower). Elderly residents were informed and encouraged to participate in the activity, but could choose whether to participate or not. Each case study resident participated in a one-on-one individual activity with the researcher that lasted approximately 45 minutes. Each resident could select the place to work on the activity. The conversation during the interview was tape recorded for conversation content analysis.

Three factors influenced the choice of the horticultural therapy activity. First, this study was conducted in a long-term care facility that did not offer a garden, so indoor activities were necessary. Second, in avoiding danger for the elderly residents with severe cognitive impairment, available activities were limited. Third, this study was conducted in winter so this activity was appropriate for that season.

"Garden Designing" involved two-dimensional garden layout, designed with consideration of the intellectual and physical ability of the participants, particularly their need for visual stimulation in order to recall memories. The elderly participants were asked to design their garden with a sheet of colored paper (green and brown) and five different colors of stickers (red, yellow, white, green, and blue). These materials were set on the table where they could be seen easily. They were encouraged to select a green or brown sheet of paper to start the garden design. They also could use as many stickers as they needed for this activity. Because of the participants' cognitive and physical disadvantage, the researcher helped them to put the color stickers on the paper if requested. With this activity, subjects were questioned about their gardening experiences and plants and encouraged to talk about their experiences. The number of verbal expressions for each elderly resident was recorded as single words or completed sentences. The conversation contents were analyzed by using the Autobiographical Memory Coding Tool developed by Kovach (1993). The content of references to horticulture was categorized as flower, garden, or vegetable.

## RESULTS

The frequency of words related to horticulture during the individual horticultural therapy activity with the four residents is presented in Table 18.1. Words related to flowers were most frequently mentioned by all the elderly residents. Recalling the name of the plants seemed to be very difficult. However, names of flowers and vegetables would suddenly be remembered in the middle of a conversation. All the flowers named in the interview were relatively common to the Midwestern prairie states. A list of the flower and vegetable names mentioned in the activity is presented in Table 18.2. Of the names mentioned, 63% were flowers, 33% were vegetables, and 4% were other garden plants. Rose was recalled 16 times (29.6% of the flower names mentioned). Carnation was recalled six times (11.1%). Among the vegetable names, onion and tomato were recalled most frequently.

The colors each elderly resident mentioned in the individual activity are presented in Table 18.3. Warm yellow and red colors were mentioned more frequently than neutral white or cold blue colors. The frequency of the "past" and "present" types of conversation from the individual activity and elderly residents' MMSE score are presented in Table 18.4. All the comments made by subjects during the individual activity were evaluated as past or present, and counted by sentence. Conversations with elderly residents who had a higher MMSE score were related to the present more than to the past. The recalled reminiscences were mostly pleasant stories. On the contrary, the conversation contents of present moments were mostly questions about what they were doing because they were forgetful and confused.

Table 18.1.  The number of word associations related to horticulture mentioned by four elderly residents with cognitive impairment in an individual activity

| Subject | A | B | C | D | Mean (%) |
|---|---|---|---|---|---|
| Flowers | 30 | 7 | 66 | 25 | 72.7 |
| Gardens | 10 | 1 | 10 | 8 | 16.5 |
| Vegetables | 4 | 0 | 10 | 5 | 10.8 |

Table 18.2.  Frequency and percentage of garden plant names mentioned by four elderly residents with cognitive impairment in an individual activity

| Flowers | | Percent |
|---|---|---|
| 16 | Rose | 29.6 |
| 6 | Carnation | 11.1 |
| 4 | Sunflower / Kansas State flower | 7.4 |
| 2 | Tulip | 3.7 |
| 1 | Daisy, Geraniums, Gladiolus, Pansy, Zinnia | 9.3 |
| Vegetables | | |
| 5 | Onion | 9.3 |
| 4 | Tomato | 7.4 |
| 2 | Corn, Potato, Beans | 11.1 |
| 1 | Peas, Carrots, Red beets, Cabbage | 7.4 |
| Other garden plants | | |
| 1 | Rosemary | 1.85 |
| 1 | Christmas tree | 1.85 |

Table 18.3.   The frequency and percentage of colors mentioned by elderly residents with cognitive impairment in an individual activity

| Subject | A | B | C | D | Total No. | % |
|---|---|---|---|---|---|---|
| Yellow | 9 | 5 | 5 | 9 | 28 | 26 |
| Red | 8 | 1 | 6 | 7 | 22 | 21 |
| White | 5 | 3 | 5 | 5 | 18 | 17 |
| Green | 2 | 5 | 6 | 5 | 18 | 17 |
| Pink | | | 10 | 2 | 12 | 11 |
| Blue | | 2 | 1 | 3 | 6 | 6 |
| Purple | | | 1 | 1 | 2 | 2 |
| Total | 24 | 16 | 34 | 32 | 106 | 100 |

Table 18.4.   Frequency of past and present conversations and Mini Mental State Exam (MMSE) scores of four elderly residents with cognitive impairment in an individual activity

| Subject | A No. | % | B No. | % | C No. | % | D No. | % |
|---|---|---|---|---|---|---|---|---|
| Past | 43 | 31 | 17 | 15 | 62 | 39 | 58 | 50 |
| Present | 103 | 69 | 99 | 85 | 99 | 61 | 58 | 0 |
| MMSE score | 12 | | not available | | 8 | | 7 | |

## CASE STUDY PARTICIPANTS

The behavior and conversation between the subjects and researcher in individual activities were described. The information on each elderly resident's characteristics and behavior outside the activity sessions was provided by staff members in the facility.

## Subject A

She was usually in a happy mood and liked to talk. She had severe problems with her short-term memory and easily forgot where she was and what she was doing. Typically, she talked often about her childhood memories but never talked about the memories of her middle age. In the activity, she started with making a border with yellow and red stickers. When asked about the meaning of the border, she said, "I do not know but a garden needs to have a border." She wanted to have the border completed before she figured out what to plant inside the border. She also said she was trying to design a vegetable garden. The conversation of gardening experience brought back memories of her childhood. She talked about the memory of poverty when she was a child. When she was in elementary school, she picked up some clothing from the trash and her mother washed and resewed them for her. She said, "Those were still good enough for me to wear and also it was fun to find the clothing. We have never wasted anything. We grew vegetables in our backyard." When asked about memories related to certain colors, she always mentioned vegetables such as red for tomatoes, green for cabbages, and white for onions. When we finished making the border, she forgot that she was designing a garden. She wanted to use the designed paper as a picture frame for a photograph of her grandchildren (Figure 18.1).

## Subject B

She usually attends activity programs, but she was always passively quiet and isolated in a social setting. She appeared to be confused that she could not recognize her family members. When confused, she became apathetic, then angry. Whenever she was confused, she said, "I cannot remember many things because I was sick for a while and it took many things away." She liked and appreciated artwork because she was an artist. In terms of garden experience, she did not have her own garden but she helped in her mother's garden in her childhood. She was not positively involved in the activity, and said, "I don't feel like it." She also did not answer questions appropri-

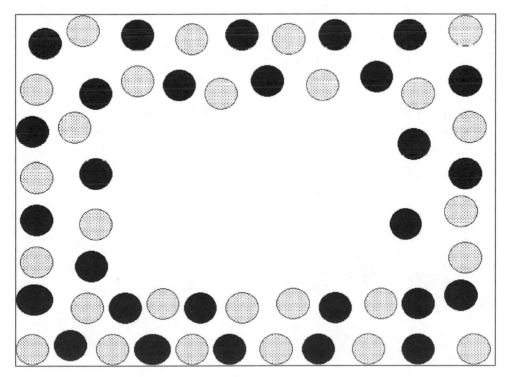

Figure 18.1.   (Subject A) This garden design by an 87-year-old female illustrates a vegetable garden border designed with red and yellow stickers.

ately. However, her attitude gradually changed through the conversation about the yellow flowers that were her favorite. She seemed to have a design idea in her mind. She chose the color, and asked for help in putting the stickers on the paper (Figure 18.2). She gradually showed more interest doing the design activity. Moreover, she did not want to stop working on this design until she was satisfied with it. Her verbal communication was limited. She made short sentences, and many of them were not completed. In our conversation, there were flowers she wanted to mention, but she was not able to remember the names. The only name of a flower she could recall was "rose."

## Subject C

She was not able to attend many activities in the facility because of her severe cognitive impairment. She was friendly and liked to talk to people, but when she was frustrated, she showed anger. She was so confused that she could not recognize her family member's faces and thought everybody around her was her relative. She had gardened in her life and raised flowers and vegetables for her church. In the activity, she talked about her plants in her garden and flower arrangements. She seemed to be confused at the beginning of the activity. She seemed to be frustrated because she did not understand what she was supposed to do. She complained about the shape of the stickers and why she had to do this activity. When we decided to talk about her garden experiences instead of designing, she started talking about all the flowers and plants she used to grow. She was easily confused during conversation and communication was difficult. However, when she talked about her gardening experiences, she remembered many details about flowers and plants. Moreover, she gave instructions in growing and preserving vegetables from the garden. "You can sow seeds at the appropriate time and season. Those will grow, if you have your flat ground, and you make rows. This is for vegetables. Let's say, you want to grow carrots, and put the rows of onions here, and put in red beets. Do you know what the red beet is? And you will put a row of red beets and . . . that's how you build your garden what you want. And if

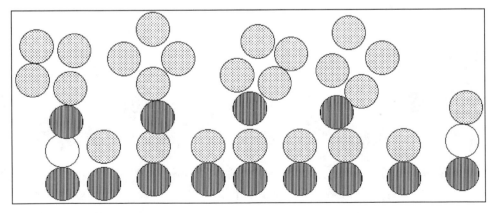

Figure 18.2. (Subject B) This garden is designed by an 88-year-old female and illustrates yellow flowers with green and white stems.

you can get good stems, you can even . . . I can a lot of stuff. Beans and peas those vegetables you can pick them and can them. I always canned a lot of stuff every year."

She was proud of herself when she talked about her gardening experience and flower arrangements for her church. The feeling of accomplishment seemed to be remembered during this activity. While she was talking about her garden, she was also putting colored stickers on the paper. Her design looked like the garden in the backyard of her house (Figure 18.3).

## Subject D

She was so confused that she did not know where she was. She frequently asked people "where do I need to sleep today?" She was always quiet and did not have behavioral problems. She recognized the familiar people, but did not remember their names. Because of her cognitive impairment, she was not confident participating in activities. The only activity she regularly participated in was church service. She always recited poems she had learned while teaching children in elementary school. Her parents were farmers, and she had gardened frequently. She frequently talked about her marriage and husband. She was not willing to participate in the activity at the beginning because she did not think she could do the design activity and answer the questions.

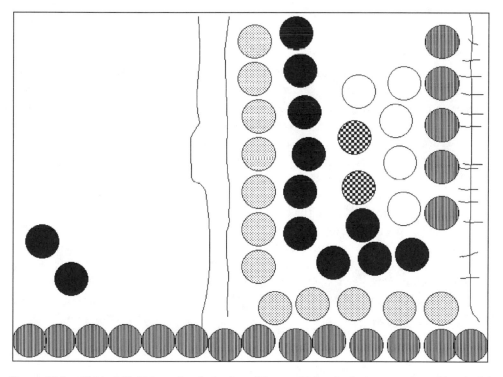

Figure 18.3.   (Subject C) This garden design by a 91-year-old female depicts a garden with colorful flowers, a pathway, and fence.

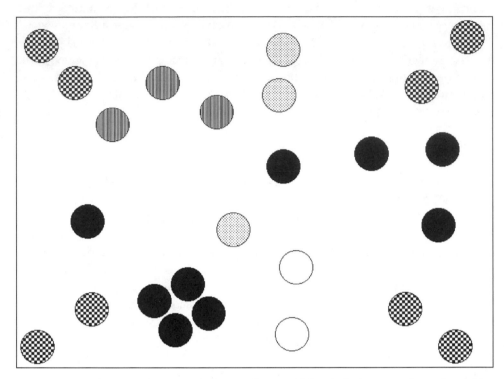

Figure 18.4.   (Subject D) This garden design by a 92-year-old female depicts many flowers in a field.

She kept saying "I am too old to remember it." She was confused about what she needed to do in the beginning. The conversation about the garden design encouraged discussion of experiences. She repeated poems from her elementary school teaching experiences. When asked about flowers and colors, poems were recited spontaneously. These poems seemed to be the answer to questions. She seemed to be visualizing a garden her father used to have. She said her family had many flowers, and mentioned names of many flowers and vegetables. Sometimes her reminiscences occurred regardless of the conversation content. During the activity, she asked to hold the paper and looked at it from a distance for a while, and then started designing her garden on the paper. She was mumbling the phrases of a poem such as "When I look down on the garden green," or "I will fly and fly . . ." and kept putting colored stickers on the paper. She did not want to talk when she was working. When asked some questions, she would ignore them and keep working. When she finished her work, she seemed satisfied with her design. She looked at her design and said, "I don't know. I guess I will fly again and fly. . . ." She mentioned that she designed the garden after her marriage. Her garden design looked like flowers in a field (Figure 18.4).

## CONCLUSION

Positive attitudes were observed during the activities. Horticultural therapy can be beneficial by providing sensory stimuli for elderly residents with severe cognitive impairment to obtain the benefit from both horticultural and reminiscence therapies. Through this research, the diversity of potential abilities of elderly residents with cognitive impairment was observed.

It is important that an activity has relevance to the elderly residents with cognitive impairment, is voluntary, and offers a reasonable chance for success. In addition, for an activity to be meaningful for this population, it must meet certain important criteria:

- Creativity: The horticultural therapy activity seemed to stimulate the elderly residents' sense of creativity. The accomplishment in designing a landscape for living materials provides the opportunity of self-expression and self-esteem maintenances.

- Visual and verbal stimulation: During the activity, subjects needed to be provided both visual and verbal cues to evoke memories. When asked similar questions without garden design activities, elderly residents showed boredom or frustration making it difficult to get information. It would be an advantage to the elderly residents in the horticultural therapy activity if plant materials were provided having visual, taste, aromatic, and tactile stimulation. Verbal cues need to be provided to gain the benefit from multiple senses.

- Relevance/feasibility: The activity should be relevant and feasible. Achievement in the activity must be within the elderly resident's physical, intellectual, and perceptual abilities. If the activities are not relevant for this population, volunteers or therapists would do most of the work. Elderly residents would be observing much of the time because it is too difficult for them, physically and intellectually. In that situation, fewer smiles and conversations would happen.

- Control: Through the activity, desire for control was observed. In a previous study, Namazi and Johnson (1992) also indicated, "autonomy and sense of control are especially important for elderly with Alzheimer's disease whose decision-making abilities are increasingly impaired." Elderly residents liked to have the researcher help during the activity, but still they wanted to control the situation. They showed frustration or disinterest when they felt that the researcher helped more than they wanted.

- Familiarity: All the subjects talked about memories related to "sunflowers." The feeling of familiarity with the flowers they have seen might bring back memories. Providing familiar material for a floral activity may stimulate long-term memories. The familiar flowers may provide a low-stress environment thus easing confusion and anxiety.

- Confidence and self-esteem: None of the elderly residents was confident about participating in the activity, and their comments were negative at the beginning. As they gradually recalled memories and become more comfortable in the activity, they started asking for opinions about their design. At the end of the activity, all of them were satisfied with what they had done and expressed positive comments.

- Short-term memory loss: They did not remember the names of the plants grown previously in their childhood or more recent past. When they were asked a question, such as "what or how" they were easily confused and frustrated because they could not remember the name. All of them forgot what they were doing during the individual activity and asked questions about what they were doing. The way of asking questions in the activity should be carefully considered for elderly residents with severe cognitive impairment because they already know they have problems with their memories. Inappropriate questions may make them frustrated and confused.

● Participation: Encouragement for elderly residents with cognitive impairment to participate in an activity needs to be considered carefully. They should be informed as often as possible and encouraged to find the activity room. The time of the activity is also important. It is important for the horticultural therapists to know the elderly resident's behavior to increase attendance and participation.

One of the advantages of horticultural therapy is its "flexibility" to adjust to many different physical and intellectual abilities. The therapist should know as much information as possible about each client with cognitive impairment for programming activities.

## LITERATURE CITED

Beckwith, E. B., and D. S. Gilster. 1997. The paradise garden: A model garden design for those with Alzheimer's disease. *Activities, Adaptation, and Aging* 22:3-16.

Butler, R. N. 1963. The life review: An interpretation of reminiscence in the aged. *Journal of Psychiatry* 26:65-76.

Cockrell, J. R., and M. F. Folstein. 1988. Mini Mental State Examination (MMSE). *Psychopharmacology* 24:689-692.

Folstein, M. F., S. E. Folstein, and P. R. McHugh. 1975. Mini mental state: A practical method for grading the cognitive state of patients for the clinician. *Journal of Psychiatric Research* 12: 189-198.

Hoover, R. C. 1995. Healing gardens and Alzheimer's Disease. *American Journal of Alzheimer's Disease* 10(2):1-9.

Kovach, C. 1993. A qualitative look at reminiscing: Using the autobiographical memory-coding tool. In *The Art and Science of Reminiscing*, eds. B. K. Haight and J. D. Webster, pp. 103-122.

Kovach, C. R., and H. Henschel. 1996. Behavior and participation during therapeutic activities on special care units. *Adaptation and Aging* 20(4):35-45.

Mooney, P. F., and S. L. Mulstein. 1994. Assessing the benefits of a therapeutic horticulture program for seniors in intermediate care. In *Healing dimensions of people-plant relations: Proceedings of a research symposium*, eds. M. Francis, P. Lindsey, and J. Stone Rice. Davis California: Center for Design Research, UC Davis, pp. 173-194.

Namazi, K. H., and S. R. Haynes. 1994. Sensory stimuli reminiscence for patients with Alzheimer's disease. Relevance and implications. *Clinical Gerontologist* 14(4):29-46.

Namazi, K. H., and B. D. Johnson. 1992. Pertinent autonomy for residents with dementia: Modification of the physical environment to enhance independence. *American Journal of Alzheimer's Disease and Related Disorders & Research* 7(1):16-21.

Tyson, M. 1987. Memories of grandma's backyard. *Journal of Therapeutic Horticulture* 8:29-35.

# 19

# Using Bonsai as a Component of Life Enhancement for Older Adults

Brian H. Santos
John Tristan
Leda McKenry

There are times during our life span when an unexpected event occurs and causes a great deal of emotional pain. There are people in our society who cannot shift their paradigm from suffering to a more positive outlook, and thus inevitably sink into depression (which is usually accompanied by an overall negative outlook on their life). They are trapped in a situation that is conceptually *prison-like*, often with little or no way to make a change for the better.

There are also people throughout our society who believe that one way to provide a pathway to healing is through the connection that can be established in nourishing one's soul by experiencing the beauty of nature. After a person can begin to nourish their soul, their soul will nurture them, and that is when things start to change. It is not the world that changes but rather the person's perception of how they view the world and the problems that they are experiencing (Carlson and Shield 1997).

Another idea that is becoming more clear in its importance with the passage of time is that the profession of nursing can provide care for people in a holistic manner (meaning that a person is cared for with the focus on the balance between the spiritual being and the physical needs of the body), and should do so when appropriate to the situation. By doing this, nurses can assist clients to become involved with those pieces of the client that call to their own sense of spirituality, love, and whatever else that may be important to them. It has been discovered that clients want these changes made for their benefit: they do not want to be prayed for by their primary healthcare provider, but rather be joined in prayer alongside their primary healthcare provider (Dossey and Dossey 1998).

Older adults often experience loneliness, helplessness, and boredom in their lives whether they are living in institutional facilities or living on their own, which can lead to depression and a quality of life that would be regarded as poor by most standards of living (Thomas 1996). Factors such as chronic physical disability, cognitive impairment, institutional confinement, diminished life expectancy, loss by death of friends/loved ones, and heavy healthcare utilization are not unique to elders yet are more commonly associated with them.

Older adults who can reduce their levels of stress and increase their perceived quality of life live healthier and happier lives (Pender 1996). Because people are unique individuals, it seems reasonable to have a variety of activities (i.e., bonsai classes and other types of therapeutic alternatives) available to be used for the appropriate individual situation, with the end result being that stress and depression are diminished and the quality of life is enhanced.

Bonsai as a horticultural therapy activity was evaluated to determine if it could provide a stimulus of focus for positive change for a group of older adults in a holistic manner. When the term *horticultural therapy* is used, the images that quickly come to mind are of people who are experiencing healing by establishing their connection with nature through the act of caring for plants. It is from this relationship that the benefits, which at first are elusive, later become apparent to the person engaged in the activity (Gough 1986). There are situations that have been documented that involve elderly clients who are impaired in their quality of life due to their physical condition, but are still engaged in relationships with their gardens and receive therapeutic rewards (Whitson 1995; Frozena 1994).

Horticultural therapy activities have facilitated personal growth for various populations. For example, these activities may include the use of Ikebana (a Japanese art that uses flower and plant arrangement to express personal stories) in combination with conversation with a therapist (Sneh and Tristan 1991) and the use of tropical conservatory greenhouses to ease the resettlement of Asian refugees into a new country of patriation (Tristan and Nguyen-Hong-Nhiem 1989). This paper reports on the therapeutic healing effects over a short length of time of bonsai as a horticultural therapy activity.

## THE REASON FOR USING BONSAI

Bonsai is an art form that is more than 2000 years old and has been carried on by many practitioners who have experienced its healing power. Currently, bonsai is very much alive as an art form and is growing in popularity in the United States. It is the authors' theory that bonsai may possess certain characteristics, inherent as an art form activity, which make it ideal for use as a therapeutic modality with all population groups, specifically older adults in this case. Among the characteristics of bonsai that contribute to its therapeutic value are patience, nurturance, creativity, nature aesthetics, and spirituality. These qualities may be experienced by anyone practicing this art. Bonsai would seem to be an ideal intervention for older adults in that it requires minimal space and is portable because of the small container in which a bonsai lives. A powerful concept that is innate to the art is its longevity, which can have two meanings. A bonsai is a small tree that lives in a small pot and with proper care can be expected to live a long life. The bonsai concept of longevity can involve the physical transference of a living tree from one caretaker to another. The older adult can pass on to a surviving family member (or acquaintance) the continuing care of a bonsai. The bonsai then takes on a very symbolic role, i.e., it can become a living link between the two people and a focal point of past memories with a particular person.

## METHODOLOGY

A bonsai workshop was offered at the Greenfield Senior Center in Greenfield, Massachusetts, serving the first 30 seniors (50 + years) on a voluntary registration basis. The number of participants was limited to 10 per workshop to allow more individualized attention and group interaction. The workshop was conducted three times, on consecutive days. Each workshop, of 60 to 90 minutes, consisted of a verbal presentation, printed handouts, and hands-on experience. A pilot study with eight participants of comparable age to the test group was conducted to evaluate the content and process of the study workshops.

Each workshop session consisted of three elements: (1) a brief history of bonsai, (2) a demonstration of bonsai creation, and (3) the creation of individual bonsai by the participants. Preworkshop and postworkshop questionnaire surveys were conducted to assess their self-reported level of knowledge of bonsai, stress reduction, overall satisfaction with the intervention for the older adult, and demographics. The questionnaires, using the Likert scale, were modified as to print size, complexity, and the number of questions.

## RESULTS

An analysis of the data from the participant responses demonstrated that the majority of the 27 participants (85%) learned a significant amount about bonsai, while a minority (15%) reported that their knowledge about bonsai had not increased or decreased. The participants' main reasons for participating were curiosity about bonsai (74%) and because it was an activity that the Senior Center had scheduled for them (63%). This suggests that these older adults trusted the staff of the Senior Center to provide them quality activities and that they were willing to try new offerings. The willingness of the seniors to participate demonstrated an interest for stimulating activities (i.e., hands-on-based activities). The stimulating activities can be a powerful motivator to engage them in activities and may indicate continued motivation for active interventions.

Results from the preworkshop questionnaire indicated that 41% of the participants were experiencing some to a lot of stress or depression, and 81% expected the workshop to be relaxing, yet only 4% selected stress as a reason for attending the workshop. Results from the postworkshop questionnaire indicated that 89% found the workshop relaxing.

Each group was asked which portion of the workshop—bonsai history, demonstration, hands-on experience—they had enjoyed the most, and the participants indicated that they had enjoyed two portions more than the third. Some participants provided more than one response; however, the majority of the participants (78%) had found the actual creation of their individual bonsai to be the most enjoyable part of the overall experience. Forty-one percent of the participants reported that the presentation on the history of bonsai had been the most enjoyable portion of the workshop. Ninety-six percent of the participants indicated they had enjoyed the workshop.

## Discussion

These results suggest that bonsai has the potential to be used as a therapeutic modality to assist older adults to develop new interests in horticultural topics and reduce their stress. It may also offer an outlet for creativity and nurturance. The participants in this project also exhibited a high degree of satisfaction with their bonsai upon completion of the workshop, as reported by the staff of the Senior Center and the results of the questionnaires.This study has helped to demonstrate that a bonsai activity may yield favorable results in a short time. Similar projects could focus on comparing the short-term effects to the long-term effects (in terms of the overall therapeutic value). Measurements might include stress levels, perceptions of stress reduction, workshop attendance over a specified time frame, and the frequency of required health care. Activities based upon horticultural arts have the potential to be used for other population groups with restricted quality of life, such as those with chronic illness, institutionalized individuals, and those who are both physically and cognitively challenged.

Perhaps this type of activity could be used as an adjunct to a larger horticultural therapy program with an audience that has a basic understanding of horticultural principles. Yet another possibility is that it can be used in an ongoing class, having weekly meetings for a period of 3 to 5 weeks. This would allow participants time to thoroughly understand all aspects of what is being taught to them and foster introspection, stress reduction, and deliberate care.

## Literature Cited

Carlson, Richard, and Benjamin Shield, eds. 1997. *Handbook for the soul/Handbook for the heart.* New York, One Spirit.

Dossey, Barbara, and Larry Dossey. 1998. Attending to holistic care. *American Journal of Nursing* 8:35-38.

Frozena, Cynthia. 1994. What flowers are for: Wisdom was the yield of a backyard for this ailing stubborn nonagenarian and her nurse. *American Journal of Nursing* 9:68-69, 71.

Gough, William. 1986. A growing interest. *American Journal of Nursing* 2:165-166.

Pender, Nola. 1996. *Health promotion in nursing practice.* 3rd ed. Stamford, CT, Appleton & Lange.

Sneh, Nili, and John Tristan. 1991. Plant material arrangement in therapy. *Journal of Therapeutic Horticulture* 6:16-20.

Thomas, William. 1996. *Life worth living: How someone you love can still enjoy life in a nursing home.* Acton, MA, VanderWyk & Burnham.

Tristan, John, and Lucy Nguyen-Hong-Nhiem. 1989. Horticultural therapy and Asian refugee resettlement. *Journal of Therapeutic Horticulture* 4:15-20.

Whitson, Anne. 1995. Seeing forever. *American Journal of Nursing* 6:80.

## Recommended Reading

Bloomer, Peter, and Mary Bloomer. 1986. *Timeless trees.* Flagstaff, AZ, Horizons West.

Palmer, Elizabeth. 1988. *The art of Japanese flower arranging.* London, Quantum.

# 20

# Horticultural Intervention as a Stress Management Technique among University of Massachusetts/Amherst Students

John Tristan
Mary Anne Bright
Michael Sutherland
Chantal Duguay

## INTRODUCTION

Horticultural activities have been helpful as therapy intervention among the physically and mentally disabled, the elderly, displaced refugees, substance abusers, public offenders, and the socially disadvantaged. Studies show that working with plants promotes health and well-being in these and other groups (Johnson 1999). College students, a population vulnerable to significant stress, have not been a focus of previous horticultural therapy research. This study was undertaken to evaluate the effectiveness of a horticultural therapy intervention to determine its effect on students' perception of their stress levels. The horticultural therapy intervention was a therapeutic tour that used visual stimulation, guided imaging, and aromatherapy in a greenhouse/garden setting.

## STRESS AND THE COLLEGE STUDENT

Stress is an arousal of both mind and body in response to the demands of life. Stress is inherent in the college student role. Taking tests, meeting deadlines, managing complex class schedules, as well as addressing roommate issues, dating relationship conflicts, and family disruptions are intrinsic to college life. Zitzow (1984), in a study of 1,146 students from four colleges across the country, reported the top six rated sources of stress reported by the students were all academic in nature, with self-induced pressure to earn good grades ranking number one and studying for tests being number two. Students can respond to the stress of college, academic pressure, time management issues, and personal conflicts in various ways, including depression and decreased health (Bush et al. 1985; Murphy 1996).

On the Amherst campus of the University of Massachusetts (UMass/Amherst), many of the students seeking support from mental health practitioners at the Univer-

179

sity Health Services do so for stress-related problems of adjustment disorder and anxiety. Referrals that require mental health services are needed by approximately 7 to 8% of the student population on an annual basis (Teixeira 1999).

The effects of stress can be manifested in numerous ways including depression, substance abuse, eating disorders, violence, and overall diminished health. When the totalities of the effects of stress are considered, the impact on a college campus is significant and the requirement on support staff is demanding. Many students also deny assistance or are reluctant to seek professional help even though they are experiencing difficulty handling stress.

## RESEARCH POPULATION

The student population of UMass/Amherst was the sample population for this study. The UMass/Amherst enrollment is currently at 24,000 students with 1,200 full-time faculty. Student subjects were directly recruited by campus advertisements in residence halls and the Student Center that advertised a free tour of the Durfee Conservatory at a variety of dates and times. Participating students were also referred from assorted academic classes, the School of Nursing, and UMass/Amherst mental health services. Participation was voluntary, anonymous, and required a waiver of consent. Consent for the study required the review and approval of a University Human Subject Review Committee. The students understood that they would participate in a tour and at the conclusion of the tour, complete a mood assessment survey.

The study occurred over three consecutive weeks during the spring 1999 semester. The final test population was 137 although more than 200 students participated in the horticulture intervention. Exclusionary criteria to qualify as a research subject excluded participants that were nonstudents, horticultural therapy students, Durfee student employees and volunteers, and Durfee visiting regulars. The prime objective was to acquire an unbiased selection of student participants that had minimal previous exposure to the Durfee Conservatory.

## RESEARCH SITE

The horticultural intervention was a specially designed greenhouse tour at the Durfee Conservatory and Gardens. Established in 1867, the Durfee Conservatory is a public, nonprofit, educational institution on the UMass/Amherst campus. Since 1980, a horticultural therapy program has serviced various populations referred by local human service organizations (Tristan & Nguyen-Hong-Nhiem 1989; Tristan 1992). These populations have included students, socially disadvantaged youths, Asian refugees, older adults, special needs youths, and public offenders.

The Durfee Conservatory is located at the center of campus and is readily accessible to the student population. A wide variety of material is available (more than 500 species) and includes ornamentals, medicinal plants, a meditation Japanese-style bamboo/stone garden, a pool with aquatic life, a sculpture formation, and an assortment of full-sized tropical trees including banana, papaya, palms, coffee, star fruit, and cocoa. This beautiful setting (complete with water sounds, lush foliage, and flowers) was considered an ideal setting to undertake a student stress study.

## METHODOLOGY

The tour consisted of a commentary that blended garden history and plant lore with assorted visual and other sensory activities. The activities included aromatherapy, which used fragrances at specific intervals throughout the tour, tactile stimulation, and visual contemplation. The students were guided by a horticultural therapist. The tour guide blended these activities together with dialogue of relevant details on horticulture, botanic knowledge, and garden history. The intention was to capture the interest and attention of the students.

Plants with aromatic qualities that were featured during the tour included *Rosmarinus officinalis* (Rosemary), *Pelargonium citrosum* (Citronella), *P. tomentosum* (Peppermint geranium), *Osmanthus fragrans* (Fragrant olive), *Jasminum officinale* (Common jasmine), *Gardenia jasminoides* (Common gardenia), *Citrus limon* (Lemon), *C. mitis* (Panama orange), *Oncidium* sp. (Orchid), *Illicium anisatum* (Chinese anise), *Plumeria rubra* (Nosegay frangipani), *Hymenocallis latifolia* (Spider lily), *Lantana camara* (Yellow sage), *Carissa macrocarpa* (Natal plum), *Mentha* sp. (mint), and *Datura immoxia* (Angels' trumpets). Plants with tactile qualities that were featured during the tour included *Mimosa pudica* (Sensitive plant), *Streptocarpus saxorum* (Cape primrose), *Episcia cupreata* (Flame violet), *Adiantum* sp. (Maidenhair fern), *Helxine soleirolii* (Baby's tears), *Asparagus densiflorus* 'Meyerii' (Asparagus fern), *Monstera deliciosa* (Mexican breadfruit), *Selaginella* sp. (Spikemoss), *Bambusa vulgaris* (Common bamboo), and *Tillandsia usneoides* (Spanish moss). Visual contemplation was facilitated through flowering displays, sculpture, and bonsai specimens.

The 30- to 45-minute tour required active participation, and the routine did not vary between groups. Group size was between 8 to 12 students per tour to ensure student involvement and supervision. Multiple tours, led by the same horticultural therapist, were conducted to service more than 200 student participants.

Immediately at the conclusion of the tour, students were required to respond to a mood assessment survey. The questionnaires were distributed to the students at the exit of Durfee Conservatory.

The mood assessment survey was designed as a posttest self-evaluation. The survey consisted of 22 questions that allowed students to rate their perceived stress and relaxation level before and after the intervention tour on a Likert scale from 1 to 5. The wording of the questions (either positively or negatively directed) was used as a check for potential response bias for those that always tend to agree (or disagree) and for those not fully reading the question. Correlation of answers helped ensure that answers were logical, consistent, and genuine.

## RESULTS AND DISCUSSION

Table 20.1 shows the results of 14 questions from the posttest self-evaluation specifically about stress and the tour intervention. Agreement or disagreement is directly related to the wording of each question (either positive or negative). This reveals correlation and trend accuracy (i.e., "tour made me less tense" [strong agreement] versus "tour visit raised stress" [strong disagreement]). Although most participants in this study (92%) felt they were *under stress* (high to low) prior to their tour intervention, a uniformly high proportion (96%) indicated that they felt *more relaxed and less stressed* after the tour (Table 20.1). Participants who reported low pretour stress also reported being more relaxed and less stressed after the tour intervention; thus, stress is not only reduced for those feeling high amounts of stress but can help maintain or improve one's sense of relaxation before stress can be perceived. This indicates that such tours could help ward off stressful feelings before they cause discomfort or become problematic. It is important to note that regardless of the reported state of stress, the intervention does no harm, and only promotes positive effects. No participant reported disagreement with this or agreed that the tour raised their stress level. In addition to stress relief, essentially everyone reported that the tour was informative and a beneficial experience for them, and that they felt a horticultural therapy program would help students better manage stress.

The impact of the tour on self-report of stress reduction was no different between students who had been to Durfee Conservatory before (86%) and first-time visitors (14%) (96% and 98%, respectively). That both groups similarly reported that the tour reduced their stress indicates that familiarity with the setting does not diminish the stress reduction effect of a tour intervention. Statistical analysis verified that the tour

Table 20.1. Participants' responses to 14 items from the 22-item post-tour self-evaluation survey

| Topic | Strongly Agree | Agree | Undecided | Disagree | Strongly Disagree |
|---|---|---|---|---|---|
| Having average amount stress | 42 | 73 | 5 | 15 | 2 |
| Have less stress than average | 1 | 13 | 18 | 76 | 29 |
| School adds to stress | 53 | 71 | 7 | 6 | 0 |
| Stress not related to school | 11 | 36 | 7 | 59 | 24 |
| Pret-tour stress was high | 19 | 66 | 18 | 29 | 5 |
| Pret-tour stress was low | 2 | 40 | 17 | 61 | 17 |
| Tour was enjoyable | 90 | 43 | 4 | 0 | 0 |
| Tour was informative | 72 | 63 | 1 | 1 | 0 |
| Tour did not benefit me | 0 | 1 | 1 | 46 | 89 |
| Tour made me less tense | 67 | 65 | 5 | 0 | 0 |
| Visit did not affect stress | 0 | 0 | 9 | 86 | 42 |
| Visit raised stress | 0 | 0 | 1 | 42 | 94 |
| Think helpful for stress | 91 | 40 | 5 | 1 | 0 |
| Program not beneficial | 3 | 4 | 5 | 37 | 88 |

effects were not due to chance but were directly and significantly the result of the guided tour interventions. Since chance improvement should occur but half of the time (improve versus not improve), we would expect half of the total 137 participants to agree and half not to agree. A chi square with one degree of freedom gives p<.001. In fact, we can see in Tables 20.2 and 20.3 that neither prior stress status nor prior visitation status had any effect on the overwhelming positive aspects of the tour.

Dual-sided questions (positively and negatively directed) were also analyzed for response bias by observing the level of expected data reversion for responses to be valid and reflective of their true opinion (Table 20.4). Correlation of answers helped ensure that answers were logical, consistent, and genuine responses. An analysis of the four reversed questions indicated an almost perfect reversal of responses, and no

Table 20.2.  Participants' percceived stress levels after a conservatory tour related to pre-tour stress

| Before tour felt under stress | Response to the statement "the tour made me feel more relaxed or less stressed" | | Subject Totals |
|---|---|---|---|
| | Strongly Agree or Agree | Undecided | |
| Yes | 103 (99%) | 1 (1%) | 104 |
| No | 30 (91%) | 3 (9%) | 33 |
| Total | 133 (97%) | 4 (3%) | 137 |

Table 20.3.  Participants' perceived stress levels after a conservatory tour related to familiarity with the site

| Have you been to Durfee Conservatory before? | Response to the statement "the tour made me feel more relaxed or less stressed" | | Subject Totals |
|---|---|---|---|
| | Strongly Agree or Agree | Undecided | |
| Yes | 87 (96%) | 4 (4%) | 91 |
| No | 45 (98%) | 1 (2%) | 46 |
| Subject Total | 132 | 5 | 137 |

response bias was evident. Chi square for reverse symmetry for the following positive/negative questions:

- On average stress was 8.2 ([p=.017] Table 20.4A)
- On school stress was 39.8 ([p<.001] Table 20.4B)
- On pretour stress was 1.2 ([p=.55] Table 20.4C)
- Not necessary for stress from the tour (Table 20.4D)

The questions on average stress (Table 20.4A) show strong symmetry, but it is the surplus of undecided in the reversed question that yields the significant chi square. The questions on school stress and stress from the tour offer no evidence of response bias and show strong symmetry of reversal (Table 20.4C and 20.4D). The questions on school stress are an exception. Upon reflection, we realized the questions themselves were not clearly reversed. A participant could agree that school adds to stress and agree that stress is not solely related to their school stress if experiencing a personal stressful situation, such as difficulties in a relationship. Significantly, however, a high proportion of students reported that school does add to their stress (90%). This correlates with previous studies that identify sources of student stress (Bush et al. 1985; Murphy 1996).

## CONCLUSION

Based on these data, it can be concluded that a horticultural intervention of a tour of Durfee Conservatory involving all five senses and led by a horticultural therapist was

Table 20.4.   Response to positive/negative paired questions

| Question | Strongly Agree/ Agree | Undecided | Disagree/ Strongly Disagree |
|---|---|---|---|
| **A.** | | | |
| Have average amount of stress | 115 | 5 | 17 |
| Have less stress than average | 14 | 18 | 105 |
| **B.** | | | |
| School adds to stress | 124 | 7 | 6 |
| Stress is not related to school | 47 | 7 | 83 |
| **C.** | | | |
| Pretour stress was high | 85 | 18 | 34 |
| Pretour stress was low | 42 | 17 | 78 |
| **D.** | | | |
| Tour made me less tense | 132 | 5 | 0 |
| Visit raised stress | 0 | 1 | 136 |

highly effective in reducing students' perceptions of stress. It can also be concluded that regardless of whether the subject felt under stress before the intervention or whether the subject had been to Durfee Conservatory prior to the intervention, the intervention is still effective for the overwhelming majority of the subjects. It is important to note that having been to Durfee before did not increase or decrease the effect of the intervention. This supports the idea that the power of the intervention does not diminish with exposure and negates the notion that those who were exposed to the conservatory previously were biased toward its effectiveness.

The data also suggested that the students of UMass/Amherst do in fact suffer from academic-related stress, much like other reported research. The majority of the students who took part in the survey agreed that academic-related matters do add to their stress and that, in general, they have an average amount of stress in their lives as a college student, if not more. However, it was found that having academic-related stress, or any stress in general, does not significantly affect the outcomes of the horticultural intervention. The reduction of perceived stress occurred whether or not the students identified themselves as stressed prior to the intervention.

The finding that having or not having stress before the horticultural intervention had no significant impact on stress reduction, and has important implications for horticultural intervention and managing students' stress. Students who came to Durfee Conservatory with low or undecided levels of stress still had a self-reported reduction in stress. This suggests that such an intervention is not only successful in alleviating already stressed students but also helps them manage their stress before it becomes a problem. The majority of the participating students was receptive to having a stress management program involving horticultural therapy at the UMass/Amherst to help its students deal with their stress, and indicated the need for innovative stress reduction methods for students such as horticultural therapy.

This study may also suggest the importance of the horticultural therapist in addition to the plants themselves. The therapist's involvement allows for expert facilitation through the intervention adding skill and knowledge as well as the human element to the subject's experience. Without the horticultural therapist, the student's experience might certainly have been a different one, possibly with less reduction of stress. Maximizing the benefit of an effective interactive tour will support healing, teach alternative coping skills, reconnect students with the serenity of nature, and generally improve life quality. As a complement to collegiate mental health services, horticultural therapy can yield positive results on the college campus.

## LITERATURE CITED

Bush, H. S., M. Thompson, and N. Van Tubergen. 1985. Personal assessment of stress factors for college students. *Journal of School Health* 55(9):370-375.

Johnson, William T. 1999. Horticultural therapy: A bibliographic essay for today's health care practitioner. *Alternative Health Practitioner* 5(3):225-232.

Murphy, M. C. 1996. Stressors on the college campus: A comparison of 1985 and 1993. *Journal of College Student Development* 37(1):20-28.

Teixeira, Natercia. 1999. Personal communication, University of Massachusetts Health Services, Amherst, MA, Spring.

Tristan, John. 1992. *A History of Durfee Conservatory*. Amherst, MA, Sara Publishing.

Tristan, John, and Lucy Nguyen-Hong-Nhiem. 1989. Horticultural therapy and Asian refugee resettlement. *Journal of Therapeutic Horticulture* 4:15-20.

Zitzow, D. 1984. The college adjustment rating scale. *Journal of College Student Personnel* 25:160-164.

# 21

# Fertile Ground in Long-Term Care Medical Facilities

Carole Staley Collins

## INTRODUCTION

The purpose of this paper is to review the Eden Alternative as a therapeutic aspect of elder health care and a research opportunity in human issues in horticulture. In the early 1990s, Dr. William Thomas introduced the Eden Alternative, an experiment in long-term health care. His goal was to change the dreaded stereotypical nursing home environments into home-like healthy habitats through an infusion of plants, animals, and children. More than 200 medical facilities now use the Eden Alternative.

With 76 million baby boomers approaching retirement, the outlook for individuals who may need long-term medical care is under scrutiny (Tavormina 1999; Uhlenberg 1997). A review of published literature from database searches including print media provided the content for this paper. These documents were primarily anecdotal and offered observational summaries of work in progress rather than completed studies. Personal visits and interviews at two facilities implementing this healthcare initiative provided the data that form the basis for this commentary from a researcher's perspective.

## ABOUT THE EDEN ALTERNATIVE

The Eden Alternative is more than a program; it is an attitude and a philosophy that can only be called a revolution in the institutionalized nursing home environment. It is the creation of a Harvard Medical School graduate, William Thomas, MD, and Judith Meyers-Thomas. Dr. Thomas's two texts outline the development of his ideas and present the philosophy behind the movement (Thomas 1996, 1998). Although his goal was not to re-create Eden but to develop healthy human environments that resemble homes, nonetheless, the title, the Eden Alternative, befits an organizational transformation that makes plants, gardens, children, and animals central to everyday life in these institutions. Indeed, the advantage of a home like the Garden of Eden is referred to in this passage from *The Herbal* by John Gerard (1597) as cited in *Gardener's Delight: Gardening Books from 1560 to 1960* by Martin Hoyles:

> Talk of perfect happiness or pleasure, and what place was so fit for that as the garden place wherein Adam was set to be the herbalist? (1994)

Thomas insists plants, pets, and children are vital components of the habitat he is promoting. His wish is to inspire "elder gardens, places where not only do green growing plants take nourishment and give sustenance, but human beings grow as well" (Vitez 1998). The Eden Alternative proposes "residents should have close and continuing contact with as much of the human habitat as they choose to embrace" (Thomas 1998), emphasizing the residents have a choice to interact but that the opportunities must be encouraged and fostered.

Residents, families, and staff work together to achieve a balance between implementing the Eden Alternative and regulatory requirements. It is important to note that state surveyors' (health inspectors') responses have been favorable overall (Thomas 1996; Bruck 1997). For example, residents might select appropriate animals from the local Humane Society or pet store and a veterinarian would certify the pet's health status as acceptable before the pet is allowed resident contact at the nursing home. If the trial period of adjustment goes well, these animals are gradually introduced into the home as full-time pets. This activity differs from having a therapeutic pet visitation program where an animal visits on certain days or hours.

Similarly, in the case of children, having local school groups visit is quite different from the kind of relationships engendered when a day care or after school program resides on site. The result of such distinctions is that spontaneity abounds. When children and animals are part of the home in a nonprogrammatic way, the atmosphere in the institution changes. The attention to integrating plants and/or gardens, the third vital component, will be discussed in greater depth in a later section.

## HISTORICAL VIEW

The Eden Alternative began in 1992 as an experiment in long-term healthcare, specifically nursing homes, and the program has mushroomed over the past decade. Why this has occurred is not very clear, but cost could be considered a major incentive, with improved public image as a close second. Its founder estimated there is potential recouping of $1.25 billion in annual government savings in healthcare expenditures (Griffin 2000).

Its founder's initial work in the early 1990s at the New York 80-bed Chase Memorial Home and a comparison facility remains impressive. His data suggested a 15% decrease in mortality, a 50% reduction in certain infections (Thomas 1996), a reduction in patients' use of antianxiety and antidepression medications, which lessened costs, and 26% fewer certified nursing aides left the institution after the Eden Alternative started. Three other New York State facilities followed suit (Bruck 1997). Since the details of his early studies are well described in his book (Thomas 1996) and these outcomes are available in the literature (Weinstein 1998; Tavormina 1999), they will not be repeated here.

Perhaps this initiative has been welcomed all the more because this sector of the healthcare industry had admitted a need for image reform and transformation of the nation's approximately 17,600 facilities to "places for living; not dying" (Wolfe 1999). The quality of care of these "feeble old people" (Uhlenberg 1997) receiving care in nursing homes has been the subject of strong criticism for decades, but now the number receiving care has reached 1.6 million (Pear 2000) at a cost of $45,000 per resident per year (Uhlenberg 1997). In the midst of this climate, the Eden Alternative was born and appears to be one of the first long-term care reform movements to capture attention at state and national levels, whatever the reasons (Scott 2000; Bruck 1997; Anonymous 1996).

Clearly, the consensus is that this experiment has taken hold. The Council of State Governments reported Missouri became the first to adopt the Eden Alternative philosophy statewide but other states offering funding or developing coalitions to support the model included Michigan, Minnesota, New Jersey, North and South Carolina (Griffin 2000), as well as Alabama, Indiana, and Texas (Anonymous 1996). According to the *Philadelphia Inquirer*, one of the first nursing homes in America to "Edenize" was the Methodist Nursing Home in Charlotte, North Carolina, and this occurred over a 2-1/2-year period (Vitez 1998).

Interestingly, the expansion of this movement has occurred with a modest number of published articles (fewer than 25 located at the time of this presentation) to support its implementation. No large-scale traditional scientific studies of outcomes have been published in research journals to date except that of the founder, William Thomas. A few publications reported very specific study outcomes that were inconclusive, as in comparing a culture of "leadership versus renovation in changing staff values" (Bond and Fiedler 1999), or were difficult to interpret. For example, physical therapists at one institution examined domains of residents' functional status, then classified observed interactions, and concluded, "therapists can be proactive in recommending environmental changes that may maximize residents' function in multiple domains" (Hinman and Hey 1999).

Some studies are under way at several institutions: Southwest Texas State University, University of Minnesota (Wolfe 1999), Christian Church Homes of Kentucky, Inc., with the University of Louisville (Griffin 2000), and the University of Texas Medical Branch in Galveston (Drew and Brooke 1999). Southwest Texas State University's Institute for Quality Improvement in Long-Term Care has preliminary data from a 2-year study in six nursing homes that mirror the original work by Thomas in New York facilities. These data suggest important reductions in residents' need for antidepressant and antianxiety drugs and safety restraints, as well as a reduction in pressure sores (Griffin 2000). Despite ongoing research, there is a lack of published studies validating effectiveness of the Eden Alternative. Nonetheless, this transformative approach has been adopted in more than 200 medical facilities (Scott 2000; Vitez 1998). As such, the Eden Alternative provides fertile ground as a research opportunity.

## IMPLEMENTING THE EDEN ALTERNATIVE

After Eden Alternative training for a few key individuals is completed, ideas are gradually introduced to residents, staff, and visitors. Each agency personalizes how it will become "Edenized," that is, how it will evolve, how fast it will happen, and who wants to be responsible for each aspect. The entire healthcare organization begins to examine ways to challenge predictability and sameness because of the belief that the vital components of the habitat provide the antidote for "loneliness, boredom and helplessness" (Thomas 1996).

Thomas's training and texts outline the particulars of the Ten Principles of the Eden philosophy to be enacted. How exactly this program is to be carried out, however, is purposely not prescribed. Instead, the way to enact these principles is interpreted by the residents, visitors, staff, and administrators in each facility. For example, although the goal is to create an atmosphere that simulates the rhythms of the natural world, facilities may interpret their capabilities differently and bring about change in different ways. Some settings began by adding gardens and horticulture activities; other agencies have put the emphasis on playgrounds, colocating day care centers or other community-based integrative programs.

In the individualization of how the program is carried out in each setting lies the potential for research. Ideally, as new programs are proposed and introduced, research to track their progress should be initiated. The evaluations should include measurements to follow changes in patient outcomes. Aside from patient outcomes and staff satisfaction, it follows that cost becomes a major consideration in what programs continue and what is researched. A logical question at the outset includes, "How much does this program cost the institution?"

*Executive Solutions for Healthcare Management* reported one care center had an up-front training cost of $75,000, excluding ongoing orientation, and maintaining hundreds of plants and animals cost $10,000–15,000/year (Anonymous 1999). At the time of this paper, Thomas's training program used 16 coordinators nationally (Evans 2000), and it still costs approximately $1,000 to train the individual(s) who, in turn, will be responsible for subsequent education within the facility (Swartz 2000). In Utica, New York, St. Luke's Memorial Hospital (160 beds) reported $30,000 in initial costs for plants and care of the animals (Evans 2000).

To put this review of the Eden Alternative in perspective, this author visited two facilities new to "Edenizing." The first setting, Levindale Hebrew Geriatric Center and Hospital in Baltimore, Maryland, is the first of its kind in Maryland. Goodwin House West in Virginia became registered in January 2000, having completed the training programs and showed commitment to the 10 guiding principles (Scott 2000).

## SITE VISITS

As announced by Levindale President and Chief Operations Officer Ron Rothstein in a recent press release, Levindale, a 292-bed facility, began its journey in September 1999. Rothstein added that "we strive to bring meaning to everyday life here because, no matter how limited we are physically or otherwise, we all need to have a sense of purpose in our lives and have fun at the same time" (Minkove 2000). Rothstein "expects Eden to add $100,000–$150,000 a year onto the home's $30 million operating expenses" (Katz-Stone 1999) but is expecting "it will serve to destigmatize the nursing home environment, to create a place that reflects the community, rather than a place where elderly are cut off from society and isolated."

Twenty local Jewish schools and synagogues have already joined their "Adopt a Resident" effort to connect volunteers from schools and synagogues with the residents who have few visitors. The fauna brought in have been primarily birds—all were declared free of psittacosis. One staff member has had a family pet checked and now the dog spends the day at the home when this individual is working. Residents, administrators, and staff are adjusting to this new environment described as "much more than fur and feathers" (Swartz 2000).

Introducing hundreds of plants, numerous pets, and children has not been problematic in terms of allergies (Thomas 1996). Allergies have not emerged as a problem at Asbury Care Center with 104 beds (Stermer 1998), Levindale (Swartz 2000), or nationally, according to one of the several Internet links to sources on the Eden Alternative website (*http://www.edenalt.com*). Rather, the Eden Alternative seems to present itself as a profound organizational restructuring that changes practice. It successfully applied key elements of the total quality movement and has made organizations rethink about the word "environment" in a totally novel way.

Metaphors about gardens and growing are everywhere. The language is written into the analogies in Thomas's texts, taught in the training, and spills out in the words the employees choose. Eden Associates training from Thomas's group suggest "warming the soil," or readying the organization for change. Participants talk about "frosts reaching a stalemate" at the same time as reiterating the Golden Rule, getting rid of top-down bureaucracy, and finding their own path.

No lesson plans are handed out; each organization must find its own way—no standard template exists on how "Edenizing" should occur. At Levindale, a second phase of training in-house helped empower 50 individuals to continue implementing the Eden Alternative (Swartz 2000). Evidence of integrating nature, flora in this example, is present at these two facilities. Integrating nature is one of the tenets of the Eden Alternative, and some of this came from within the organizational change in both nursing homes visited.

The focus on this priority can emerge from planning committees. At Levindale, the nursing units that chose to "Edenize" eliminated the traditional nursing station. Where once there was a chest-high counter separating staff and patients, there is now a table surrounded by chairs. Fragrant herbs grow in pots where staff and residents can sit and touch them. Levindale has outside garden areas tucked around the buildings (but in full view) and is gradually expanding its capabilities beyond plant cuttings in residents' rooms. A greenhouse addition has been designed to extend horticultural therapy activities. It will reach out into the already established playground area for intergenerational fun. An enabling garden with easy access for disabled participants is also planned.

Goodwin House West is a continuing care retirement home in Falls Church, Virginia. Nursing units open onto new "outside gardens" and patio areas where moveable planters were added along with comfortable seating that can be shaded or not. A facility tour proudly included a large area of distinctly separate raised beds managed by the residents in independent living. A series of garden paths are lined with benches, and the paths wander along the front perimeter but are near enough to the main entrance to be inviting. Bird feeders, sculptures, and memorial plaques can be seen just off the shaded paths. These sometimes-meditative spaces encourage staff, residents, and visitors to enjoy and be inspired in the natural setting of birds and small wildlife and the many floral beds. During the appropriate seasons, the floral beds adjacent to the putting green provided essentially all the materials for their arrangements and craft sessions. The Eden Alternative principles foster such efforts and present gardens both for resident gardeners and as sources for wildlife, food, respite, and social gatherings.

## RESEARCH ON THE FLORA ASPECT OF THE EDEN ALTERNATIVE

The horticultural aspect of the Eden Alternative appears to be less studied than the fauna aspect with its connection to pet therapy. Research specific to flora, or interaction of residents or staff with flora, was not evident in published work. Thomas dedicated about 20 pages to plants and gardens in his 1996 text and briefly addressed the health-enhancing benefits of plants as air purifiers. He recommended "two or three plants per hundred square feet of living space" (Thomas 1996) and offered that indoor air quality in "Edenized" nursing homes has not been studied. Plant care and maintenance pointers follow, along with precautions about toxic plants.

Both facilities the author visited had a horticultural therapist either on site or employed part-time. The climate was in place for researching many facets of this experiment to adopt a habitat "where plants, animals, and children promote growth among residents" (Thomas and Stermer 1999). Although no formal research had begun, residents in both settings had access to horticultural activities, and a strong commitment to improved care through the Eden Alternative was visibly present.

## IMPLICATIONS AND CONCLUSION

The Eden Alternative organization has published a free Quality of Life Survey for assessing readiness for "Edenizing" (Thomas and Stermer 1999). Agencies already registered may send their summary data for analysis (Swartz 2000). At this point, however, both facilities are using internal surveys and traditional benchmarks as evaluation measures.

This reform initiative in long-term care seems to have continued despite unanswered questions—at least the questions have not been answered in print. Perhaps what is published is the testimonial that usually precedes scientific reports in the literature. This initiative seems to have many professionals convinced of its value. These "Edenizing" agencies firmly believe they have found a philosophy that is a precedent to holistic geriatric care. Additionally, there is the realization that this movement is a call to action to use this alternative as a quality of life initiative.

Although formal research in quality outcomes and residents' functional status is the initial direction of early investigations, the impact of an "Edenized" institutional transformation on the psychological well-being of all who live and work there has not been explored. The intention of this paper was to inform and motivate others to research the Eden Alternative as a part of human issues in horticulture. After reviewing the published work documenting this paradigm in long-term care, this author is convinced the horticultural component is underresearched and inadequately understood. The Eden Alternative provides fertile ground for significant research.

## LITERATURE CITED

Anonymous. 1996. Building Eden: A process, not a program. *The Brown University Long-Term Care Quality Advisor* 8(22):6.

Anonymous. 1999. A long-term care paradise: Eden Alternative lowers drug costs and turnover; improves payer mix and revenue. *Executive Solutions for Healthcare Management* 2(9):13-19.

Bond, G. E., and F. E. Fiedler. 1999. A comparison of leadership versus renovation in changing staff values. *Nursing Economics* 17(1):37-43.

Bruck, Laura. 1997. Welcome to Eden. *Nursing Homes* 46(1):28-32.

Drew, J. C., and V. Brooke. 1999. Changing a legacy: The Eden Alternative. *Annals of Long Term Care* 7(3):115-121.

Evans, Heidi. 2000. The "Eden Alternative": Remaking the nursing home as a place to live, not die. *Newsday*:1-4.

Griffin, Lisa. 2000. Eden in old age. *State Government News* May:10-13.

Hinman, M. R., and D. Hey. 1999. Influence of the Eden Alternative on the functional status of nursing home residents. *Physical Therapy* 79(5):S20 (Conference Abstract).

Hoyles, Martin. 1994. *Gardeners delight: Gardening books from 1560 to 1960*. Boulder, CO: Pluto Press, p. 100.

Katz-Stone, Adam. 1999. Garden variety: A new program called Eden Alternative aims to alleviate loneliness among Levindale residents. *Baltimore Jewish Times*, September 3:45-46.

Minkove, Judy. 2000. Levindale becomes first "Eden Alternative" facility in Maryland. Baltimore, MD: *Levindale Hebrew Geriatric Center and Hospital*, June:1-2.

Pear, Robert. 2000. U. S. recommending strict new rules at nursing homes. *New York Times*, July 22:1-5.

Scott, Abigail. 2000. The flora and fauna alternative: The Eden Alternative enhances the lives of long-term care residents with plants, pets and children. *Advance for Nurses* 2(2):8-10.

Stermer, Miriam. 1998. Not-for-profit report. Notes from an Eden Alternative pioneer. *Nursing Homes* 47(11):35-36.

Swartz, Carol. 2000. Nursing Director, Levindale Hebrew Geriatric Center and Hospital, Baltimore, MD. Personal interview, June.

Tavormina, Candace E. 1999. Embracing the Eden Alternative™ in long-term care environments. *Geriatric Nursing* 20(3):158-161.

Thomas, William H. 1996. *Life worth living: How someone you love can still enjoy life in a nursing home*. New York: VanderWyk and Burnham.

Thomas, William H. 1998. *Open hearts open minds: The journey of a lifetime*. New York: Summer Hill Company, Inc.

Thomas, William H., and Miriam Stermer. 1999. Eden Alternatives hold promise for the future of long-term care. *Balance* 3(4):14-17.

Uhlenberg, Peter. 1997. Replacing the nursing home. *Public Interest* 128 (Summer):73-84.

Vitez, Michael. 1998. New meaning to "growing old." *Philadelphia Inquirer*, July 13:D1+.

Weinstein, L. B. 1998. The Eden Alternative: A new paradigm for nursing homes. *Activities, Adaptation and Aging* 22(4):1-8.

Wolfe, Warren. 1999. A place for living, not dying. *Star Tribune*, April 11:E1+.

# 22

# Sensory Garden Tours at Denver Botanic Gardens

Janet Laminack

"The ultimate goal of farming is not the growing of crops, but the cultivation and perfection of human beings" (Fukuoka 1978). What Masanobu Fukuoka, philosopher and farmer, said can be applied to sensory gardens. Sensory gardens cannot be designed without considering the human element. Unlike traditional display gardens that are meant to be observed from a distance, sensory gardens draw the visitor in to touch, smell, and actively experience the garden with all senses.

## THE FACILITY

The Morrison Horticultural Demonstration Center at Denver Botanic Gardens was built in 1983 to be used for horticultural therapy. It includes office space, a classroom, greenhouse, and the Sensory Garden. The Sensory Garden demonstrates small space container gardening and accessible gardening techniques, such as raised bed planters. The garden also displays plants that appeal to all five senses, and the primary use of the space is for sensory tours. In addition, the garden is used as a learning environment and reference for the students of the Center for Horticultural Therapy Studies. Themed plantings, such as a raised bed of old-fashioned flowers or everlasting flowers, are popular with horticultural therapy students who are working in long-term care facilities. The planting provides prompts for reminiscing as well as supplying the facility with materials for winter activities. Other themes include plants of the Mediterranean (which typically is known as the "Pizza Bed"), Neapolitan (plants that look, smell, or taste like chocolate, vanilla, or strawberry), unusually colored vegetables, and moonlight gardens. These themes are designed to have multiple levels of appeal and to engage a wide range of audiences.

## SENSORY TOURS

Sensory tours provide highly interactive garden experiences to people with special needs. Participants can be children with visual impairments, adolescents in foster care, adults with developmental disabilities, and seniors with dementia. Each group

195

experiences different challenges, and, of course, rewards. The purpose of sensory tours is to bring a meaningful garden experience to persons regardless of their ability. A single focus on colorful bedding plant displays does not offer much significance for persons who are blind, but adding plants that are fragrant, edible, and touchable translates into a garden with wider interest. Likewise, persons with dementia or severe developmental disabilities may be stimulated by fragrant and tactile plants, while a more ordinary garden setting may not initiate active responses.

Volunteer sensory guides make each garden visit unique and personal. Using an example from our own corps, a volunteer who is a landscape architect approaches the world of plants differently than the former elementary educator. The landscape architect focuses on structure of plants, rare specimens, and the total design, while the educator leans toward adaptations of plants, the interaction of flowers and pollinators, and human uses of plants. Because the guide determines the agenda, tours can be tailored to the interests and preferences of the group. This is built into training, with emphasis put on the sensory qualities of plants and the joy of being in a garden. Plant names and horticultural information are of secondary importance in the context of a sensory tour. Each guide shares his/her delight of plants in his/her own way with the tour group. Participants are encouraged to rub the lamb's ear on their face, taste the chocolate mint, and take some flowers home with them. There is no other area at Denver Botanic Gardens where a visitor is allowed to do this, and certainly not encouraged!

## LEADERSHIP GROUP OF VOLUNTEERS

The key to the success of this program is the volunteers. Sensory guides are highly motivated and dedicated. Since 1998, a leadership group of guides has been instrumental in improving tours, promotion, recruitment, and expansion of the program. The leadership group meets year-round, with meetings as often as twice a month in the early spring. This group consists of six members including the scheduler, co-chairs, the Sensory Garden's staff person, and two other experienced guides.

## RECRUITMENT AND TRAINING OF VOLUNTEERS

The majority of volunteer guides are recruited by press releases, a volunteer open house, and through the garden's own newsletter. Training consists of three sessions, each lasting 3 hours. The first day is basic orientation to Denver Botanic Gardens and the horticultural therapy program, as well as the sensory program. Additionally, we cover safety procedures and disability awareness, which entails the "people first" philosophy where the emphasis is put on relating to the person as an individual rather than focusing on the disability. However, we do provide basic disability information and conduct simulation experiences so that the guides get a taste of what it is like to be in a wheelchair or have impaired vision. Also on the first day, guides receive notebooks that contain all the information included in training: disability awareness, safety, guiding techniques, plant lists, and folklore.

The second day of training is spent with three experienced guides. Each has a unique way of interpreting the Sensory Garden, and they present their style to the newcomers. This is to encourage the new guides to bring their own passions and interest to their own tours. The final day of training is our plant "refresher" day and potluck. All sensory guides attend this day to refresh their knowledge on the permanent collection, orient to the annual planting themes, and enjoy some delicious food, catch up with old friends, and meet the new volunteers. The trainees are encouraged to "shadow" experienced guides for the first few weeks, or for as long as it takes to become comfortable leading on their own. On occasion, there are educational opportunities offered throughout the year, some unique to the sensory program and others include all of the volunteer guides at the Gardens. In addition, our season wrap-up party is a time to discuss successes and failures and learn from each other in an informal setting.

## THE FORMAT AND PROCEDURES OF TOURS

Sensory tours are held on Tuesday, Wednesday, and Friday, June through September. They are scheduled for 1-hour time slots, 9 am to 3 pm, and Tuesday evenings are scheduled June through August, 6 pm to 8 pm. The maximum number of participants for one tour is 15, with one staff person for each wheelchair user (if the wheelchair will need to be pushed). To ensure an intimate garden experience, we assign one guide per five participants.

To schedule a tour, a facility or group contacts the registrar of Denver Botanic Gardens, receives basic information, such as size limits and location, and is later sent a confirmation letter. The registrar is responsible for recording on a calendar the contact information and group specifics (such as the type of disabilities). The guide scheduler, a volunteer, works from this information and assigns guides, based on their availability sheets (Figure 22.1). Once scheduled, guides are responsible for finding their own substitutes if they are not able to conduct a tour. For each group, one volunteer is assigned as tour leader who has responsibility to contact the facility beforehand to confirm the date and time, remind people to bring sunblock and drinking water, answer any questions, give them directions to our garden, and also give them a contact name in case they need to cancel at the last minute (Figure 22.2). The tour leader is also responsible for documentation in our record books and contacting the other guides if there is a cancellation.

## PROMOTION OF TOURS

In spring, we send a sensory tour flier to the facilities, such as group homes, Alzheimer's units, and health associations, on our mailing list. This list has been developed over the years, mostly through giving talks in the community. Others have been added to the list through word of mouth and reading about our program in newsletters. We send out press releases to general news sources and to associations on our

**Name:**
**Phone number:**

### Sensory Garden Tours 2000
### Guide Availability for June through September

Please provide as much detail as possible; your availability determines how many tours you get.

| When are you available? | Morning | Afternoon | Evening |
|---|---|---|---|
| Tuesday | yes or no | yes or no | yes or no |
| Wednesday | yes or no | yes or no | N/A |
| Friday | yes or no | yes or no | N/A |

How many tours would you like to give per month?

| | |
|---|---|
| 2 tours _____ | 3 tours _____ |
| 4 tours _____ | 5 tours _____ |
| Other # _____ | |

Would you like to have your tours scheduled back to back?   yes    no    occasionally

Would you be a substitute? _____

When is your vacation scheduled?   June _____ July _____ Aug _____ Sept _____

Do you have any other special considerations (for instance: black-out times, prefer a full day of tours, any scheduling considerations)?

Anything else you want us to know?

Please note:  schedules will be prepared using this information, so it is vital you complete it as soon as possible and forward it to our scheduler, Jane Smith, 1 Main St, Denver CO 80014, 303.333.3333. Contact Jane with any changes to your availability.

**Schedule is made and put in the mail to guides by the 20th of the preceding month.** When you are scheduled for a tour, it is your responsibility to find a replacement if you are unable to do the tour!

You might want to make a copy of this for your records.
If you have any questions, concerns, or just need to talk, you can contact the Sensory Guide Chairman (Joy and Terry) at any time. Thank you for your participation in the BEST guide program at DBG!

Figure 22.1.   Availability form completed by all Denver Botanic Gardens Sensory Garden tour guides.

## SENSORY GARDEN TOUR INFORMATION

*Tour Leader*_____     *te & time of tour*_____ *Date confirmed*_____

*Name of facility*_____*Phone*_____ *Contact person*_____

*Address*_____

*No. clients expected*_____*No. staff expected*_____*Types of disabilities*_____

*********************************************************************************

**PRIOR TO TOUR - 5 TO 7 DAYS: Call group contact person to confirm**
_____date, time, number attending
_____status of clients and number of attendants needed  (need 1 attendant/4 clients if mobile;
        1 attendant/1 client if in wheelchair)
_____recommend sunblock, hats, sun protective clothing or other specific needs for  outdoors
_____confirm that contact person knows how to get to the Sensory Gardens and give directions (below)
_____give group contact person your name and telephone number and ask that you be contacted if tour
        is canceled; explain why this is important
_____if client cancels, encourage s/he to reschedule by calling 303-370-8020 or  8019

**Directions to Sensory Garden:** The Sensory Garden is located at 2320 East 11th Avenue, on the corner of York Street and 11th Avenue, across from the main campus of the Denver Botanic Gardens. Parking for visitors with mobility impairments is just east of the building; enter from 11th Avenue, between York and Josephine Streets.

**After speaking to contact person, tour _not_ canceled**
_____determine if number of guides is appropriate; adjust number by adding or canceling guides
_____call scheduled guides to confirm their attendance
**If tour is canceled**
_____notify Janet Laminack 303-370-8098 and other guides

### DAY BEFORE TOUR

_____reconfirm with contact person
_____ask if there are any questions
_____if tour canceled, notify Janet and guides, as above

### DAY OF TOUR
**Prior to tour** (tour leader or guides)
_____make certain that drinking water, hats etc. are available
_____greet client; explain about tour
 **At end of tour**
_____ask client to complete evaluation form
_____complete Sensory Tour Information Sheet as accurately as possible; this is the only statistical data
        source
_____remind guides to complete individual time logs
_____be certain that water cooler, hats etc are put away if this is last tour of the day (tour leader or guides)
*********************************************************************************

### TOUR LEADER – TO BE COMPLETED AFTER TOUR
Additional tour guides_____
Actual number of participants_____          Number of staff_____
Comments following tour (problems, participant feedback, quotes, suggestions, etc.) Use reverse if needed.
_____

Figure 22.2.   Denver Botanic Gardens Sensory Garden tour information.

mailing list to publish in their newsletters. One season, we were very successful with a "call-a-thon." We called groups who had visited the previous few years, and with schedule book in hand, signed them up for a tour on the spot.

## EVALUATION

The simplest way a sensory tour is evaluated is with a short form that the guides give the participants at the conclusion of the tour. This evaluation is designed to capture the "feel-good" comments that the actual participants have as soon as they have finished the tour. The second evaluation is longer and is mailed to the facility at the end of the season. This form asks for feedback in the areas of what their favorite part was, what we could do better, what challenges they faced, and what we could add to the experience. The disadvantage is that it is hard to ensure that the person who fills out the form was present on the tour. Moreover, if he/she was on a tour, he/she will probably rate the experience solely based upon his/her experience rather than solicit input from colleagues who were with a different guide at the same time. The final evaluation is given to the guides themselves to find out how the scheduling process worked, how they felt about the number of tours they gave, plants that were especially popular, and any ideas for the future.

All of these evaluations are valuable. The positive remarks are used to promote the tours, recruit new guides, and thank the current guides. The few negative remarks we receive are used to help us improve the program or better meet the needs of the populations we serve. We receive several new ideas from these evaluations and gain not only a great sense of what we are doing right but also where we could head in the future. We learned that the little things make a difference when we got this comment from a participant: "I liked the big hats and the guide knew my name!" We are reminded why we do tours in the first place: "We come because it's such a good sensory experience for all our clients; where else can you taste, touch, smell, and learn about the world of plants? It's the highlight of our summer!" In addition, the guides themselves offer suggestions about what we need to work on: "Consider shadowing for more experienced guides to enhance their skills; we can all learn from one another."

## CHALLENGES AND SOLUTIONS

Our biggest challenge is that the cancellation rate for our tours has been at 30% for the past few years. One reason this rate is so high is the unpredictability of the populations: one behavior problem and the whole trip is off, or an emergency at the facility requires staff to cancel off-site trips. Another issue tends to be transportation problems: it seems as though the bus breaks down more often than it runs! The last contributor to the cancellation rate is the high turnover of staff at these facilities. A new activity director will not have access to the schedule set up by the previous director. The phone calls that the tour leader does the week or day before will sometimes catch this problem, so that at least the volunteers can be aware of the cancellation before driving in to the Gardens. Unfortunately, the high cancellation rate seems to be a perennial problem.

A related challenge faced by the sensory program and the biggest complaint from the volunteers is that they want more tours! To accommodate this need, and to compensate for our cancellation rate, we added additional tours on Tuesday evenings and Wednesdays in 1999 and began to book on the hour, every hour. We were able to increase the number of tours from 53 in 1998 to 104 in 1999. Furthermore, we began to use the volunteers better in the training process of new guides, thus giving them another opportunity to share the garden, and it's a great way to influence the program. Other challenges that will probably always remain with us are the heat and sun. We provide hats, sunblock, and water for tour participants. Our guides are adept at keeping the group in the shade and bringing the garden to them. However, most of our visitors have not been out of doors often and may experience sensory overload just by being outside. Medications will also contribute to problems encountered by heat and sun. The guides inform the facilities before they arrive what the conditions are so that they can come prepared, and the guides accommodate the situation as best they can, but in the summer, heat and sun are inevitable!

## PLANS OF THE PROGRAM

The Sensory Garden is undergoing a renovation that will provide new opportunities for sensory experience with a fresh and cohesive garden design. The program has also received a grant to begin an outreach program, taking the Sensory Garden to senior centers and assisted-living facilities. This will be an excellent way to keep the volunteer force involved year-round, as well as to better serve the community and those who cannot come to the gardens.

## LITERATURE CITED

Fukuoka, Masanobu. 1978. *The One-Straw Revolution*. Emmaus, PA: Rodale Press.

# 23

# The Role of Botanical Gardens in Research, Design, and Program Development: A Case Study of Botanica, the Wichita Gardens

Patricia J. Owen

## INTRODUCTION

Although some research efforts by botanical gardens are readily identified for their contribution to horticultural research, others are remembered for different reasons. For example, the 1787 voyage led by Captain William Bligh on the HMS Bounty is recalled because of the mutiny that occurred rather than for the original purpose, which was for research commissioned by the Royal Botanical Gardens, Kew. Captain Bligh was on an expedition to collect breadfruit from Tahiti to grow in the West Indies as a food source for the slaves working on the plantations (Simmons 1982).

Extensive collection expeditions as well as research into plants' medicinal and ornamental characteristics remain hallmarks of botanical garden research efforts. Visitor studies (why people visit a garden, who they are, where they are from, etc.) and research on experiential learning are also common at botanical gardens. However, research on the innate relationship between people and plants has rarely occurred. This case study discusses physiological research conducted within a botanical garden; the design and development of a new garden area; and the ensuing programs developed to use this garden for horticultural therapy activities. These events occurred at Botanica, a botanical garden located in Wichita, Kansas.

Botanica contains less than 10 acres and is young for a botanical garden. The garden was established in 1987. Because a botanical garden is an outdoor environment landscaped with a variety of plants that may include trees, shrubs, and flowers, the environmental design provides a milieu of the sights, sounds, and smells of nature. A botanical garden is a unique outdoor setting that lends itself to research because access can be controlled and monitored. At the same time, outdoor settings present certain research challenges. Ambient air temperatures, moisture levels, wind, sunlight, and cloud cover are some of the variables under nature's control when the research laboratory is an outdoor environment. Despite the inherent challenges, the need exists to study humans within gardens and other natural settings. This study investigated the physiological and psychological effects of a botanical garden experience on people. The research was an exploratory attempt to measure people's responses and perceptions within a botanical garden environment.

## METHODOLOGY

Participants in the study were visitors to the garden who responded to a sign displayed near the admissions desk requesting help for an environmental research study being conducted by a graduate student from Kansas State University. Each person who decided to participate in the study was required to read and sign an informed consent form. The study was conducted on four Sundays in September and October 1991.

The protocol for the study follows:

1. *Physiological measurements.* Blood pressure and heart rate were measured before participants entered the garden. Three sets of physiological measurements were recorded for each participant with an automated blood pressure cuff, programmed to measure both blood pressure and heart rate at 1-minute intervals.
2. *Previsit survey.* A previsit survey to determine the participant's expectations of the visit was completed by each participant.
3. *Visit the garden.* After the participants completed the physiological measurements and the survey, they left the building and entered the garden. No directions were given to participants nor were any suggestions offered concerning the time spent in the garden.
4. *Physiological measurements.* After participants visited the garden, their blood pressure and heart rate were measured again.
5. *Postvisit survey.* Participants completed a postvisit survey that included evaluations of the visit, and were offered a copy of their previsit and postvisit blood pressure and heart rate measurements. The final form completed by each participant was a health survey.

The previsit and postvisit survey results and the health survey results will not be addressed in this paper.

## RESULTS

Of the 127 people who participated in the study, 57 were men and 70 were women. The average age was 46. Participants spent an average of 51 minutes in the gardens. The median time since they had last eaten was 2 hours and 43 minutes.

Participants' mean systolic blood pressure (the pressure while the heart is beating) measurements are presented before and after visiting the garden in Table 23.1. A Newman-Keuls Multiple Comparison Test of previsit and postvisit blood pressure measurements found significant decreases ($p < 0.01$) in systolic blood pressure on 3 days of the study. As shown in Table 23.2, on the first and fourth days of the study, participants' diastolic blood pressure (the pressure when the heart is resting between

Table  23.1.    Mean systolic blood pressure before and after visiting a botanical garden

| Week | Previsit (mm Hg) | Postvisit (mm Hg) | Change (mm Hg) |
|------|------------------|-------------------|----------------|
| 1 | 119.46 | 112.79 | (−)  6.67* |
| 2 | 112.19 | 113.30 | (+)  1.11 |
| 3 | 126.79 | 120.92 | (−)  5.87* |
| 4 | 128.59 | 122.65 | (−)  5.94* |

Note: mm Hg = millimeters of mercury.
* $p < 0.01$.

Table  23.2.    Mean diastolic blood pressure before and after visiting a botanical garden

| Week | Previsit (mm Hg) | Postvisit (mm Hg) | Change (mm Hg) |
|------|------------------|-------------------|----------------|
| 1 | 70.66 | 67.50 | (−)  3.16* |
| 2 | 66.42 | 67.37 | (+)  0.95 |
| 3 | 71.16 | 70.91 | (−)  0.25 |
| 4 | 77.36 | 71.29 | (−)  6.07* |

Note: mm Hg = millimeters of mercury.
* $p < 0.01$.

beats) means decreased significantly ($p < 0.01$). Significant systolic blood pressure decreases were recorded on these 2 days as well.

On the first 3 days of the study, participants' heart rate measurements before and after visiting the garden registered increases (Table 23.3). On the second day of the study, the mean heart rate increased by 4.09 beats per minute. A Newman-Keuls Multiple Comparison Test of previsit and postvisit heart rate measurements found significant increases in heart rate ($p < 0.05$) on the second day of the study. The second day was the only day of the study when heart rate increased significantly.

Table  23.3.    Mean heart rate before and after visiting a botanical garden

| Week | Previsit (bpm) | Postvisit (bpm) | Change (bpm) |
|------|----------------|-----------------|--------------|
| 1 | 71.49 | 72.43 | (+) 0.94 |
| 2 | 69.99 | 74.08 | (+) 4.09* |
| 3 | 70.64 | 71.70 | (+) 1.06 |
| 4 | 76.76 | 75.40 | (−) 1.36 |

Note: bpm = beats per minute.
* $p < 0.05$.

The changes on the second day in heart rate and blood pressure are contrary to the changes on the other 3 days of the study. Two factors that could account for the differences on the second day are the ambient air temperature and the sample size. The high temperature for the second day of the study was 37.2°C at 3:30 pm while the study was in progress (Table 23.4). As a comparison, on the fourth day of the study, the ambient temperature was 21.7°C at 3:30 pm (Table 23.4). Because of the heat and humidity on the second day, only 56 people visited the garden, and 15 people participated in the study (Table 23.4).

## DISCUSSION

Current research suggests that systolic blood pressure should be the major criterion in the diagnosis and treatment of hypertension. This study documented decreases of approximately 6 millimeters mercury (mm Hg) in the mean systolic blood pressure for participants after spending time in a botanical garden environment (Table 23.1). Recognition of the relevance of systolic blood pressure to human health puts in perspective the potential value of the systolic blood pressure decrease recorded in this study (Kannel 1996).

A growing body of new research indicates that moderate levels of physical activity can lead to health benefits (Pratt 1999). Information from the Centers for Disease Control and Prevention and the American College of Sports Medicine defines moderate physical activities as bodily movement sufficient to use about 150 calories while engaged in activities, such as walking briskly for a 30-minute period. This amount of activity, on a regular basis, decreases the risk of coronary heart disease and hypertension as well as symptoms of anxiety and depression. Participants in this study walked through the garden, which is considered light physical activity (Pate et al. 1995). Whether the decreases in blood pressure were a result of participants' physical activity, a result of the garden environment, or a combination of both factors are topics worthy of future exploration.

This exploratory study measured people's physiological responses to a garden environment and found significant decreases in systolic blood pressure. Kaplan and Kaplan (1989) explain that when research is conducted within natural settings, one

Table 23.4.   Mean ambient air temperature, number of visitors to the garden study site, and number of visitors that participated in the study

| Week | Temperature (degrees C) | Number of Visitors | Number in Study |
|---|---|---|---|
| 1 | 35.6 | 71 | 18 |
| 2 | 37.2 | 56 | 15 |
| 3 | 26.7 | 134 | 41 |
| 4 | 21.7 | 255 | 53 |

Note: Ambient air temperatures were recorded between 3 and 4 pm while study was in progress.

must give up some of the control. It is through the accumulation of data acquired from numerous studies that knowledge is acquired. The value of this research may be that it indicates and justifies the need for further investigation into the significance of garden environments to people.

## GARDEN DEVELOPMENT

When this study was conducted, Botanica did not have a horticultural therapy program. However, because of this research, Botanica realized that the addition of a garden for horticultural therapy would be beneficial. Herbs were considered essential for the garden because they are resilient, emit wonderful fragrances while being handled, and they can be used in a myriad of activities. Several years after the research project was completed, funds were donated to Botanica for the construction of a memorial herb garden.

A landscape architect was hired to create a tentative design for the Sally Stone Sensory Garden. Although this garden was to be wheelchair accessible, the original designs by the architect included steps at the entry area. Improvements were made to this design as the result of collaboration among a landscape architect, a horticultural therapist, and Botanica's Director of Horticulture. This ensured the final design had no steps but rather an entrance with a lesser grade that was wide enough for two wheelchairs to traverse.

The Sally Stone Sensory Garden was created as a place for the public to enjoy herbs and to learn more about accessibility for gardeners with physical challenges, as well as an area for Botanica to implement a new horticultural therapy program. Funding was required to support this new program.

## BEGINNING A HORTICULTURAL THERAPY PROGRAM

The fact that Botanica could not afford to fund a horticultural therapy program is a problem common to many botanical gardens. However, many granting agencies will fund programs for special populations. Grant applications were made requesting support for a horticultural therapist, interns in horticultural therapy from Kansas State University, and programmatic operating costs. As a result, Botanica received funding to provide horticultural therapy programming for 750 people from various local agencies during the summer and fall of 2000, as well as support for horticultural therapy staff and supplies.

Training sessions for staff were organized prior to the start of the horticultural therapy program. The training introduced basic principles of horticultural therapy and sensitized staff to some of the challenges that clients participating in the program might face. As part of this training, a gerontologist introduced Botanica's staff to a variety of physical challenges that client groups might have. For example, hearing impairments were presented with an audio tape that illustrated how people with different levels of hearing might perceive the same sentence.

Training sessions were also held for volunteers who wanted to participate in activities with client groups. Volunteers learned about horticultural therapy, became familiar with different challenges the client groups might exhibit, role-played potential difficulties clients might experience, and participated in representative activities. After the training, each volunteer was required to observe several activity sessions. Volunteers were then allowed to assist the horticultural therapist and interns with horticultural therapy activities.

The number of volunteers required for each activity was determined by the nature of the client group and activity. For example, a group of extremely low-functioning clients might require volunteers to work one-on-one with each client, so for seven clients, five volunteers, the horticultural therapist, and the intern would be needed to provide a meaningful experience for each person. If the group were composed of persons who did not require one-on-one involvement, a lesser number of volunteers would be scheduled, along with the horticultural therapist and intern.

In today's society, botanical gardens are realizing their role as institutions designated to share horticulture with the entire community. Horticultural therapy is an excellent vehicle that expands the viability of horticulture to many special populations. During the summer of 2000, groups of abused and neglected children, mentally challenged adults and children, and those with physical handicaps, as well as aging populations were among the client groups who participated in activities in the Sally Stone Sensory Garden.

The horticultural therapy program at Botanica was funded through local granting agencies. As Frazel (1998) explains, once a pilot program in horticultural therapy begins at a botanical garden, local awareness of the value and need for such services grows. The horticultural therapy program continues at Botanica because local funding agencies recognize the value of the program and continue to provide financial support. In fact, in 2001, the program was expanded. Outreach programs were offered at local long-term care facilities and more than 100 individuals participated in vocational training opportunities.

The development of a horticultural therapy program is not an easy task; however, the rewards are tangible. They can be seen in the faces of children who create an herbal bouquet, adults who pot a plant, or a person with cerebral palsy sitting in a wheelchair surrounded by butterflies enjoying the wonder and beauty of gardens and nature.

## LITERATURE CITED

Frazel, Matthew. 1998. Botanical gardening: Design, techniques, and tools. In *Horticulture as therapy: Principles and practice*, eds. S. Simson and M. Straus. Binghamton, NY: Haworth Press, pp. 355–376.

Kannel, W. B. 1996. Blood pressure as a cardiovascular risk factor: Prevention and treatment. *Journal of the American Medical Association* 275(20):1571-1576.

Kaplan, R., and S. Kaplan. 1989. *The experience of nature: A psychological perspective*. New York: Cambridge University Press.

Pate, R. R., M. Pratt, S. N. Blair, W. L. Haskell, C. A. Macera, C. Bouchard, D. Buchner, W. Ettinger, G. W. Heath, A. C. King, A. Kriska, A. S. Leon, B. H. Marcus, J. Morris, R. S. Paffenbarger, K. Patrick, M. L. Pollock, J. M. Rippe, J. Sallis, and J. H. Wilmore. 1995. Physical activity and public health—A recommendation from the Centers for Disease Control and Prevention and the American College of Sports Medicine. *Journal of the American Medical Association* 273(5):402-407.

Pratt, M. 1999. Benefits of lifestyle activity versus structured exercise. *Journal of the American Medical Association* 281(4):375-376.

Simmons, John B. E. 1982. *Kew: Gardens for science and pleasure.* Edited by F. Nigel Hepper. Owings Mills, MD: Stemmer House.

# 24

# Welfare and Medical Institutions in Kyushu, Japan, Engaged in Horticulture: A Survey

E. Matsuo
Y. Fujiki
H. Takafuji
H. Kweon
F. A. Miyake
K. Masuda
K. Mekaru

## INTRODUCTION

Horticulture has been used to some extent at psychiatric hospitals in Japan. It is not regarded, however, as a real therapeutic tool but a means of occupying clients' time (Matsuo and Fujiki 1996). Since horticultural therapy was reintroduced into Japan in the early 1990s, Japanese people have become increasingly interested in it, and horticultural therapy became fashionable in the late 1990s. Regardless of the fashion among the public, the welfare and/or medical institutions' administration are not always interested in horticultural therapy, nor have they used horticulture as a therapeutic tool. The level of interest and experience, of course, varies with different kinds of institutions and/or locations across the country.

To clarify the current status of the use of horticulture in welfare and/or medical institutions and to promote the development of horticultural therapy in Japan, a nationwide research project team was organized in 1998. This interdisciplinary project team consisted of horticulturists, a sociohorticulturist, a physician, a nurse, and an educator, all of whom work at universities and are located across all of Japan. This paper reports results obtained in four prefectures: Fukuoka, Kumamoto, Saga, and Kagoshima prefectures on Kyushu Island in the southern part of Japan.

## METHODOLOGY

This project extends preliminary research conducted in Fukuoka Prefecture (Matsuo 1997; Matsuo et al. 1977) in 1995 to a nationwide survey in Japan. The questionnaire used in the preliminary research was revised for use in the nationwide survey. The questionnaire included questions on the following topics:

1. Location of horticulture within the institution
2. Experiencing horticulture
3. Items of horticulture such as kind of activities, cultivated plants, supervisor, etc.
4. Aims
5. Effects or benefits
6. Problems
7. If they knew what horticultural therapy was

Questionnaires were sent to welfare and medical institutions by mail with a stamped, addressed envelope. A total of 1,344 questionnaires were mailed to the heads of the institutions. Questionnaires returned within 3 weeks of mailing were analyzed.

## RESULTS AND DISCUSSION

Questionnaire return rate was 57%. Fifty-eight percent of the institutions that responded were engaged in horticulture (Table 24.1). The type of institution most engaged in horticulture was psychiatric hospitals (84%). The type of institution that had the fewest engaged in horticulture was institutions that served the physically disabled (40%). This difference between institutions may be due to differences between occupational therapy programs and physical therapy programs. In general, psychiatric hospitals hire occupational therapists while institutions serving the physically dis-

Table 24.1.  Percent of welfare and medical institutions in Fukuoka, Kumamoto, Saga, and Kagoshima Prefecture, Kyushu Island, Japan, engaged in horticulture

| Institution Type | Nursing Home | Elderly Health Care | Physically Disabled | Mentally Disabled | Psychiatric Hospital | Total |
|---|---|---|---|---|---|---|
| Number of questionnaires sent (S) | 465 | 230 | 139 | 329 | 181 | 1,344 |
| Number of questionnaires returned (R) | 228 | 126 | 96 | 209 | 108 | 767 |
| R/S (%) | 49 | 55 | 69 | 63 | 60 | 57 |
| Number of institutions engaged in horticulture (H) | 105 | 64 | 38 | 146 | 91 | 444 |
| H/R (%) | 46 | 61 | 40 | 70 | 84 | 58 |

abled hire physical therapists. Further investigation is needed to understand this significant variance in the use of horticulture at these different types of institutions.

Ninety percent of the psychiatric hospitals regarded horticulture as a therapeutic or rehabilitative tool (highest among the five groups), while less than 30% of the nursing homes did so (lowest of the five groups) (Figure 24.1).

More than 70% of the institutions grew vegetables and/or flowers, and less than 20% grew fruit trees, herbs, garden trees, crops, and others (Figure 24.2). Although herb growing is in fashion in Japan, herbs were grown at only 11% of the institutions. Growing primarily vegetables and flowers may be influenced by the lack of professional horticulture supervisors in these institutions and by the fact that the staff who work with the clients in horticulture are not horticultural professionals (Matsuo and Fijuki 1996). Results from this survey support this fact. More than 37% of institutions responded that the primary problem for engaging in horticulture is that there are no professionals with both medical/welfare and horticultural knowledge and techniques.

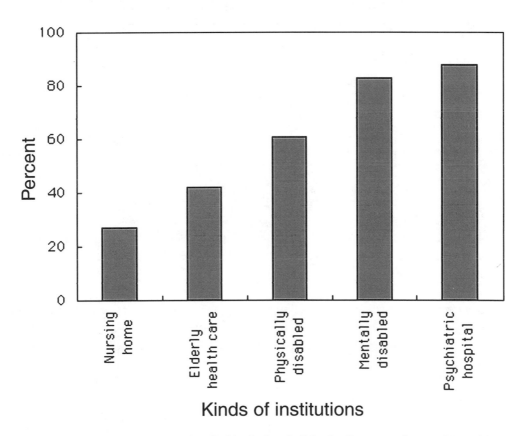

Figure 24.1.   Percent of welfare and medical institutions in Fukuoka, Kumamoto, Saga, and Kagoshima Prefecture, Kyushu Island, Japan, that regarded horticulture as a therapeutic or rehabilitative tool.

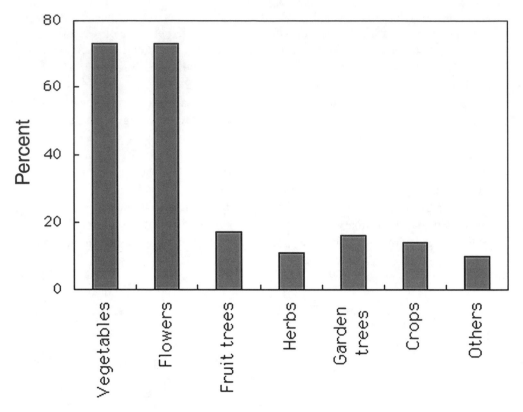

Figure 24.2.   Types of plants grown by welfare and medical institutions in Fukuoka, Kumamoto, Saga, and Kagoshima Prefecture, Kyushu Island, Japan.

Other top problems pointed out follow:

1. There are not enough staff for supervising horticulture (33%).
2. There are no facilities such as a greenhouse to be used when weather conditions are undesirable (30%).
3. Evaluation of horticulture is difficult (28%).

Less than 30% of the institutions responded that they knew what horticultural therapy is (Figure 24.3). Our research survey is under way, but the results obtained so far suggest that horticulture is not always recognized as a useful therapeutic tool. Thus, enlightening the institutions about horticultural therapy is needed to promote and develop its appreciation in Japan.

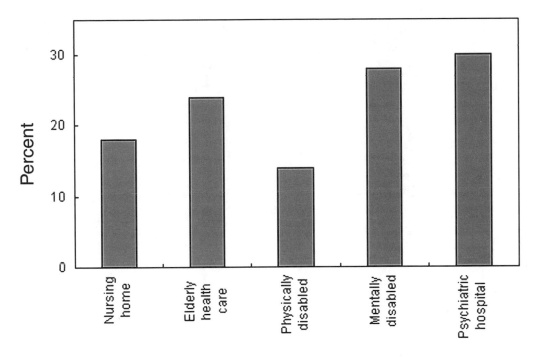

Figure 24.3.   Percent of welfare and medical institutions in Fukuoka, Kumamoto, Saga, and Kagoshima Prefecture, Kyushu Island, Japan, that knew what horticultural therapy is.

## LITERATURE CITED

Matsuo, E., ed. 1997. *Kokoro no Kenkou ni taisuru Engei Katsudou to sono Seisanbutsu no Kouka (Effect of gardening and its products on physical/mental health)*. Ministry of Education and Science Report, Japan.

Matsuo, E., and Y. Fujiki. 1996. Therapeutic use of horticulture in Japan. In *Proceedings people-plant interactions in urban areas—A research and education symposium*, eds. P. Williams and J. Zajicek. Dept. of Horticultural Sciences, Texas A&M University, TX.

Matsuo, E., Y. Fujiki, Y. Miyajima, Y. Ozaki, and K. Fujiwara. 1997. Therapeutic use of horticulture at the welfare facilities and the mental hospitals in Fukuoka Prefecture. *Journal of the Japanese Society for Horticultural Science Supplement* 2:694-695 (in Japanese).

## AUTHORS' NOTE:

This survey was partially supported by the Grant in Aid for the Scientific Promotion (No. 11460015) from the Japan Society for Scientific Promotion.

# IV

# Research

# 25

# Postoccupancy Evaluation and the Design of Hospital Gardens

**Clare Cooper Marcus**

For those professionals who design outdoor spaces in medical settings, for those who provide programs within such gardens, and for those seeking to promote the restorative benefits of nature, it is essential that we understand how to evaluate existing spaces. Postoccupancy evaluation (POE) with a standardized methodology that can be replicated is one such tool. With more POE studies conducted on similar sites, their results published and disseminated, we can incrementally improve the efficacy of designed settings. The discussion that follows refers principally to hospital gardens for passive use or nonprogrammed spaces—i.e., not horticultural therapy gardens— but the methods described would be equally suitable for use in the latter.

## OVERVIEW OF POSTOCCUPANCY EVALUATION METHOD

A POE is a systematic evaluation of a designed and occupied setting from the perspective of those who use it. Such evaluations are rarely carried out by the designers themselves since they lack the training, and contracts with a client never provide for what should be an essential conclusion to the design process: How does the setting work for people?

The methods of POE, although not complex, are too lengthy to describe in detail in a brief paper. Hence, what follows is an outline; for further details, readers are referred to Chapter 8, "Post-occupancy evaluation," in Clare Cooper Marcus and Carolyn A. Francis, eds., *People Places* (1998); John Zeisel, *Inquiry by Design: Tools for Environment-Behavior Research* (2001); and Robert Bechtel, Robert Marans, and William Michelson, eds., *Methods in Environmental and Behavioral Research* (1987). The important steps in a POE are summarized below.

1. Walk-through: Walk through the site with the designer and/or the client, and/or staff who work in or maintain the space. All of these individuals are significant informants who can tell you the "story" of how this place came about, what works and what does not from their perspective. This will set the stage for further steps in the evaluation.

2. "Hanging out": It is important before engaging in systematic data gathering to spend some time in the setting unobtrusively noticing what is going on, who is doing what, which areas seem to be heavily used or unused, and so on. At this time, make

notes on what you might want to later observe systematically, or ask people about in interviews. It is important, too, to record your *own* experience of this space, to sense the essence of this setting, how it is for you and, perhaps, for others.

3. Activity mapping: This is a procedure for systematically recording

— *Who* is using the site? (men? women? staff? patients? children? elderly? etc.)

— *Where* do they tend to gravitate? (sun? shade? a particular form of seating? close to water? etc.)

— *What* are people doing? (gardening? looking? smelling? dozing? talking? etc.)

At a minimum, a site needs to be observed on at least four separate occasions, on different days and different times of day. On each visit, walk through the site so that each part of it can be observed. With a site plan and notebook in hand, on each visit, record everyone in the space by locating them as dots on the site plan, numbering each dot, and in a notebook, record that number plus gender, approximate age, activity and, if possible, the role of the person observed (visitor, patient, employee, staff member, etc.). Repeat this procedure with a new site plan on each visit. With this relatively simple procedure, you have data that can be aggregated in terms of an overall activity map (all observations on one map); graphs of use by time, by gender, by activity, etc.; activity maps by gender, time, etc. This provides a very detailed assessment of how the space is used and may identify underrepresented groups; subareas not used; areas preferred by older people; and so on.

All of the above is much more accurately recorded via systematic observation than it would be, say, by asking people what they do. However, what this approach *cannot* tell you is why people come to the setting, what they may like changed, and so on.

4. Interview survey: Observations during the previous phases should result in questions you will want to ask. With a well-composed interview survey, you will be able to gather data on how far people have traveled to this place, how often they come, what they like best about it, what they would like to see changed, how they feel different after being in the space, and so on.

It is critical that an expert on social research or a definitive text (e.g., Zeisel 2001; Bechtel et al. 1987) be consulted at this stage so that the sample size and the data gathered will provide you with reliable information.

5. Archival research: Parallel with the above procedures, it may be useful to study any archival records on the site, such as its history, design goals (if recorded), security records, accident reports, etc.

6. Final report: A final report on the POE should be well illustrated with activity maps and bar graphs (from data in step 3); tables (from data in step 4); and an analysis drawing on data from all the steps in the process and answering questions such as the following:

— Who is using this place?

— How are they using the site and its subareas?

— What potential users are not represented?

— Which areas seem over- or underused?

— Do any areas seem misused, damaged, or poorly maintained?
— Are there any nonconformities between what the designer and/or administration apparently intended as suitable activities on the site and what is actually happening there?
— Were the design cues not clear enough to guide users?
— Are administrative or staff messages (verbal or symbolic) about use weak or inappropriate?
— Are the intended uses irrelevant to current users?
— Finally, what works well on this site?
— What are people and staff happy with?
— What design features facilitate this and why?

If the POE has been conducted with the purpose of making changes or modifications at this site, the report should conclude with a list of recommendations. If the POE can be compared with others of similar sites, it might be appropriate to conclude with a set of potential guidelines that would assist the clients and designer of future settings.

## A POE STUDY OF FOUR CALIFORNIA HOSPITAL GARDENS

In 1994, a POE study of four hospital gardens in the San Francisco Bay Area was carried out by this author and Marni Barnes (Cooper Marcus and Barnes 1995). The methods used were similar to those described above. Two gardens were in acute care hospitals; one in a chronic care facility; and one served a number of outpatient clinics at a county hospital. Some of the pertinent findings included the fact that staff (59%) were the most frequent users of these gardens, compared to patients (26%), and visitors (15%). In interviews, the staff spoke enthusiastically, sometimes emotionally, about how much it meant to them to come outdoors to an environment quite unlike the interior of the hospital. Many used the gardens daily, or several times a day.

The most frequent activities for all users were relaxing (94% of total at all four sites), strolling (61%), "outdoor therapy" (53%); also important were eating and talking.

Significantly, all but 5% felt positive change of mood after spending time in the hospital garden; people reported feeling calmer, more relaxed, refreshed, stronger, and rejuvenated. Of particular importance in terms of staff retention, many staff reported returning to work feeling more relaxed, more able to cope, concentrate, and work more efficiently.

When asked what garden qualities seemed to assist in their change of mood, most users cited plants and trees (69%); other sensory stimulants (fresh air, birdsong, sounds of water, fragrances, etc.) (58%); and settings in which one could feel a sense of "escape" and privacy, places to be with friends, etc. (50%). (More details about the results at each of the four gardens can be found in the monograph cited above and in the book *Healing Gardens* [Cooper Marcus and Barnes 1999].)

To summarize, current research indicates the following:

- Some studies suggest that people under stress often turn to nature for a sense of relief (Francis and Cooper Marcus 1991, 1992; Cooper Marcus 1997).
- When nature is present in a hospital setting—usually in the form of a garden—patients, staff, and visitors will voluntarily go into such a space (sometimes several times a day) knowing that the experience will help them feel better (Cooper Marcus and Barnes 1995).
- Research confirms that actual physiological changes, commensurate with stress reduction, take place when people are exposed to nature (Ulrich 1999).

Now it is pertinent to ask what elements and qualities are necessary in a garden for it to have a restorative effect. Ten such elements and qualities will be briefly discussed. For the first four—opportunities for exercise, opportunities to express choice, opportunities for social support, and engagement with nature—research results indicate that each is associated with a reduction in stress, and thus it is postulated that including such elements in a garden will help to create a healing and restorative environment (Ulrich 1999). The subsequent six elements and qualities are necessary for a garden to be accessible and usable, and are gleaned from the four POE studies of hospital gardens that exist and from observational research by this author at more than 70 acute care hospitals in three countries (the United States, the United Kingdom, and Canada).

## DESIGN GUIDELINES FOR HOSPITAL GARDENS

1. Opportunities for exercise—Exercise is associated with a whole spectrum of significant health benefits, particularly for the elderly, postsurgery patients, and those with psychiatric problems, and people are more likely to walk if there is an attractive, accessible, visible place to walk in (Brannon and Feist 1997; Ruuskanen and Parketti 1994; Emery and Blumenthal 1991).
2. Opportunities to make choices, seek privacy, and experience a sense of control—Spending time as an inpatient in a hospital often takes away a sense of control (little choice about what to wear, what to eat, when to eat, etc.), and this can be a major source of stress. Enabling people to go outdoors and choose where to walk, where to sit, what to look at, etc., can help them regain a sense of autonomy.
3. Opportunities for people to gather together and experience social support—A large body of research indicates that those with higher levels of social support are less stressed and have better health than those who are more socially isolated (Ulrich 1999; Cohen and Syme 1985; Sarason and Sarason 1985; Schwarzer and Leppin 1989). Hence, hospitals have given greater attention to encouraging family visits, providing more attractive waiting rooms, and allowing parents to sleep in the room with their sick child. Finding a private place to converse in a

garden is especially important for visitors and ambulatory patients; parents and hospitalized children; family members visiting a hospice; staff with colleagues; and staff working with patients (for example, in sessions of physical therapy, psychotherapy, and so on).

4. Engagement with nature—Research results indicate that viewing elements of green nature, even for a short time, helps to reduce symptoms of stress (Ulrich 1984; Ulrich 1979; Ulrich et al. 1991; Hartig et al. 1996). Hence, it is essential that a garden intended as healing contain a plentiful variety of trees, shrubs, and perennial plants; planting choices that provide seasonal variations; plants that appeal to multiple senses (smell, taste, sound, touch, sight). It is also important to provide still or moving water features since water evokes a contemplative mood, and its sounds can screen out nearby conversation. Finally, wildlife in the form of butterflies, birds, and small mammals reminds people in a stressful state that life goes on.

Although the above guidelines represent the most critical design elements, there are others that are essential in, so to speak, a "support" role.

5. Visibility—If you do not know a garden is available, you will not be able to use it! In more than 70 hospitals visited by this author, only three had signs indicating the presence of a garden, and (it is speculated) few provide information about gardens in patient or staff orientation material. Hence, it is essential that outdoor spaces are visible from the main foyer or access corridor, and from as many waiting areas and patient rooms as possible. If people can see a garden is available they will be more likely to use it. Moreover, for those inpatients that cannot get outside, a view out to a garden can help to reduce the stress of hospitalization.

6. Accessibility—It is not enough to know that a garden is available. People need to gain access to it. This means that the access door is unlocked (unfortunately, many potentially restorative courtyard spaces in hospitals are kept permanently locked); that there are appropriate walking surfaces (for example, people who are elderly or infirm need smooth, nonreflective surfaces); and that those using wheelchairs or pulling intravenous poles do not have to negotiate an entry "lip" or steps.

7. Sense of security—Once in a garden, patients need to feel a sense of security and safety. Depending on the circumstances, this may mean that the garden is enclosed and not accessible to outsiders and/or that it is visible from a nurses' station.

8. Sense of comfort—It is essential that people feel comfortable in both a physiological and a psychological sense. Some patients on chemotherapy drugs or AIDS medication need to stay out of the direct sun, hence a choice of sitting in the shade is important. If ambulatory inpatients are to be encouraged to use a garden, it is important that it is a space where they will feel comfortable in a

hospital gown; hence, an entry garden facing onto a road would not be appropriate.

9. Quiet—It should go without saying that a garden needs to be quiet except for the sounds of nature, such as water and birdsong. However, all too often, hospital gardens are located within auditory range of an emergency helicopter-landing pad, or a potentially restful fountain is sited too close to a noisy air conditioning unit. Mechanical noises clearly disrupt a restorative garden experience.

10. Design elements that are unambiguously positive—When under stress, people tend to project their fears and negative emotions onto other people and onto the physical environment. Hence, a piece of outdoor sculpture that might be greeted with interest or curiosity in a museum garden may be seen by stressed patients as threatening or confusing. Great care must be taken in selecting potentially ambiguous art, both inside and outside the hospital buildings. (See Ulrich 1999).

## CONCLUSION

The notion of the "therapeutic" or "healing" garden has received considerable attention in the past decade (Mintner 1993; Tyson, 1998; Gerlach-Spriggs et al. 1998; McDowell and McDowell 1998; Jay 1998). For the client or designer of such a garden to create a space that is truly therapeutic, it is essential that they are aware of relevant research on the restorative benefits of nature and POEs of existing hospital gardens in order not to reinvent the wheel or repeat past mistakes. Since the number of POEs of gardens specifically labeled as "healing" is minimal, it is critical that many more such studies are conducted. The methods discussed above provide a procedure used successfully in many other settings that can be applied to the evaluation of hospital outdoor space in general, healing gardens, or gardens used for horticultural therapy.

## LITERATURE CITED

Bechtel, Robert B., Robert W. Marans, and William Michelson, eds. 1987. *Methods in environmental and behavioral research.* Van Nostrand Reinhold, New York.

Brannon, L., and J. Feist. 1997. *Health psychology* (3rd ed.). Brooks/Cole, Pacific Grove, CA.

Cohen, S., and S. L. Syme, eds. 1985. *Social support and health.* Academic Press, New York.

Cooper Marcus, Clare. 1997. Nature as healer: Therapeutic benefits in outdoor places. *Nordic Journal of Architectural Research* 10(1):9-20.

Cooper Marcus, Clare, and Marni Barnes. 1995. *Gardens in health care facilities; uses, therapeutic benefits, and design recommendations.* The Center for Health Design, Inc., Martinez, CA.

Cooper Marcus, Clare, and Marni Barnes, eds. 1999. *Healing gardens: Therapeutic benefits and design recommendations.* John Wiley & Sons, New York.

Cooper Marcus, Clare, and Carolyn A. Francis. 1998. Postoccupancy evaluation. In *People*

*Places: Design guidelines for urban open space,* 2d edition, ed. Clare Cooper Marcus and Carolyn A. Francis, ch. 8. John Wiley & Sons, New York.

Emery, C. F., and J. A. Blumenthal. 1991. Effects of physical exercise on psychological and cognitive functioning of older adults. *Annals of Behavioral Medicine* 13:99-107.

Francis, Carolyn A., and Clare Cooper Marcus. 1991. Places people take their problems. In *Proceedings of the 22nd Annual Conference of the Environmental Design Research Association.* Mexico.

Francis, Carolyn A., and Clare Cooper Marcus. 1992. Restorative places: Environment and emotional well-being. In *Proceedings of the 23rd Annual Conference of the Environmental Design Research Association.* Boulder, CO.

Gerlach-Spriggs, Nancy, R. Kaufman, and S. B. Warner. 1998. *Restorative gardens: The healing landscape.* Yale University Press, New Haven.

Hartig, T., A. Böök, J. Garvill, T. Olsson, and T. Gärling. 1996. Environmental influences on psychological restoration. *Scandinavian Journal of Psychology* 37:378-393.

Jay, R. 1998. *Sacred gardens: Creating a space for meditation and contemplation.* Thorsons, London.

McDowell, C. F., and T. Clark McDowell. 1998. *The sanctuary garden.* Simon and Schuster, New York.

Mintner, S. 1993. *The healing garden.* Headline Book Publishing, London.

Ruuskanen, J. M., and T. Parketti. 1994. Physical activity and related factors among nursing home residents. *Journal of the American Geriatric Society* 42:987-991.

Sarason, I. G., and B. R. Sarason, eds. 1985. *Social support: Theory, research, and applications.* Nijhoff, The Hague, Netherlands.

Schwarzer, R., and A. Leppin. 1989. Social support and health: A meta-analysis. *Psychology and Health* 3:1-15.

Tyson, Martha. 1998. *The healing landscape: Therapeutic outdoor environments.* McGraw Hill, New York.

Ulrich, R. S. 1979. Visual landscapes and psychological well-being. *Landscape Research* 4(1):17-23.

Ulrich, R. S. 1984. View through a window may influence recovery from surgery. *Science* 224:420-421.

Ulrich, Roger S. 1999. Effects of gardens on health outcomes: Theory and research. In *Healing gardens: Therapeutic benefits and design recommendations,* Ch. 2, ed. Clare Cooper Marcus and Marni Barnes. John Wiley & Sons, New York.

Ulrich, R. S., R. F. Simons, B. D. Losito, E. Fiorito, M. A. Miles, and M. Zelson. 1991. Stress recovery during exposure to natural and urban environments. *Journal of Environmental Psychology* 11:201-230.

Zeisel, John. 2001. *Inquiry by design: Tools for environment-behavior research.* Revised ed. Cambridge University Press, New York.

# 26

# Research Methodologies for Studying Human Responses to Horticulture

**Candice A. Shoemaker**

Formal research to document and isolate many of the impacts of plants on people began to appear in the 1970s with a few research studies from social and medical scientists (Kaplan 1973; Talbott et al. 1976). In 1990, the first national People-Plant Symposium was held bringing together for the first time researchers from the fine arts, sociology, psychology, urban planning, forestry, environmental psychology, and history. In her introduction at the symposium, Paula Diane Relf of Virginia Tech University (1992) stated that the ultimate goal of the symposium was "to establish a research initiative on Human Issues in Horticulture and, with that initiative, a system to communicate the research to a comprehensive audience and to work toward the application of the research findings." She continued by saying that "a research initiative on Human Issues in Horticulture is needed to document clearly through research the value to people of active and passive experiences with plants in order to implement actions that will allow the greatest number of people to garner these benefits." At each subsequent People-Plant Symposium, there were similar calls for conducting research, some of which follow.

At the 1992 symposium, Russ Parsons (Parsons et al. 1994) stated, "Unfortunately, documented evidence of the therapeutic benefits of gardening is equivocal at best, and virtually no research has been reported concerning the effects of such passive interactions as merely viewing plants." From the 1994 symposium proceedings, Mark Francis, Patricia Lindsey, and Jay Stone Rice (1994) said, "As a young science, people-plant research must develop more comprehensive approaches and research designs that will help to expand our understanding of the multifaceted nature of people-plant interactions." At the symposium in 1996, Jack Kerrigan (1996) stated, "A research base for the field of People-Plant Interactions is growing, but there is much more that could and should be researched and documented for the field to have a greater impact on the attitudes and understanding of the public and decision makers."

And in 1998 at the first international people-plant symposium, Paula Diane Relf (1999) stated, "Through research, we must document the essential role of plants in our daily life and identify and describe the optimum types of plants, their configuration in the landscape, and methods of caring for these plants to gain the greatest benefit to each person regardless of age, sex, disabilities, cultural, or geographic differences." She continued by saying, "Horticulture has long addressed the production of plants for their numerous benefits to people. We must apply that same quality of research to understanding the role of cultivating a garden in human health and well-being. The questions have multiplied many fold over the last 10 years; we must find and share the answers and inspire others to join us in this endeavor." Moreover, here, in 2000, we once again discuss the fundamentals of research in this rapidly expanding field.

Determining sound methodologies for studying human responses to horticulture, conducting research, and communicating the findings have always been an important component of the People-Plant Symposia and the People-Plant Council (www.hort.vt.edu/human/ppcmenu.html). Research has been a focus of these symposia because historically, most of the information supporting many of the benefits of plants has been anecdotal (Cotter et al. 1978). The benefits that an individual can derive from plants have been discussed for thousands of years (Stein 1990). Psychiatrists and other mental health workers have been using horticulture as a therapeutic tool for more than 100 years (McCurry 1963; O'Connor 1958), with benefits cited including intellectual and emotional growth, as well as improved social and motor skills (Hefley 1973). Additionally, beliefs in the curative effects of gardening for the mentally ill are at least several hundred years old (Watson and Burlingame 1960). Yet, much more remains to be documented.

In many ways, it seems incredible that we need to find evidence of how people depend upon and value plants. The reality is obvious all around us—we use plants to surround our homes, to celebrate holidays and special occasions, as gestures of love and more. Yet, at the same time, our modern lifestyle has distanced us from nature and plant life, often by our own choice.

The information gathered through basic and applied research in this area is critical in directing the growth and development of the professions of horticulture, horticultural therapy, and landscape architecture, just to name a few. As Charles Lewis (1988) stated about the horticulture industry, "It seems obvious that an industry whose sole survival depends on the purchase of plants should understand the meaning plants may hold and the kinds of needs they satisfy in the people who purchase them."

Roger Ulrich and Russ Parsons (1992) clearly justify the need for further research at the 1992 symposium when stating

> Unfortunately, intuitive arguments in favor of plants usually make little impression on financially pressed local or state governments, or on developers concerned with the bottom line. Politicians, faced with urgent problems such as homelessness or drugs, may dismiss plants as unwarranted luxuries. The lack of research on plant benefits also has tended to reduce spending for plants in other important settings, such as workplaces, healthcare facilities, and outdoor areas of apartment complexes.

With this repeated call for research, many wonder how they can contribute to this new, growing body of information. The remainder of this paper will review some common methodological approaches currently used to study the physiological, psychological, and sociological responses of humans to horticulture and their environment.

## RESEARCH METHODOLOGY

### *What Is Research?*

To begin with, we first need to agree on what exactly research is. The word research can mean many different things. We "research" various cars before buying a new one. We "research" vacation spots when planning our vacation. However, when I talk about research I mean the discovery of new knowledge in a systematic and unbiased way. The key points to this definition are "new knowledge," "systematic," and "unbiased." Being systematic and objective will lead to the development of generalizations, principles, or theories, resulting in prediction and possibly ultimate control of events.

### *Characteristics of Research*

- Research begins with a question: We all have asked "Why?" or "What's the cause of that?" or "What does it all mean?" The question or questions are the departure point for research.
- Research requires a plan, direction, and design: This is a critical component of research. The more planning we do, the more certain we will be of our conclusions. Research is an orderly procedure, planned and logical in design.
- Research demands a clear statement of the problem: A clear statement of the problem means that you understand the problem and have looked at it objectively. We must see clearly what it is we are attempting to research.
- Research deals with the main problem through subproblems: Most researchable problems have within them various other problem areas of lesser breadth and importance. By defining the subproblems, the project becomes more manageable.
- Research seeks direction through appropriate hypotheses or educated guesses to assist you in discovering the solution and in giving you direction in looking for the facts.
- Research deals with facts and their meaning.
- Research is circular: We begin with a question and through the process of research, we solve the problem and return to the question with the answer.

### *Defining the Problem*

Clearly defining the problem from the onset will increase the likelihood of obtaining useful, relevant information. A suggestion for getting started is to look at another's research. Researchers are often inspired by another's research. Expanding on previous research or modifying it will provide new insights. This has certainly been the case in people-plant research. Recent research has indeed built on and expanded the early research on the psychological and physiological responses of people to plants.

For example, in Roger Ulrich's classic research paper (1984), he reported on the health benefits to hospital patients from having a room with a view of trees rather than a view of a brick wall. He showed that these patients spent less time in the hospital (7.96 versus 8.70 days), used fewer doses of strong pain relievers, and received fewer negative comments from hospital staff on their charts. Several studies have followed Ulrich's, investigating the same thing—a view from a room—but with different populations. West (1985) found that prison cell window views of nature, compared to views of prison walls, buildings, or other prisoners, were associated with lower frequencies of health-related stress symptoms such as headaches and digestive upsets. Tennessen and Cimprich (1995) tested college students in their own dormitory rooms—some with window views dominated by nature, others with views dominated by hardscape (i.e., parking lots, sidewalks, walls, other buildings)—and learned that students with nature views were better able to focus their attention to the desired tasks given them. A recent study by Kuo and Sullivan (1996) asked apartment complex residents about domestic violence. Some lived in apartments surrounded by trees, others lived in apartments without green surroundings and were not able to select what apartment they would live in. Kuo and Sullivan learned that incidents of domestic violence were significantly lower for residents in apartments surrounded by greenery—22% versus 13% had engaged in violence, and 14% versus 3% had hit their children, respectively.

A more difficult application of this principle would be in studying the therapeutic effects of horticultural therapy. For example, Song and Sim (1999) studied the effects of a horticultural therapy program on schizophrenics with negative symptoms. Two likely modifications would be to change the population studied or change the therapeutic application used while all other variables remained the same—this last part is what makes this difficult. However, this does not mean that previously reported research on therapeutic effects of horticulture cannot be used to guide your research. Also, consider modifying occupational therapy and physical therapy research—in this case, the modification could be the therapeutic application.

When defining the problem, we must state operational definitions to allow quantification of the results. These definitions help to establish the frame of reference with which the researcher approaches the problem. For example, when we talk about "quality of life" or "psychological well-being" what are we talking about? What is "quality of life"? What is "psychological well-being"? When writing operational definitions, consider things such as physical health, behavioral repertoire, reaction to stress, and competence to define these terms.

We must also state assumptions implicit in the study. For example, say we wanted to study the relationship between landscape architecture and therapeutic garden landscapes. An implicit assumption to this problem is that there is a relationship between the two types of landscape. It may seem obvious but it is important to state. We must also state limitations, which are those conditions beyond the control of the researcher that may place restrictions on the conclusions of the study and their application to other situations. Examples of limitations are a data-gathering instrument that has not been validated or the inability to randomly select and assign subjects to experimental and control groups. In addition, we must state delimitations, which are the boundaries beyond which the study is not concerned. Simply put, delimitations are premeditated limitations. For example, the researcher deliberately narrows down, excludes, and is selective with all subjects.

Also, consider the following questions:

- Is this the type of problem that can be effectively solved through the process of research?
- Is the problem significant? Does it provide fresh insights for your profession?
- Is the research on the problem feasible? Do you have sufficient time to carry it out? Do you have access to subjects? Do you have access to the proper equipment or research tools?

### Stating the Problem

When you have defined the problem, you can then state the problem in either question form or as a declarative statement. State the problem fully and precisely. It must be limited enough in scope to make a definite conclusion possible. Statements such as "psychological implications of horticulture in urban environments" or "health benefits of gardening for older adults" are not statements of problems. They are broad areas of concern from which problems may be selected. A problem suggests a specific answer or conclusion.

### The Hypothesis

The research hypothesis is a reasonable guess, an educated conjecture that may give direction to thinking with respect to the problem, and thus, aid in solving it. It limits the focus of the investigation to a definite target and determines what observations are to be made. A good hypothesis is reasonable, consistent with known facts or theories, and is stated in such a way that it can be tested and found to be probably true or probably false.

### Sampling

Proper sampling is a critical issue to allow study findings to be generalized to an entire population, whether the population is one of plants or people. In people-plant research, researchers must determine in advance which people will be the appropriate subjects and then a sampling procedure may be used to select the subjects. The basic premise is that results may be generalized only to those individuals who in principle had an equal chance of being included in the survey sample (Vining and Stevens 1986). For example, if we wanted to study the role of sympathy flowers in the funeral ritual, we might survey all those who recently received sympathy flowers or all those who had recently sent flowers as a sympathy gift. If we chose to study those who had recently sent flowers as a sympathy gift, we could not conclude anything about people who do not use flowers as part of the funeral ritual.

## Types of Research

Methods for studying people's responses to plants can generally be grouped into two approaches. The first approach seeks to measure people's responses to plants in their environment objectively. This is done either by recording subjects' preferences and values on numerical scales in a manner comparable to the measurement of physical benefits such as air and water quality (Schroeder 1987) or by using physiological measures, such as blood pressure readings, to document physical changes in people in response to plants (Ulrich and Parsons 1992). This approach is quantitative. The second approach is qualitative and seeks to identify the meaning or significance of plants in people's subjective experiences of their environments. This approach recognizes the importance of emotion, imagination, and intuition in people's experience of the natural world. It does not seek to quantify value, but to describe how people interpret their surroundings relative to their own experiences (Schroeder 1987).

Although it is common to distinguish between quantitative and qualitative methods, I prefer to not consider them as dichotomous, but rather complementary. For example, in conducting a survey, there can be both quantitative ("How many times did you purchase a cut-flower arrangement in January?") and qualitative ("How did you feel when receiving a foliage plant as a gift?") questions used. Qualitative methods, such as observations, can also be used for collecting quantitative data. For example, the number of items within a workstation, such as photographs of family members, posters, and plants, not only describes and quantifies a particular attribute of the workstation, but also characterizes the behavior of its occupant. For a thorough review of quantitative and qualitative research methods used in human issues in horticulture research see Shoemaker, Relf, and Lohr (2000).

### Quantitative Research

Quantitative research design can be classified as experimental, quasiexperimental, and survey (Marans and Ahrentzen 1987). The strength of quantitative research lies in its effectiveness for testing deductive theories and generalizing the findings.

The nature of experimental design is that a presumed "treatment" or independent variable is introduced to one group of participants but not to others, and the conditions and characteristics of those receiving and not receiving the treatments are assumed equivalent. In other words, experimental research deals with the phenomenon of cause and effect. Equivalency is established in human issues in horticulture research through the random assignment of participants to treatments.

Quasiexperimental designs are of two types: nonequivalent control group designs and interrupted time-series designs (Marans and Ahrentzen 1987). A distinguishing feature of quasiexperimental studies is that, unlike experimental design, treatment groups are not randomly assigned. The nonequivalent control group design allows for comparisons between the effects of a treatment and its absence to be made using pretest and posttest comparison between nonequivalent groups. These designs are similar to those for experimental designs, except that the members in the control and treatment groups are constrained by forces beyond the experimenter's control, and thus are not randomly assigned. Interrupted time-series designs are used when comparisons are made among the same subjects before and after a treatment (Marans and Ahrentzen 1987). The time-series design consists of taking a series of evaluations and then introducing a variable or new dynamic into the system after which another series of evaluations is made. If a substantial change results in the second series of evaluations, we may assume with reasonable experimental logic that the cause of the difference was because of the factor introduced into the system.

Survey designs use questionnaires, structured interviews, and observation. They are deceptively easy to compose and distribute; however, correct use demands consideration of many issues. Dillman's book *Mail and telephone surveys: The total design method* (1978) has been the standard in this field for many years and is highly recommended; a revision of this classic reference, incorporating Internet surveys, has just been published (Dillman 2000).

## Qualitative Research

The most common methods of gathering qualitative data are interviews, focus groups, observations, and questionnaires. Other qualitative methods not yet commonly used in people-plant interaction research, but that do have application, include phenomenological methodologies, comparative historical methodologies, ethnographic methodologies, and discourse analysis.

The qualitative nature of personal interviews makes them ideal to explore people's thoughts in depth, as they provide a unique opportunity to probe for more information when an answer is ambiguous or when it appears to provide unique insights or observations. To maintain objectivity in the collection of data, however, it is critical that the survey administrators be well trained to avoid the biases that they can elicit by injecting personal interpretations to questions or providing positive feedback to specific lines of answering.

Krueger (1988) defines a focus group as a special type of group in terms of purpose, size, composition, and procedures. Focus groups are useful either as a self-contained means of collecting data or as a supplement to both quantitative and other qualitative methods. The strength of focus groups is the explicit use of the group interaction to produce data and insights that would be less accessible without the interaction found in a group. The major advantage of focus groups is that they offer the chance to observe participants engaging in interaction that is concentrated on attitudes and experiences that are of interest to the researcher. Experienced interviewers should be employed to run focus groups to ensure unbiased results.

When studying human behavior, there is a concern with both preference and survey research: the link between preference (attitude-opinion) and actual behavior may not be explicitly evaluated. Observing or measuring behavior in the actual setting, although not feasible in many cases, is a useful means for obtaining this information. Recording people's behavior provides a direct indicator of human activities in a particular environment. Behavioral measures are a broad class of methods for directly observing and measuring human activities that have many advantages, the most obvious of which is their face validity (Vining and Stevens 1986).

## COLLABORATION

Another useful approach to generating successful research studies in human issues in horticulture is through collaboration. The interdisciplinary research team approach is an excellent model for studying people-plant relationships since it impacts so many different disciplines—horticulture, design, sociology, education, geography, medicine, and psychology to name a few. The interdisciplinary research team approach offers many benefits:

- It offers opportunities for cross-fertilization of ideas as individuals in seemingly unrelated areas recognize shared interests.
- It highlights and builds on the expertise of each individual on the team.
- It provides differing perspectives when approaching a research problem based on the understanding of the discipline of each individual on the team.
- It serves as a focus or support for other activities such as development of workshops.

Collaboration is also useful for those of us wanting to do research but who are not in the academic world. We can look within our own institutions for opportunities for collaborations or consider other institutions. Quite often, horticultural therapists must defend the value of what they are doing to administrators and funders yet may not know how to determine the value. Understanding how to do research is critical for the horticultural therapist and the profession of horticultural therapy.

## COMMUNICATION

A guiding principle that all human issues in horticulture researchers should heed is communication. What is the point in doing research if you are not going to communicate your findings? Communicating the results of research must be planned as an integral part of a comprehensive research initiative. Communicating negative or incomplete results is as important as positive results, as they can serve as guides to other researchers. Several levels of communication need to be considered from publishing in refereed journals to communicating with the public. Paula Diane Relf (1992), in the proceedings of the first people-plant symposium, discusses the need for communicating the findings as well as the pros and cons of the various levels of communication that are needed.

## CONCLUSION

Our understanding of the value of plants to people has progressed significantly over the past 10 years. Yet there are still so many questions to be answered about the impact of plants on people that there is room for all types of research strategies to be used.

I hope I have demonstrated the wide array of research methods available for studying human responses to horticulture and their environment. I also hope I have inspired you to help document what we all know intuitively and hear anecdotally. Examine what you believe about how plants are affecting you. Consider what others are saying about how plants are affecting them. Generate researchable questions and then communicate your findings.

To conclude, horticulture benefits us economically, psychologically, mentally, emotionally, socially, and educationally. We stand to gain in many areas if we prove this in substantial terms through research because it will direct the growth and development of the profession.

## LITERATURE CITED

Cotter, D. J., R. E. Gomez, and V. I. Lohr. 1978. Enhancing ASHS efforts at the plant-people interface. *HortScience* 13:216.

Dillman, D. A. 1978. *Mail and telephone surveys: The total design method.* Wiley, New York.

Dillman, D. A. 2000. *Mail and internet surveys: The tailored design method.* Wiley, New York.

Francis, M., P. Lindsey, and J. S. Rice. 1994. Introduction. In *The healing dimensions of people-plant relations,* eds. Francis, M., P. Lindsey, and J. S. Rice. Ctr. for Design Res., Univ. Cal., Davis. p 1.

Hefley, P. D. 1973. Horticulture: A therapeutic tool. *Journal of Rehabilitation* 39(1): 27-29.

Kaplan, R. 1973. Some psychological benefits of gardening. *Environment and Behavior* 5:145-162.

Kerrigan, J. 1996. Anatomy of a people-plant research project connecting resources. In *People-plant interactions in urban areas,* eds. Williams, P., and J. Zajicek. Texas A&M University, Department of Horticulture, College Station, TX. p. 56-58.

Krueger, R. A. 1988. *Focus groups: A practical guide for applied research.* Sage Publ., Newbury Park, CA.

Kuo, F. E., and W. C. Sullivan. 1996. *Do trees strengthen urban communities, reduce domestic violence?* For Rpt. R8-FR55, Tech. Bul. No. 4. USDA For. Serv. Southern Reg., Athens, GA.

Lewis, C. 1988. Hidden value. A look at the often overlooked profits plants generate for our peace of mind. *American Nurseryman* 168 (Aug 15):111.

Marans, R. W., and S. Ahrentzen. 1987. Developments in research design, data collection, and analysis: Quantitative methods. In *Advances in environment, behavior, and design,* eds. Zube, E. H., and G. T. Moore. Vol. 1. Plenum, New York. p. 251-277.

McCurry, E. 1963. *Flowers and gardens—Therapy unlimited.* Pontiac: Pontiac State Hospital.

O'Connor, A. H. 1958. *Horticulture as a curative.* Cornell Plantation, Ithaca, 14(3):42. As cited in Hefley (1973).

Parsons, R., R. S. Ulrich, and L. F. Tassinary. 1994. Experimental approaches to the study of people-plant relationships. In *People-plant relationships: setting research priorities,* eds. Flagler, J., and R. P. Poinselot. Haworth Press, Binghampton, NY. pp. 347-372.

Relf, D. 1992. Introduction. In *The role of horticulture in human well-being and social development,* ed. D. Relf. Timber Press, Portland, OR. pp. 13-15.

Relf, D. 1999. Moving toward a new millennium in people-plant relations. In *Towards a new millennium in people-plant relationships,* eds. Burchett, M. D., J. Tarran, and R. Wood. University of Technology, Sydney, Printing Services, Australia. pp. 1-7.

Schroeder, H. W. 1987. Psychological value of urban trees: Measurement, meaning and imagination. In Proc. *3rd Natl. Urban For. Conf. Amer. For. Assn.,* Wash., DC. pp. 55-60.

Shoemaker, C. A., P. D. Relf, and V. I. Lohr. 2000. Social science methodologies for studying individuals' responses in human issues in horticulture research. *HortTechnology* 10(1):87-93.

Song, J., and Sim, W. 1999. An experimental study on the effects of horticultural therapy—With special reference to negative symptoms of schizophrenia. In *Towards a new millennium in people-plant relationships,* eds. Burchett, M. D., J. Tarran, and R. Wood. University of Technology, Sydney, Printing Services, Australia. pp. 292-300.

Stein, A. B. 1990. Thoughts occasioned by the Old Testament. In *The meaning of gardens,* eds. Francis, Mark, and Randolph T. Hester Jr. MIT Press, Cambridge, MA. pp. 38-45.

Talbott, J. A., D. Stern, J. Ross, and C. Gillen. 1976. Flowering plants as a therapeutic/environmental agent in a psychiatric hospital. *HortScience* 11:365-366.

Tennessen, C. M., and B. Cimprich. 1995. View to nature: Effects on attention. *J. Environ. Psychol.* 15:77-85.

Ulrich, R. S. 1984. View through a window may influence recovery from surgery. *Science* 224:420-421.

Ulrich, R. S., and R. Parsons. 1992. Influences of passive experiences with plants on individual well-being and health. In *The role of horticulture in human well-being and social development,* ed. D. Relf. Timber Press, Portland, OR. pp. 93-105.

Vining, J., and J. J. Stevens. 1986. The assessment of landscape quality: Major methodological considerations. In *Foundations for visual project analysis,* eds, Smardon, R. C., J. R. Plamer, and J. P. Felleman. Wiley, New York. pp. 167-186.

Watson, D. P., and A. W. Burlingame. 1960. *Therapy through horticulture.* New York: MacMillan Company.

West, M. J. 1985. Landscape views and stress response in the prison environment. Unpublished Master's Thesis. Department of Landscape Architecture, University of Washington, Seattle.

# 27

# The Effect of Cut Flowers in the Japanese Tea Ceremony

Kenji Yamane
Miwa Umezawa
Seiya Uchida
Nobuaki Fujishige
Masao Yoshida
Masayoshi Katagiri

## INTRODUCTION

Tea ceremony is a Japanese traditional custom and contributes to better human relations and relaxation. Cut flowers are generally used in tea ceremonies, but their role in the ceremony has not been elucidated.

Several physiological or medical methods have been used to assess human responses to plants (Shoemaker et al. 2000). Ulrich (1981) reported that alpha amplitudes were consistently higher when individuals viewed natural scenes rather than urban scenes. Another study showed that viewing cut rose flowers after stress significantly promoted alpha wave amplitude (Yamane et al. 1999). Kim and Mattson (1999), measuring brain waves (electroencephalogram, EEG), reported that red-flowering geraniums significantly enhanced recovery from stress. Electromyograpy (EMG) can be used for biofeedback therapy to reduce excess tension and headaches (Budzynski et al. 1973). Skin conductance (electrodermal activity, EDA) has been used to detect mental conditions of people (Lee and Sim 1999). Stern et al. (1984) showed that eye-blinking rate could be used to assess mental conditions. Adachi et al. (2000) used Profile of Mood States (POMS) to assess moods and feeling.

The objective of this study was to evaluate the effects of cut flowers on emotions and physiological parameters after drinking green tea. EEG, EMG, EDA, eye-blink rate, and POMS were measured before and after drinking green tea with or without the presence of cut flowers.

## MATERIALS AND METHODS

### Subjects

The subjects were 20 students (10 males, 10 females) in faculty of agriculture, education, and international studies of Utsunomiya University. They were 19 to 24 years old, and the average age was 21 years old. The subjects were chosen for the control group (no cut flowers) or the treatment group (with cut flowers). Subjects were tested individually. The subjects were told the purpose of the study was to measure the effect of green tea in a tea ceremony. They did not know the true objective of the experiment. After the tests were completed, the subjects were told the real purpose of the experiment and asked not to tell the untested subjects. This experiment was performed as a single blind test.

## Experimental Setting

Experiments were performed from the 8th to the 10th of February 2000. Experiments were performed in a tea ceremony room (3.6 m × 4.5 m) in Utsunomiya University. The subjects were seated in the room that was environmentally controlled at 18°C, with a relative humidity of 40 to 50%, and a light level of about 100 lux. Three cut flowers of camellia (cv. *Akawabisuke*), with red corollas, were arranged in a vase. The cut camellia flowers of this variety have almost no fragrance. A tray with the tea bowl and with or without the vase with cut flowers was put on the table where each group could see it.

## Procedures and Measurement Items

The schedule of procedures for one experimental session is shown in Table 27.1. EDA was measured with a device, which the researchers assembled according to the recommendation by the Society for Psychophysiological Research (USA) (Niimi 1983). Two Ag-AgCl electrodes (NS-335S, *Nihon Koden* Ltd.) were fixed at second finger joints of the left forefinger and the middle finger. EDA was measured throughout the experimental schedule.

Table 27.1. The schedule of the procedure for one experimental session

| Duration | Activity |
| --- | --- |
| 2--3 min | Set up EDA and video; Instruction |
| 1 min | EEG and EMG (30 s with eyes closed, 30 s open) |
| 3 min | POMS, Eating sweets |
| 5 min | Quiz regarding tea ceremony |
| 1 min | EEG and EMG (30 s with eyes closed, 30 s open) |
| 1 min | Drinking tea (with or without flowers) |
| 1 min | EEG and EMG (30 s with eyes closed, 30 s open) |
| 3 min | POMS |

Note: EDA = electrodermal activity; EEG = electroencephalogram; EMG = electromyography; POMS = Profile of Mood States.

EEG and EMG at the forehead position (Fp1) were measured by means of a biofeedback instrument (mindNAVI, Hitachi ULSI Systems Corp.) three times, 1 minute each time, during the experimental session. EEG data were transformed to fast fourier transform (FFT) data and analyzed as power spectral ($\mu$V2). The alpha/beta ratio (alpha wave = 8–12Hz, beta wave = 12–25Hz) was calculated as an index expressing the degree of amenity (Nishina et al. 1998). The measurements recorded after completing the quiz on tea ceremonies and drinking tea were standardized using the initial measurements.

Throughout the experiment, the facial expressions of subjects were videotaped with a digital video camera (DV700, Victor). Eye-blink rate during the EEG and EMG measurements with eyes open was counted using the videotapes.

After the first and third EEG and EMG measurements, the subjects were asked to answer the 32-question POMS (Japanese version II, Kaneko Shobo Ltd.). T scores were calculated from 32 questions. T score results of six feelings (Tension-Anxiety, Depression-Dejection, Vigor, Anger-Hostility, Fatigue, and Confused) before and after drinking tea were compared.

Following completion of the first POMS, the subjects were asked to complete a quiz consisting of 100 original questions about tea ceremonies. The quiz was intended to induce a light stress. After completion of the quiz on tea ceremonies, the second EEG and EMG were measured, and then the subjects drank tea.

One-way analysis of variance and paired $t$ tests were run on all data using SuperANOVA (Abacus Concepts, California, U.S.A.).

## RESULTS AND DISCUSSION

Although EDA patterns of each treatment group were parallel, the variance between subjects was too great to use for comparison (data not shown).

Alpha/beta ratio slightly decreased after drinking tea (Figure 27.1). There were no significant differences in alpha/beta ratio with or without cut flowers (Figure 27.1). Enhanced alpha wave and/or suppression of beta wave by plants were reported in several papers (Kim and Fujii 1999; Kim and Mattson 1999; Nishina et al. 1998; Yamane et al. 1999). However, alpha blockage was not found in this experiment even after respondents took the quiz. It may have been caused by technical problems. Since this instrument was made for biofeedback therapy, alpha amplitude was emphasized too much in the default setting. After this study, we changed the setting; thus, we were able to get better sensitivity to a human stress.

After drinking tea, the EMG level of the treatment group with eyes opened was significantly lower than the control group (Figure 27.2). The EMG of the treatment group with eyes closed was also lower than the control group. In a related study, potted cyclamen also decreased EMG levels after watching a stressful video (Yamane et al. 2000). Since the EMG level measure at the forehead indicates tension and stress (Budzynski et al. 1973), these results suggest that the presence of cut flowers may reduce tension during a tea ceremony.

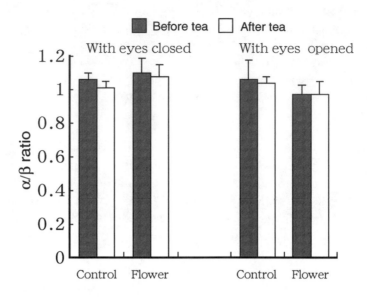

Figure 27.1. The changes in brain wave alpha/beta ratios before and after drinking tea, measured when subjects' eyes were closed and open. All data were standardized by the initial measurements. Vertical bars show SE.

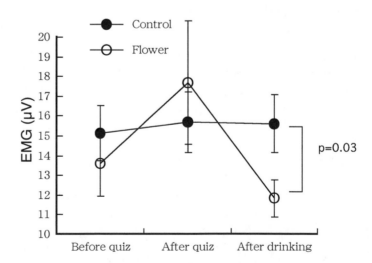

Figure 27.2. The changes in electromyography (EMG) before and after a mild stress inducer and after drinking tea with eyes open, with and without cut flowers. Mean SE.

Table 27.2. The effect of the presence of cut flowers during a tea ceremony on eye-blink rate before and after a mild stress inducer (taking a quiz) and after drinking tea (counts/30 sec from videotape)

| Subject | Before Quiz | After Quiz | After Tea |
|---------|-------------|------------|-----------|
| Control | 18.1±2.2 | 17.7±2.0 | 16.2±2.5 |
| Flower | 26.2±2.6a | 22.3±3.2ab | 18.8±1.9b |

Note: Significant at p<0.05 by paired *t* test.

Eye-blink rate decreased through the experimental session (Table 27.2). The facial expressions of subjects suggested they were nervous when the experiment started (data not shown). The decrease in eye-blink rate suggests that the subjects gradually relaxed through the experiment (Table 27.2). Other researchers have also reported this (Lohr et al. 1996). There were no significant differences between the treatment and control group throughout the experiment period. There was a significant difference in the eye-blink rate of the treatment group between before the quiz and after the tea ceremony (Table 27.2).

The presence of cut flowers decreased T scores of "Depression-Dejection," "Anger-Hostility," and "Fatigue" by 4.3 (p=0.046), 6.7 (p=0.025), and 2.7 (p=0.004), respectively (Table 27.3). Decreased T scores of "Vigor" by cut flowers was a negative response to the existence of flowers, but it may indicate mind calmness because "Fatigue" was decreased by flowers. Adachi et al. (2000) reported that a floral display had positive effects on human emotions such as "confidence." Talbott et al. (1976) reported that flowering plants had some beneficial effects on the behavior of hospitalized psychiatric patients.

Drinking tea itself decreased T scores of "Tension-Anxiety" by 7.7 (p=0.009) and

Table 27.3. The effect of the presence of cut flowers during a tea ceremony on changes in T scores from the Profile of Mood States (POMS) before and after drinking tea

| Mood | Control | Flower |
|------|---------|--------|
| Tension/Anxiety | -7.7** | -1.9 |
| Depression/Dejection | -0.2 | -4.3* |
| Anger/Hostility | -0.8 | -6.7* |
| Vigor | +9.3* | -3.2 |
| Fatigue | -0.9 | -2.7** |
| Confused | -6.7 | -1.5 |

* Significant at P=0.05 by paired *t* test, before and after drinking tea.
** Significant at P=0.01 by paired *t* test, before and after drinking tea.

increased that of "Vigor" (p=0.013) (Table 27.3). Subjects could have had a preconception that drinking tea had positive effects because we announced to them that this test was planned for checking the effect of green tea. However, these results suggest that the combination of cut flowers and drinking tea had positive effects on human emotions.

## CONCLUSION

The results of this study suggest that the presence of flowers during a tea ceremony may have physiological effects, noted by decreased tension levels of foreheads, and may have positive effects on human emotions, such as "Depression-Dejection," "Anger-Hostility," and "Fatigue." Methods of measurement for EEG and EDA need to be improved, and more work is needed to reveal the complex human responses to different flower colors and species.

## ACKNOWLEDGMENT

We thank Mr. S. Ishizaka, who provided the camellia flowers, and Miss M. Kawashima, who helped us analyze the data. We also thank Dr. N. Kirihara and Mr. H. Suga, who helped us assemble the EDA circuit. This research was supported by MOA Health Science Foundation.

## LITERATURE CITED

Adachi, M., C. L. E. Rohde, and A. D. Kendle. 2000. Effects of floral and foliage displays on human emotions. *HortTechnology* 10(1):59-63.

Budzynski, T. H., J. M. Stoyva, C. S. Alder, and D. J. Mullaney. 1973. EMG biofeedback and tension headache: A controlled outcome study. *Psychosomatic Medicine* 35:484-496.

Kim, E. I., and E. Fujii. 1999. A study of sight-psychological effects by a color area of green space. In *Towards a New Millennium in People-Plant Relationships,* eds. M. Burchett, J. Tarran, and R. Wood. University of Technology, Sydney: Printing Services. pp. 301-311.

Kim, E. H., and R. H. Mattson. 1999. Human psychophysiological responses to red- and non-flowering geraniums during recovery from an induced stress. In *Towards a New Millennium in People-Plant Relationships,* eds. M. Burchett, J. Tarran, and R. Wood. University of Technology, Sydney: Printing Services. p 269.

Lee, J., and W. Sim. 1999. The effects on recovery from psychological stress by indoor plants. In *Towards a New Millennium in People-Plant Relationships,* eds. M. Burchett, J. Tarran, and R. Wood. University of Technology, Sydney: Printing Services. pp. 323-327.

Lohr, V. I., Pearson-Mims, C. H., and Goodwin, G. K. 1996. Interior plants may improve worker productivity and reduce stress in a windowless environment. *Journal of Environmental Horticulture* 14(2):97-100.

Niimi, Y. 1983. A trial equipment to measure skin conductance, based on the recommendations of the society for psychophysiological research. *Japanese Journal of Psychology* 53(5):325-327.

Nishina, H., Y. Nakamoto, S. Watamori, N. Masui, and Y. Hashimoto. 1998. Analysis of amenity effect of ornamental foliage plants on human psychology by means of brain waves and semantic differential technique. *Journal of Society of High Technology in Agriculture* 10(2):65-69.

Shoemaker, C. A., P. D. Relf, and V. I. Lohr. 2000. Social science methodologies for studying individuals' responses in human issues in horticulture research. *HortTechnology* 10(1):87-93.

Stern, J. A., L. C. Walrath, and R. Goldstein. 1984. The endogenous eyeblink. *Psychophysiology* 21:22-33.

Talbott, J. A., Stern, D., Ross, J., and Gillen, C. 1976. Flowering plants as a therapeutic/environmental agent in a psychiatric hospital. *HortScience* 11:365-366.

Ulrich, R. S. 1981. Natural versus urban scenes. Some psychophysiological effects. *Environment and Behavior* 13(5):523-556.

Yamane, K., S. Fukaya, N. Fujishige, K. Yoshino, and M. Katagiri. 1999. Effects of cut flowers on physiological and psychological parameters of human beings under stress. In *Towards a New Millennium in People-Plant Relationships,* eds. M. Burchett, J. Tarran, and R. Wood. University of Technology, Sydney: Printing Services. pp. 328-334.

Yamane, K., M. Umezawa, S. Uchida, N. Fujishige, M. Yoshida, and M. Katagiri. 2000. Effects of flowers on human physiology and psychology. *Journal of the Japanese Society for Horticultural Science* 69 (Suppl. 1):381.

# 28

# Impact of Cut Roses at Different Flower-Opening Stages on Customers' Perceptions of a Restaurant Environment

Megumi Adachi
Yasushi Takano
Anthony D. Kendle

## INTRODUCTION

Cut flowers are often harvested at bud stages so that consumers can appreciate the best opening stage after purchase. Cut roses, for instance, are usually harvested at bud stage (Durkin 1980) so that consumers can appreciate them from bud to opening. The ornamental quality, appearance, and physiological condition (e.g., size, shape, and freshness) will change as the flower opens. It is important for the floriculture industry to know which stage of opening of cut flowers is the best as it relates to customer satisfaction. Although there are many reports on the physiological changes of cut roses at postharvest (Marrisen, and Brijn 1995), there are no reports on how these changes of flower opening may affect peoples' perceptions of the flower itself.

This research used the semantic differential method (Osgood and Suci 1955; Doyle et al. 1994; Kim and Fujii 1996; Lohr and Pearson-Mims 2000) to ask subjects to evaluate the total interior environment of a restaurant so that the effects of cut roses as one of all factors in an interior landscape could be evaluated. Cut roses at bud stage, half-open, fully open, and senescence were used to decorate tables in a restaurant so that the differences in decorative effects between flower-opening stages and the best flower-opening stage for appreciation could be investigated.

## MATERIALS AND METHODS

### Plants

Cut red roses of two popular cultivars, 'Souvenir' and 'First Red,' were obtained from retail florists. Pink rose 'William Baffin' was cut from a flower bed on the University of Reading campus. Each stem had one corolla and was arranged in a small white vase on all 20 tables in the mezzanine area of the Blue Room Restaurant at the University of Reading. Flowers were placed in the restaurant at bud stage, half-open, fully open, and senescence so that the differences in decorative effects between flower-opening stages could be investigated. Moreover, two flower sizes of 'First Red'—large and normal size—were compared to study the decorative effects of the volume of the corollas. Injured and noninjured flowers were compared to determine the decorative effects of petal surface injury. Cut flowers were held in a cold room until they were used.

Flower characteristics varied at the different flower-opening stages.

'Souvenir'

- Corollas were closed and tulip shaped in the bud stage.
- Petal color gradually faded during flower opening and senescence.

'First Red' (normal size)

- Corollas were slightly opened in the bud stage.
- Outermost petals at half-open and fully opened stages were slightly injured and were removed from fully opened flowers.

'First Red' (large size)

- Corollas were larger (5 cm diameter) than normal size 'First Red' (4 cm diameter).

'William Baffin'

- Smaller buds (1.5 cm diameter) than red roses 'Souvenir' and normal size 'First Red' (4 cm diameter).
- Corollas were slightly opened in the bud stage.
- Center petals were closed in the half-open stage and opened at the fully open stage.
- Center petals withered and misshapen, and outer petals had small spots at senescence.

## Restaurant Environment

All the surveys were done during lunchtime (11:00 am to 2:30 pm) when the weather was cloudless. The flowers were placed on the tables before lunchtime and removed after lunchtime. Most of the seats were constantly occupied. The temperature, relative humidity, and light conditions were 24°C, 45 to 60% relative humidity, and 10 lux. There were green foliage potted plants located near the entrance and exit of the restaurant, painted landscape scenes were on the walls, and there were window views of outside trees and lawn from all tables.

## Questionnaire

People who visited the Blue Room Restaurant during lunchtime were asked to evaluate the entire environment by answering a questionnaire based on the Semantic Differential (SD) Method (Osgood and Suci 1955). The subjects were made up of many nationalities and ages and were randomly sampled, except that attempts were made to survey equal numbers of male and females (Table 28.1). Each survey assessed about 25 subjects, except for when there were no flowers (64 people surveyed to increase the control sample).

The questionnaire consisted of 32 bipolar adjective pairs (e.g., Beautiful-Ugly). Ten of the adjective pairs were the primary set of scales and considered the absolute minimum essential for coverage of the meanings attributable to designate environments (Hershberger and Cass 1988). The remaining 22 adjective pairs were selected from some "secondary scales" (Hershberger and Cass 1988) and "environmental descriptors" (Kasmar 1988) to clarify the environment more as supplementary factors (Hershberger and Cass 1988) and were regarded as closely related to ornamental expressions.

Each adjective pair represented an image factor and was evaluated at one of seven degrees where the left-hand adjective had a positive meaning and was expressed by a degree of one, and the right-hand adjective had a negative meaning with a degree of seven. Therefore, lower values in factor scores could increase the evaluated degree in each adjective pair. Factor analysis with 32 adjective pairs was also done.

Table 28.1. Numbers of subjects, males, females, and subjects' nationalities evaluating a restaurant environment with different cut flowers as table decorations

| Table Decoration | Red Cut Rose | | | | | | | | | No Table Decoration |
|---|---|---|---|---|---|---|---|---|---|---|
| | 'Souvenir' | | | | 'First Red' | | | | | |
| Flower-Opening Stage | Bud | Half-open | Fully opened | Senescence | Bud | Half-open (injured) | Half-open | Fully opened | Large-size | |
| Subjects | 30 | 27 | 29 | 28 | 29 | 25 | 22 | 27 | 26 | 64 |
| Males | 20 | 18 | 18 | 13 | 17 | 15 | 17 | 19 | 21 | 33 |
| Females | 10 | 9 | 11 | 15 | 12 | 9 | 5 | 8 | 4 | 30 |
| Nationalities | 9 | 5 | 8 | 9 | 12 | 8 | 8 | 9 | 6 | 15 |

| | Pink Cut Rose | | | |
|---|---|---|---|---|
| | 'William Baffin' | | | |
| Flower-Opening Stage | Bud | Half-open | Fully opened | Senescence |
| Subject | 28 | 29 | 28 | 27 |
| Males | 20 | 15 | 18 | 18 |
| Females | 7 | 14 | 10 | 9 |
| Nationalities | 13 | 5 | 12 | 6 |

One-way ANOVA analysis between flower-opening stages in each rose was done in each adjective pair to know the differences in decorative effects between flower-opening stages. Factor analysis was also carried out using the SPSS (Statistical Package for the Social Sciences) program. In the factor analysis, all of the data were analyzed together and then the mean value of factor scores in each condition was calculated. The survey at each decorative condition was done once on one day during each August and September, except for three replications of the "no table decoration treatment" as a control.

## RESULTS

There were significant differences for Clear-Ambiguous (p=0.012), Exciting-Calming (p=0.037), Friendly-Unfriendly (p=0.042), Interesting-Boring (p=0.041), Pleasing-Annoying (p=0.008), and Warm-Cool (p=0.027) in the ANOVA test among flower-opening stages in 'Souvenir,' and the mean scores were generally lower at half-open than those at other stages. In 'First Red,' there were significant differences for Beautiful-Ugly (p=0.037), Deluxe-Shabby (p=0.003), and Formal-Casual (p=0.034) in the ANOVA test among the three stages (bud, half-open, fully open). The ANOVA test among the bud, half-open, and fully open stages in 'First Red' and the large-size of 'First Red' showed significant differences for Deluxe-Shabby (p=0.015), Formal-Casual (p=0.051), and Tasteful-Tasteless (p=0.029). There was only a significant difference for Facilitating-Distracting in the ANOVA test among flower-opening stages in the pink rose; however, there were significant differences for Colorful-Colorless (0.051), Facilitating-Distracting (p=0.002), Fashionable-Unfashionable (p=0.006), Natural-Artificial (p=0.027), and Vibrant-Subdued (p=0.004) in ANOVA between the control (no plant) and the pink rose at the four flower-opening stages.

There were significant differences for Fresh-Musty (p=0.450) and Pleasant odor-Unpleasant odor (p=0.440) among roses whose petals were partly injured and no injured roses at half-open stages.

Factor loadings in factor analysis are shown in Table 28.2. Considering the high factor loadings, Factors I to VI were regarded as referring to "Aesthetic attraction," "Comfort," "Brightness," "Refreshing," "Interest," and "Elegance." Factor scores in Factor I and Factor VI were lower at fully opened stage in the three cultivars of roses (Table 28.3). Factor scores in Factor III were lower in red roses at the half-open stage (Table 28.3).

## DISCUSSION

Factor I in the factor analysis showed that red bud roses were generally satisfactory as a table decoration. Freshness was considered first for ornamental effectiveness in the cut roses at bud stage. However, pink bud roses were too small to be effective as a floral decoration, so the factor scores were higher.

This study also showed that shape and size (volume) of the roses were the most

Table 28.2. Varimax factor loadings above 0.3 from Factor I to Factor VI in factor analysis for cut flowers at different development stages used as a restaurant table decoration

| | Varimax Factor Loading | | | | | |
| | Factor I: | Factor II: | Factor III: | Factor IV: | Factor V: | Factor VI: |
| Factor Meaning | Aesthetic attraction | Comfort | Brightness | Refreshing | Interest | Elegance |
|---|---|---|---|---|---|---|
| Attractive-Unattractive | .647 | .406 | | | | |
| Beautiful-Ugly | .438 | | .508 | | | |
| Bright-Dull | | | .741 | | | |
| Clear-Ambiguous | | | .668 | | | |
| Colorful-Colorless | .393 | | .491 | | | .343 |
| Comfortable-Uncomfortable | .309 | .654 | | .313 | | |
| Complex-Simple | | | | | | .447 |
| Delicate-Rugged | | | | .526 | | |
| Deluxe-Shabby | .551 | .344 | | | | |
| Elegant-Plain | .470 | | | | | .523 |
| Exciting-Calm | | | | | | |
| Facilitating Distracting | | | | .461 | | |
| Fashionable-Unfashionable | .569 | | .306 | | | |
| Feminine-Masculine | .404 | | | | | |
| Flexible-Rigid | | | | | | |
| Formal-Casual | | | | .509 | | .435 |
| Fresh-Musty | | | | | | |
| Friendly-Unfriendly | | .493 | | .543 | .399 | |
| Humid-Dry | | | | | | |
| Interesting-Boring | .366 | | | | .642 | |
| Lively-Dull | .367 | | | | .587 | |
| Natural-Artificial | .439 | | | | | |
| Pleasant odor-Unpleasant odor | | .395 | | | | |
| Pleasing-Annoying | | .494 | | .315 | .319 | |
| Protect-Exposed | | .434 | | | | |
| Refreshing-Wearing | .389 | .440 | | | | |
| Relaxed-Tense | | | | .519 | | |
| Romantic-Unromantic | .573 | | | | | |
| Tasteful-Tasteless | .486 | .360 | | | | |
| Traditional-Contemporary | | | | | | |
| Vibrant-Subdued | .498 | | | | | |
| Warm-Cool | | | | | | |
| Percent Variance | 11.32% | 7.11% | 7.09% | 6.77% | 4.78% | 3.68% |

important factors in decoration. The factor scores for Factor VI "Elegance" at fully open stage were lower than those at the other stages, suggesting that fully opened roses were more effective for adding an elegant image to the restaurant. Since fresh weight and size increased from the bud stage and reached a maximum at the fully opened stage (data not shown), opened roses might help attract people to the restaurant. Development in the helical folds of the petals at the fully opened stage could also be aesthetically important. Double petals or frilling in the roses can be used as an

Table 28.3. Factor scores from Factor I to Factor VI in factor analysis of 32 adjective pairs evaluating a restaurant environment that included a cut flower table decoration. Mean values of factor scores in each condition (cut flower at different opening stage) were calculated after all of the data together were analyzed by factor analysis

| | Factor | | | | | |
| | Factor I: | Factor II: | Factor III: | Factor IV: | Factor V: | Factor VI: |
| Factor Score | Aesthetic attraction | Comfort | Brightness | Refreshing | Interest | Elegance |
|---|---|---|---|---|---|---|
| **Condition** | | | | | | |
| **'Souvenir' (red cut rose)** | | | | | | |
| Bud | -0.201 | -0.009 | 0.080 | -0.007 | -0.039 | 0.195 |
| Half-open | -0.319 | -0.370 | -0.388 | -0.244 | -0.277 | -0.051 |
| Fully opened | -0.111 | -0.294 | -0.036 | 0.255 | 0.209 | -0.128 |
| Senescence (outset) | 0.117 | -0.000 | 0.053 | -0.173 | 0.037 | 0.027 |
| **'First Red' (red cut rose)** | | | | | | |
| Bud | -0.087 | -0.169 | -0.156 | -0.119 | -0.240 | 0.179 |
| Half-open (injured petals) | 0.205 | 0.066 | -0.210 | 0.481 | -0.113 | 0.298 |
| Half-open (no injured petals) | 0.089 | 0.293 | -0.155 | 0.018 | 0.091 | -0.004 |
| Fully opened | -0.098 | 0.057 | 0.109 | -0.064 | -0.282 | -0.282 |
| Fully opened (large size) | -0.204 | -0.158 | 0.054 | 0.000 | 0.133 | -0.227 |
| **'William Baffin' (field pink rose)** | | | | | | |
| Bud | -0.013 | 0.373 | 0.094 | 0.201 | -0.042 | -0.016 |
| Half-open | -0.125 | 0.215 | 0.169 | -0.158 | 0.118 | 0.118 |
| Fully opened | -0.138 | 0.172 | -0.218 | 0.143 | 0.111 | -0.080 |
| Senescence (outset) | 0.006 | 0.044 | -0.089 | 0.021 | -0.131 | 0.025 |
| No Table Decoration | 0.412 | -0.073 | 0.267 | 0.148 | -0.015 | -0.015 |

effective way to add adequate complexity. Adequate complexity may retrieve preference in natural settings (Kaplan and Kaplan 1982). In 'Souvenir,' there were lower factor scores in Factors I to V at the half-open stage than those at the bud stage. Since 'Souvenir' was completely closed at the bud stage and opened at the half-open stage, the half-open roses could be aesthetically more preferable. Historically, large-sized roses with a larger number of petals have been bred as one of the most favorable type of roses (Gault and Synge 1971). This preference for large-sized roses coincided with this study: the roses with more helical folds or a larger size had better factor scores.

Brightness was also regarded as an important factor in decoration in this study with regards to Factor III. The mean scores in Factor III were generally lower in the red roses at half-open stage and the pink field rose at the fully opened stage, possibly because fresh color was kept and the volume increased to an effective level.

In general, fresh and aesthetically pleasing flowers are preferable to visually aged and injured ones (Graham and Kligman 1985). Senescent and injured cut flowers were negatively qualified in the floriculture industry (Halevy and Mayak 1979). The mean value of the factor score for Factor I was higher at senescence than at the other stages. The mean scores in Fresh-Musty and Factor IV "Refreshing" were lower for noninjured flowers than slightly injured flowers, suggesting freshness was felt more with noninjured 'First Red' flowers at the half-open stage. The results showed that senescent or injured flowers influenced aesthetic evaluation. Heerwagen and Orians (1993) reported that defoliated and injured trees were less attractive than healthy trees. The lower scores at half-open stage in 'Souvenir' may be attributed to the flower's freshness as well as the shape and volume of the flowers.

Although there were some differences among stages in the roses in factor analysis, the differences were not extreme. Possibly the existence of the cut rose itself was important if the flowers were neither injured nor senescent. Green leaves could attribute to the psychological effectiveness in cut roses at all stages.

## CONCLUSION

Factor analysis showed that red roses at bud stage were generally preferred as a flower decoration. The fully opened red and pink roses increased sophisticated impressions of the restaurant environment such as elegance and feminine image more strongly than those at the other stages. Brightness was strongly felt in the red roses at half-open stage and in the pink rose at the fully opened stage, possibly because of the fresh color and increased volume. However, changes of ornamental effectiveness at various open stages were not large. Existence of a rose on a table itself may be effective as a decoration.

## LITERATURE CITED

Doyle, K. O., A. M. Hancheck, and J. McGrew. 1994. Communication in the language of flowers. *HortTechnology* (4):211-216.

Durkin, D. J. 1980. Roses. In *Introduction to floriculture,* second edition, ed. R. A. Larsen. Academic Press, Inc., San Diego, CA. pp. 67 92.

Gault, S. M., and P. M. Synge. 1971. *The dictionary of roses in colour.* Ebury Press and Michael Joseph, England.

Graham, J. A., and A. M. Kligman. 1985. *The psychology of cosmetic treatments.* Praeger, New York.

Halevy, A. H., and S. Mayak. 1979. Senescence and postharvest physiology of cut flowers. *Horticultural Reviews* 1:204-236.

Heerwagen, J. H., and G. H. Orians. 1993. Humans, habitats and aesthetics. In *The biophilia hypothesis,* eds. S. R. Kellert and E. O. Wilson. U.S.A. pp. 138-172.

Hershberger, R. G., and R. C. Cass. 1988. Predicting user responses to buildings. In *Environmental aesthetics: Theory, research, & applications,* ed. J. L. Nasar. Cambridge University Press, U.S.A. pp. 195-211.

Kaplan, S., and R. Kaplan. 1982. *Cognition and environment.* New York: Praeger.

Kasmar, J. V. 1988. The development of a usable lexicon of environmental descriptors. In *Environmental aesthetics: Theory, research, & applications,* ed. J. L. Nasar. Cambridge University Press, U.S.A. pp. 144–155.

Kim, E., and E. Fujii. 1996. *A fundamental study of physio-psychological effects of the colour of plants* (In Japanese with English summary). JILA. 58:141-144.

Lohr, V. I., and C. H. Pearson-Mims. 2000. Physical discomfort may be reduced in the presence of interior plants. *HortTechnology*. 10:53-58.

Marrisen, N., and L. L. Brijn. 1995. Source-sink relations in cut roses during vase life. *Acta Hort*. 405:81-88.

Osgood, C. E., and G. J. Suci. 1955. Factor analysis of meaning. *Jour. Exp. Psychol*. 50:325-338.

# 29

# The Effect of Flower Color on Respondents' Physical and Psychological Responses

Chun-Yen Chang

Is the natural environment good for people's health? If so, how does the natural environment benefit people's health? Studies have begun to define the relationships between the natural environment and human health (Ulrich and Parsons 1992; Kaplan 1995), the benefits green landscapes have on people's psychological condition (Ulrich and Simons 1986; Ulrich 1981; Ulrich et al. 1991), and the effects of natural environments on stress management and relaxation (Kaplan and Kaplan 1989).

In the 1970s, research focused on the psychological effects of nature on people (Kaplan 1973; Ulrich 1979). In the 1980s, researchers also began to study the physiological responses people have to nature (Ulrich 1981, 1983, 1986; Ulrich and Simons 1986). By the 1990s, researchers understood the need to study both the psychological and physiological effects so began studies on the psychophysiological effects of plants and nature on people (Tarrant et al. 1994).

In many of these early studies, the visual effect of the overall landscape on human health and well-being were tested. However, specific factors in the landscape such as specific plants were not examined. Therefore, this study examined the effect of different flower colors on people's physical and psychological conditions. Specifically this study evaluated the influence of different flower colors on respondents' electromyography (EMG) values and their state-anxiety values.

The EMG instrument picks up the weak electrical signals generated during muscle action. Each muscle is comprised of many muscle fibers with "motor neurons" or an electrical connection to higher levels of the nervous system. Muscle contraction occurs when these motor neurons carry electrical activating signals to the muscle fibers. A small part of this electrical energy leaves the muscle and migrates through surrounding tissues. Some of this energy becomes available for monitoring at the surface of the skin. Studies have applied this factor to recording respondents' reactions under different kinds of stimulation; thus, in this study EMG will be used to record a person's physical reaction to different flower colors.

In 1966, Spielberger suggested that conceptual anxiety could be introduced to multifaceted definitions of anxiety by distinguishing trait anxiety from state anxiety. Spielberger defined trait anxiety as an individual's predisposition to respond, and state anxiety as a transitory emotion characterized by physiological arousal and consciously perceived feelings of apprehension, dread, and tension. The distinction between trait and state anxiety is analogous to the distinction between potential and kinetic energy (Endler and Kocovski 2001). Thus, state anxiety records a respondent's short-term psychological reaction, which is often used in psychological studies on anxiety.

## METHODOLOGY

### Subject Selection

A total of 105 students of National Hsing University volunteered as the study subjects.

### Physiological Measurement

An operating staff gave instructions to the research subjects concerning the testing purpose to prevent them from becoming nervous. EMG was measured using a Musclewave EMG4 (by MOE Co.). Two electrodes were placed on the skin 1-1/2 inches above the eyebrows, and the reference electrode was placed between these two. To obtain a high EMG value, subjects were asked to raise their eyebrows. This was necessary to obtain a range for each respondent that would be used to standardize the respondents' recorded data. A resting EMG value was obtained while subjects viewed a blue imageless slide. Two periods of EMG responses were recorded. During the first period, subjects were asked to express their feelings related to the flower color (spoken period). During the second period, a 15-second quiet period was observed for the subjects to sense the flower color. Respondents' EMG values were recorded every 0.1 second.

### Psychological Measurement

The state-anxiety value was determined using the self-evaluation questionnaire designed by Spielberger et al. (1983). The questionnaire has 20 statements like "I feel calm," "I feel secure," and "I am tense," which the subjects responded to by selecting "not at all," "somewhat," "moderately so," or "very much so." The total score of these items represents the subject's degree of state anxiety.

### Flower Colors

The following criteria were implemented to select the colors that were used:

- Popularity: The color needed to be popular in Taiwan.
- Availability: The color needed to be available and easily used in floral design work.

- Identity: The flower color needed to be distinct enough to not be confused with other colors.
- Recognizable: The flower needed to be recognized by nonprofessional people.

Using these criteria, seven colors of roses were selected: red, orange, champagne, white, crimson, pink, and yellow. Flower colors were simulated using PhotoImpact 4.0 image software to present all of the respondents with the same flower image and the same color effect. The simulation of flowers involved projecting colored light on a gray flower to keep its shadow. Red, blue, and green lights were combined to make the desired color.

## Experimental Setting

A Kodak DP850 Ultra LCD Projector was used to show the simulated color roses. Data were collected from March 20 to May 23, 2000, in the Lab of Landscape Design of National Chung Hsing University, Taichung, Taiwan.

Subjects were first shown general landscape slides with images like a seashore to become familiar with the light of the projector. Next, a slide with all seven flowers was shown to become familiar with the testing content. Following these two slides, an empty blue slide was displayed to separate the effect of the previous slides and to provide a sense of calm. The next two slides were the same image of one of the test flower colors. During the first flower color slide, the respondent was asked to express their feelings related to the flower color verbally. During the second flower color slide (showing the same image), the respondent was given a 15-second quiet period to sense the flower color. Next, the state-anxiety questionnaire was administered by showing each statement on a slide that included the flower color image and asked the respondent to indicate the level of agreement. Thus, an evaluation of one flower color consisted of a respondent viewing 25 slides, which included one general slide, one slide composed of all seven flower colors, one blank blue slide, two flower color slides, and followed with 20 flower slides with 20 questioning items on each slide (Figure 29.1). Each flower color was evaluated using the same procedure.

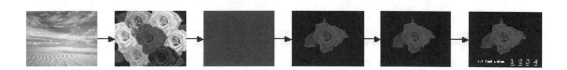

Figure 29.1. The slide sequence of the testing procedure to evaluate the psychological and physical effects of viewing flower color.

## Data Analysis

The average EMG value of each flower color was calculated to represent the subject's physical reaction. The average score was standardized with its z score. The total score of the state-anxiety items was calculated to represent the subject's psychological reaction. Repeated measures ANOVA with software SPSS (Statistical Package for the Social Sciences) 8.0 were used to analyze the physiological and psychological differences among different flower colors.

## RESULTS AND DISCUSSION

A total of 105 students participated in the study. Included were 46 males (44%) and 59 females (56%). Sixty-five (62%) of the subjects were majors in horticulture; 40 (38%) were not.

## EMG Responses

The EMG values for the spoken period and the quiet period were distributed in a similar pattern (Table 29.1). The spoken period had lower EMG values, indicating that the subjects were in a more relaxed condition. The involvement level of the spoken period might cause this result.

Results of the repeated measures ANOVA of the EMG measurements showed there were significant differences ($p < 0.05$) among different color flowers (F=2.50, p=0.02)

Table 29.1. Subjects' electromyography (EMG) values while viewing roses of different colors during periods when they were or were not making verbal statements about their feelings of the flower color

| Color | With Statement | | Without Statement | |
|---|---|---|---|---|
| | Mean | SD | Mean | SD |
| Red | 1.07 | 0.54 | 1.13 | 0.59 |
| Orange | 1.03 | 0.55 | 1.06 | 0.58 |
| Champagne | 1.04 | 0.57 | 1.06 | 0.56 |
| White | 0.99 | 0.54 | 1.07 | 0.55 |
| Crimson | 1.06 | 0.54 | 1.08 | 0.55 |
| Pink | 1.06 | 0.56 | 1.06 | 0.55 |
| Yellow | 1.06 | 0.57 | 1.09 | 0.55 |

Note: N=104; the measurement value for EMG is microvolts.

Table 29.2. The repeated measures ANOVA of different colors of roses during periods when they were and were not making verbal statements about their feelings of the flower color and the state-anxiety values

| Source | Type III SS | df | F | Signif. |
|---|---|---|---|---|
| The EMG value in the spoken period | 0.40 | 6 | 2.50 | 0.02 |
| Error | 16.63 | 618.00 | | |
| The EMG value in the quiet period | 0.39 | 6 | 2.02 | 0.06 |
| Error | 19.66 | 618.00 | | |
| State-anxiety | 20477.39 | 6 | 40.05 | 0.00 |
| Error | 53170.33 | 624 | | |

during the spoken period (Table 29.2). The red flower color was related to higher tension, while white flower color resulted in less tension. The period of quiet thinking showed no significant differences at the 95% standard but is significant at the 90% standard ($F=2.02$, $p=0.06$).

## State-Anxiety Responses

According to the data, pink and yellow flower colors were associated with higher state-anxiety values, which represent the subjects in a more relaxed condition. The white and crimson flower colors were associated with lower state-anxiety values, showing that the subjects were in a nervous condition. From the results of the repeated measures ANOVA, there were significant differences among different color flowers concerning the state-anxiety value ($F=40.05$, $p=0.00$) (Table 29.3).

Following the analysis, the subjects' statements were compared with the emotion-scale model of Russell and Pratt (1980) (Table 29.4). The model provides a four-scaling dimension of aroused, sleepy, pleased, and displeased to measure respondents' current affective state. Red and yellow flowers had more pleasure/arousal facets.

Table 29.3. Subjects' state-anxiety values while viewing roses of different colors

| Flower Color | Mean | SD |
|---|---|---|
| Red | 9.50 | 8.30 |
| Orange | 10.56 | 9.50 |
| Champagne | 10.22 | 10.99 |
| White | 0.74 | 12.60 |
| Crimson | 4.03 | 11.48 |
| Pink | 17.75 | 8.28 |
| Yellow | 13.82 | 9.59 |

Note: N=105.

Table 29.4. The facet of emotional statements related to viewing different flower colors

| | Emotional Statement | | | | | |
|---|---|---|---|---|---|---|
| | Pleased Aroused | Aroused Displeased | Sleepy Displeased | Pleased Sleepy | Others | |
| | N (%) | N (%) | N (%) | N (%) | N (%) | Total |
| Red | 118 (67) | 0 (0) | 10 (6) | 29 (17) | 18 (10) | 175 |
| Orange | 40 (4) | 10 (6) | 27 (16) | 71 (42) | 20 (12) | 168 |
| Champagne | 10 (6) | 8 (5) | 47 (27) | 88 (50) | 23 (13) | 176 |
| White | 21 (12) | 33 (18) | 93 (52) | 13 (7) | 19 (11) | 179 |
| Crimson | 43 (25) | 22 (13) | 62 (36) | 32 (18) | 15 (9) | 174 |
| Pink | 69 (37) | 3 (2) | 10 (5) | 88 (47) | 17 (9) | 187 |
| Yellow | 79 (47) | 8 (5) | 10 (6) | 52 (31) | 18 (11) | 167 |

Orange, champagne, and pink indicated more pleasure and sleepy facets. White and crimson are more in the facet of displeasure and sleepy. Yellow flowers show anxiety, tension, pleasure, and arousal while white flowers illustrate a diverse condition. In the physical reaction, it shows the relaxing condition; however, in the psychological condition, it shows a tension condition (Table 29.5). Further statistical testing should be designed and tested to realize the differences among different flower colors.

Red roses were related to feelings of tension, pleasure, and arousal (Table 29.5). Orange and champagne roses were related to feelings of pleasure and sleepiness. Crimson roses were related to feelings of tension, displeasure, and sleepiness. Viewing pink roses brought feelings of comfort, low anxiety, pleasure, and sleepiness. Feelings of comfort, pleasure, and arousal were demonstrated when viewing yellow

Table 29.5. The comparison of respondents' physical and psychological reaction with their emotional facet

| | EMG | State-Anxiety | Emotional Facet |
|---|---|---|---|
| Red | Tension | Mediate | Pleased-Aroused |
| Orange | Mediate | Mediate | Pleased-Sleepy |
| Champagne | Mediate | Mediate | Pleased-Sleepy |
| White | Relax | Anxiety-Tension | Sleepy-Displeased |
| Crimson | Mediate | Anxiety-Tension | Sleepy-Displeased |
| Pink | Mediate | Comfortable | Pleased-Sleepy |
| Yellow | Mediate | Comfortable | Pleased-Aroused |

roses. White roses illustrated diverse conditions. Physiologically (EMG), subjects were relaxed while psychologically they were tense, sleepy, and displeased when viewing white roses. Subjects, while viewing the white flower, stated feelings of calm, peace, cold, and death.

In this study, different flower colors were shown to have different effects on respondents' physical and psychological responses. This research has shown that the effect of color on humans' perception can be applied when studying plant colors. However, very few related tests have been done that limit the opportunity to compare the results with other studies. Further studies are planned, and comparisons among different culture backgrounds are also needed for a more comprehensive understanding.

## Conclusions

Although many studies (Ulrich 1979, 1983, 1986; Kaplan 1995; Parsons et al. 1998) have looked at people's perceptions of the visual landscape, little has been reported on the relationship between people's perception and the physical effects of viewing landscapes or plants. This paper tries to test the psychological and physical effects of flower color, the basic element of the landscape, to realize the influential factors in a landscape.

The effects of flower colors showed significant differences. Further studies could be done on multiflower effects, as well as the effect of mixed color flowers. The effects of other landscape elements like green cover rate in a landscape, different landform, and land cover, urban versus rural landscape, etc., and the effects of other psychological and physical indicators should be tested. Meta-analysis should be used to analyze the results of different studies. Understanding the effect of various parts of the landscape on human responses can contribute to the understanding of the effect of therapeutic landscapes.

## Literature Cited

Endler, S. N., and N. L. Kocovski. 2001. State and trait anxiety revisited. *Journal of Anxiety Disorders* 15:231-245.

Kaplan, R. 1973. Some psychological benefits of gardening. *Environment and Behavior* 5(2):145-162.

Kaplan, R., and S. Kaplan. 1989. *The experience of nature: A psychological perspective.* Cambridge University Press.

Kaplan, S. 1995. The restorative benefits of nature: Toward an integrative framework. *Journal of Environmental Psychology* 15:169-182.

Parsons, R., L. G. Tassinary, R. S. Ulrich, M. R. Hebl, and M. Grossman-Alexander. 1998. The view from the road: Implications for stress recovery and immunization. *Journal of Environmental Psychology* 18:113-140.

Russell, J. A., and G. Pratt. 1980. A description of affective quality attributed to environment. *Journal of Personality and Social Psychology* 38(2):311-322.

Spielberger, C. D., R. L. Gorsuch, P. R. Vagg, and G. A. Jacobs. 1983. *Manual for the State-Trait Anxiety Inventory.* Palo Alto, CA: Consulting Psychologists Press.

Tarrant, M. A., M. J. Manfredo, and B. L. Driver. 1994. Recollections of outdoor recreation experiences: A psychophysiological perspective. *Journal of Leisure Research* 26(4):357-371.

Ulrich, R. S. 1979. Visual landscape and psychological well-being. *Landscape Research* 4:17-23.

Ulrich, R. S. 1981. Natural versus urban scenes: Some psychophysiological effects. *Environment and Behavior* 13:523-556.

Ulrich, R. S. 1983. Aesthetic and affective response to natural environment. In *Behavior and the Natural Environment,* eds. I. Altman and J. F. Wohlwill. New York: Plenum Press. pp. 85-125.

Ulrich, R. S. 1986. Human responses to vegetation and landscape. *Landscape and Urban Planning* 13:29-44.

Ulrich, R. S., U. Dimberg, and B. L. Driver. 1991. *Psychophysiological indicators of leisure.* State College, PA: Venture Publishing, Inc.

Ulrich, R. S., and R. Parsons. 1992. Influences of passive experiences with plants on individual well-being and health. In *The role of horticulture in human well-being and social development,* ed. D. Relf. Portland: Timber Press. pp. 93-105.

Ulrich, R. S., and R. F. Simons. 1986. Recovery from stress during exposure to everyday outdoor environments. In *Proceedings of the Seventeenth Annual Conference of the Environmental Design Research Association,* eds. J. Wineman, R. Barnes, and C. Zimring. Washington DC: EDRA. pp. 115-122.

# 30

# Plant Decoration in Front of the House Entrance or Gate

H. Kweon
E. Matsuo
H. Takafuji
F. A. Miyake
K. Masuda
K. Mekaru

## INTRODUCTION

The major function of the front border of a house is not only a boundary line between the house and the road, but also an intermediate link between private and public space (Maki et al. 1980). Typically, the front border includes a wall and a gate—outside the wall is considered public and inside the wall is private. They say in Japan that the house entrance is a face of the dweller. Also, many Japanese arrange plants in front of the house entrance, even if the land space does not belong to them (Matsuo 1995). When a garden is available, however, it is very rare that flowers and/or greenery are to be seen in front of the gate and/or the wall (including hedge) of the residence (Matsuo 1995).

The purpose of the current study was to understand changes taking place in the application of plants in the public area of the front border and their significance to cultural changes in Japan. Historically, the public area did not have a garden or plants, particularly when plants or a garden was in the private area of the residence. The Japanese consider gardening a personal and private activity. Is there a change occurring to now consider gardening as a display for the public?

## METHODOLOGY

### Study 1

From 1994 to 1999, single-family one- and two-story homes in a mixed-use community (community included single-family homes, multiple-family homes, businesses, factories, etc.) were examined to see whether plants were used in the front border in 240 areas in Japan. In addition, the number of houses or buildings, such as stores, factories, and houses more than three stories, were counted to clarify the trait of the surveyed areas. The author personally assessed each site. The least number of residences with a garden and/or residences without a garden was more than 50.

## Study 2

Results from Study 1 indicated that the front border was landscaped with containers and/or planting beds. In Study 2, conducted in 1999 and 2000, the use of containers versus planting beds in the front border was compared. In addition, the construction year of the housing development was obtained at the local government office to determine if there was a correlation between construction year and use of plants in the front border. A housing development is a planned residential area in the suburbs of a city, not a mixed-use community as in Study 1. A housing development rather than a mixed-use community was chosen for this study because it was thought that residents of a housing development may have different views on plant decoration than those living in a mixed-use community.

## Study 3

The construction year of the planting bed outside the gate was surveyed in nonhousing development residential areas (areas where residential housing was built up spontaneously, usually near downtown, rather than developed systemically) in 2000 to see if there was a difference between nonhousing development residential areas and housing development areas.

## RESULTS AND DISCUSSION

Almost all houses in the mixed-use communities used plants in the front border (Figure 30.1). In addition, very few houses in the mixed-use community had gardens

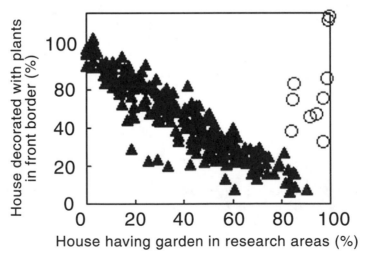

Figure 30.1.   Percent of houses in mixed-use communities with private gardens that used plants in the front border. Circles indicate housing developments.

in the private area of the residence. Houses with a garden in the private area rarely used plants in the front border (Figure 30.1). However, several areas show different attributes from other areas (the areas marked by circles in Figure 30.1). Almost all houses in housing developments had gardens in the private area of the residence. The correlation between having gardens and not using plants in the front border was not as strong for houses in housing developments. In housing developments, 80 to 100% of the houses  had gardens, with 30 to 100% of these houses also using plants in the front border. Comparatively, only 2 to 30% of 80% of the houses in the mixed-use community with gardens used plants in the front border.

In most of the housing developments studied, the percent of houses using plants in the front border was greater than 40% (Figure 30.2). There was no correlation between the construction year of the housing development and the percent of houses using plants in the front border.

Figure 30.3 illustrates the relationship between the year of housing development construction and the percent of houses with planting beds in the front border. These results suggest that the construction year of the housing development is somewhat correlated with the likelihood of finding houses with plants in the front border. This suggests that the housing developments are planned to make the residents pay attention to the landscape of their own residences. This tendency to have plants in the front border in housing developments may influence the residents in nonhousing developments, which was considered in Study 3.

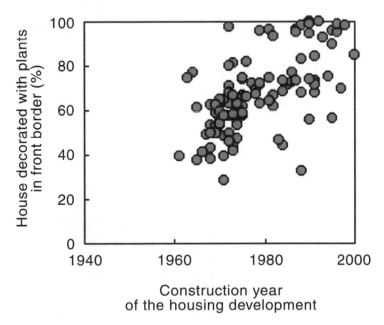

Figure 30.2.    Percent of houses in housing developments, by construction year, that used plants outside the front gate.

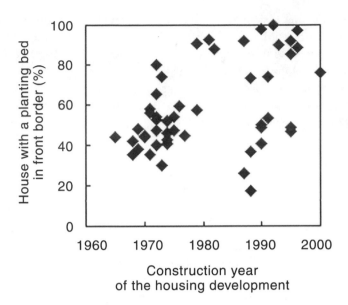

Figure 30.3. Percent of houses in housing developments, by construction year, with planting beds outside the front gate.

The number of houses with a planting bed in the front border was counted for 3 years in a row and is illustrated in Figure 30.4. Planting beds were rarely constructed before 1973, but increased in the late 1970s. This suggests that the design of the houses in mixed-use communities may have been influenced by houses in housing developments since their use of planting beds in the front border followed approximately 10 years later.

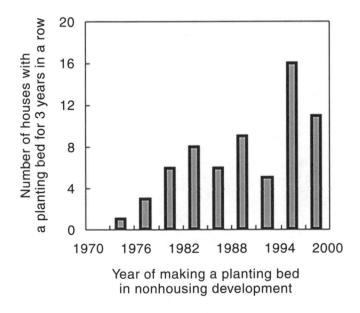

Figure 30.4. Number of houses in mixed-use communities with a planting bed outside the front gate.

The Japanese traditional idea is that inside the gate and wall is private space, and outside the gate is public space. The results of these studies showed that those who do not have a garden use the area near the entrance and the wall as a "surrogate" garden. Perhaps the environmental interest that developed during the three decades from 1970 to 2000 influenced this increased use of the front border for gardens. Finally, these results show that many Japanese pay attention to the space outside the gate and wall of their residence, turning part of the public space into a function more commonly seen in the private areas of the residence.

## LITERATURE CITED

Maki, H., Y. Wakatsuki, H. Ono, and T. Takatani. 1980. *Miegakuresuru toshi*. Kasima Press: Tokyo.

Matsuo, E. 1995. What does the plant use on the station platform give an account of? *Acta Hort.* 391:311-318.

# 31

# Childhood Contact with Nature Influences Adult Attitudes and Actions toward Trees and Gardening

Virginia I. Lohr
Caroline H. Pearson-Mims

## INTRODUCTION

Intuitively, many people feel that being around plants and nature is restorative to the human spirit, and they incorporate plants in a variety of living environments, including homes, work spaces, gardens, shopping malls, and theme parks. Why is this feeling so universal? The biophilia hypothesis maintains that humans have an innate affinity for the natural world, implying that our attraction for nature had significance for the survival of our early ancestors (Wilson 1993). There is evidence that people's responses to nature have both innate and learned components.

More than 80% of the U.S. population lives in urban areas, where chances to interact with nature are greatly reduced (U.S. Census Bureau 2000). As these areas become progressively more urbanized, more trees will be threatened to make room for new development. Will children raised in such stark surroundings fail to develop strong, positive responses to nature, because of fewer opportunities to interact with nature?

### Promoting Positive Attitudes toward Nature and the Environment in Children

Educators have long realized the importance of experiential education for children. Nature education and outdoor experiences help children gain a respect for living things, stimulate their curiosity, and provide them with meaningful life experiences that may influence adult responses to nature (Bullock 1994; Cooper Marcus 1992). Instruction in environmental education promotes positive attitudes toward the environment in elementary school children, and these positive attitudes are retained during childhood (Jaus 1984; Skelly and Zajicek 1998).

Organizers of children's tree planting programs believe that children who plant trees will develop a sense of ownership for the trees they have planted because they have invested their time and energy and, therefore, have a strong incentive to maintain the plants and assure their survival. It is hoped that these children will have a "sense of kin-ship and respect for the natural world" when they are adults (Lewis 1996).

## *Adult Attitudes: Participation in Urban Forestry Programs*

Certain positive attitudes have been identified among adults who volunteer for tree planting programs. The "deep values" that people have for the urban forest, namely, those that are aesthetic, emotional, or spiritual, are the most important reasons for par-ticipation in urban forestry programs (Westphal 1992). The utilitarian benefits of trees, such as their ability to improve air quality, promote cooling, and increase prop-erty values, are much less important to these volunteers. Others have reported similar findings (Dwyer et al. 1991; Hull 1992). Tree planting also provides social benefits for the participants, including contributing to an enhanced sense of community, empowerment of inner-city residents to improve their neighborhoods, and promotion of environmental responsibility and ethics (Dwyer et al. 1992; Sommer et al. 1994).

## OBJECTIVES

Children's experiences with nature are influenced by numerous factors, including parental attitudes, the surroundings where they are raised, and participation in gar-dening or tree planting programs. Are children who have positive or extensive expe-riences with nature more likely to understand and appreciate the values of the urban forest when they are adults?

The goal of this project was to examine the relationship between childhood contact with nature and adult attitudes toward the urban forest among residents of large met-ropolitan areas in the United States. The specific objectives were to (1) assess the pub-lic's understanding of trees and their benefits in urban areas, (2) examine the rela-tionship between childhood contact with nature and adult attitudes, and (3) determine if there are differences in these relationships based on demographic factors, such as ethnic backgrounds. Objective 3 will not be addressed in this paper.

## METHODS

A nationwide 20-minute telephone survey was conducted. It was administered by the Social and Economic Sciences Research Center (SESRC) at Washington State University. John Tarnai and Don Dillman, SESRC, cooperated on the project. Financial support was provided by the United States Department of Agriculture Forest Service and the National Urban and Community Forestry Advisory Council. Three types of questions were included in the survey.

## *Demographics*

Background information about the participants, such as age, education, and income, was gathered. People also were asked questions to evaluate their ethnic and cultural identity, such as "How often do you participate in or identify yourself with activities, groups, or events related to your ethnic or cultural heritage?"

## Adult Attitudes and Actions

Participants were surveyed regarding their current understanding and appreciation of urban trees. Questions assessed different types of values that people assign to trees, including utilitarian benefits, social benefits, and "deep values." Respondents were read a series of opinion statements about trees in urban areas; for example, "Trees should not be planted, because their roots crack sidewalks." They were asked how much they agreed or disagreed with each statement. Possible responses ranged from 1 ("strongly agree") to 4 ("strongly disagree").

Respondents also were questioned about their participation in various activities, such as community service and gardening classes. For example, they were asked, "During the past year, have you participated in any activity or program to enhance the environment, such as a clean-up on Earth Day?" These questions were designed to examine whether adult attitudes toward trees would translate into tangible actions.

## Childhood Contact with Nature

Participants were surveyed regarding their childhood memories of the surroundings where they were raised and their early experiences with nature. Examples of these questions include "Before age 11, how easy was it for you to get access to outdoor places with trees or plants?" and "Was your home or residence next to a large grass area? ...busy streets? etc.?"

## Sample

The sample population was all adult respondents (18 years and older) in households with telephones in large metropolitan areas. A sample combining randomly generated and directory-listed telephone numbers for urban households was purchased from Genesys Sampling Systems of Fort Washington, Pennsylvania. The sample consisted exclusively of households in the 112 most populated metropolitan areas in the continental United States. Respondents in each household were selected from persons 18 years and older based on who had the most recent birthday.

Several procedures were used to encourage respondents to cooperate with the survey. All cases received a minimum of 20 call attempts. If an interviewer called at an inconvenient time, they rescheduled a specific time to recontact the household for an interview. Interviewers left messages on answering machines after three attempts had been made to reach respondents. Use of refusal prevention statements was left to the discretion of the interviewer, who would use one of several prepared statements designed to address specific concerns the respondent might express. Refusal conversion calls were made by a select set of interviewers 3 weeks after the initial refusal. Approximately 11% of the sample refusals were converted into completed interviews.

Completed surveys were obtained from 2,004 randomly selected adults. The overall response rate was 51.8%, which is high for residents of large metropolitan areas (Groves and Couper 1998). The sampling error for the survey was approximately +/- 2.2% for the binomial variable questions.

## RESULTS AND DISCUSSION

Respondents ranged in age from 18 to 90 years with an average age of 42 years. Forty-four percent were male; 56% were female. A majority identified themselves as White or of European background (Table 31.1). Less than half had completed a 4-year college degree.

In general, respondents expressed positive attitudes toward trees in urban areas (Table 31.2). People appear to appreciate trees and be cognizant of the diverse benefits they provide. We anticipated this trend, but were surprised by the magnitude of the response. For example, more than 90% of those surveyed somewhat or strongly agreed with the statements on the utilitarian and social benefits of trees. Almost 98% somewhat or strongly agreed with the "deep values" statement. For even the esoteric aesthetic value question, 75% of the respondents somewhat or strongly agreed.

### Relationships between Adult Attitudes and Childhood Experiences

We will now examine more closely the responses to two specific questions—one on adult attitudes, and one on adult actions.

1. "Do trees have a particular personal, symbolic, or spiritual meaning to you?" A majority (59%) answered "yes" to this "deep values" attitude question.
2. "During the past year, have you participated in a class or program about gardening?" Ten percent responded positively to this specific action-oriented question.

The influence of childhood experiences with nature on the responses to these two questions will now be addressed.

Table 31.1. Selected demographic characteristics of respondents of a survey on attitudes about nature

| Ethnic Background | Number (%) | Educational Attainment | Number (%) |
|---|---|---|---|
| Asian-American/Pacific Islander | 2.2 | High school or less | 28.8 |
| Black/African-American | 9.1 | 2-year college degree or less | 30.7 |
| Hispanic/Latino | 5.3 | 4-year college degree or less | 22.5 |
| Multiethnic/Other | 4.7 | Some graduate school or more | 18.0 |
| Native American | 3.0 | | |
| White/European | 75.7 | | |

Table 31.2. Responses to selected statements used to assess adult attitudes toward urban forests

| Survey Statement | Strongly Agree (%) | Somewhat Agree (%) | Somewhat Disagree (%) | Strongly Disagree (%) |
|---|---|---|---|---|
| "Trees should be planted in business districts to reduce smog and dust." (Utilitarian benefit) | 61.5 | 29.0 | 6.5 | 3.0 |
| "Trees should be used in cities, because they make interesting sounds as their leaves rustle." (Aesthetic value) | 31.6 | 43.9 | 14.2 | 10.3 |
| "Trees in cities help people feel calmer." (Social benefit) | 63.0 | 31.8 | 3.0 | 2.2 |
| "You consider trees important to your quality of life." (Deep value) | 83.2 | 14.6 | 1.1 | 1.1 |

## *Childhood Experiences: Home Surroundings*

### Natural Elements

Respondents were asked questions about what surrounded their childhood homes. Of those who said that their childhood home was "next to a garden or flower beds," 61% also said that trees had spiritual meaning to them. Only 49% of those who lived in homes that were not next to gardens felt that trees had such value (Figure 31.1A). This showed that childhood surroundings could influence adult attitudes.

Being raised in a home next to a garden affected adult *actions* as well as *attitudes* (Figure 31.1B). Of the respondents who lived in homes near gardens during childhood, 11% reported taking a gardening class in the last year, while only 6% of those whose childhood home was *not* near a garden during childhood reported participating in a gardening program in the last year. These two questions clearly showed that having gardens near children could be positive.

### Nonnatural Elements

We next examined the influence of being raised near nonnatural elements, such as busy streets and parking lots, by examining responses about the proximity of large buildings. Of those who said that their childhood home was "next to large buildings," only 52% also said that trees had spiritual meaning, while 60% of those who were raised in homes that were not next to large buildings felt that trees had this value (Figure 31.2). This response indicated that being raised near urban elements could have a negative effect on adult attitudes toward trees.

Being raised in a home next to large buildings did not, however, negatively affect adult participation in gardening programs (data not shown). Adults whose childhood homes were near large buildings were as likely to have taken a gardening class in the last year as were adults whose childhood homes were not near large buildings.

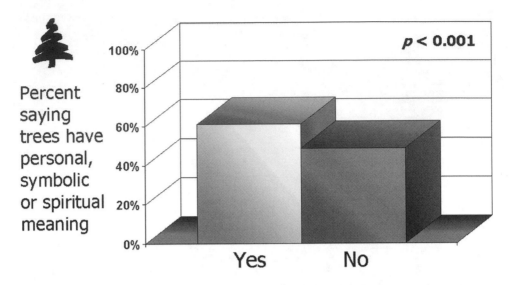

**A**

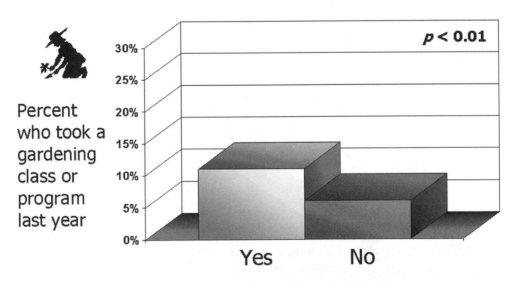

**B**

Figure 31.1.   Effect of being raised next to a garden on adult attitudes and actions.

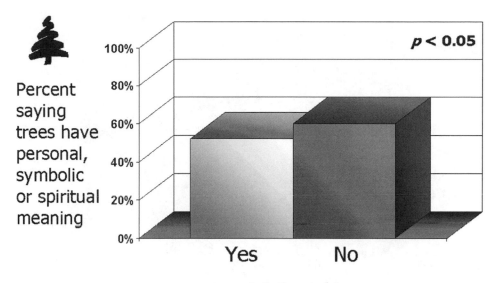

Figure 31.2.    Effect of being raised next to large buildings on adult attitudes.

## Childhood Experiences: Nature or Environmental Education

Most of us assume that childhood education on nature or gardening positively affects adults' attitudes toward plants. We looked at this relationship by asking people if they had participated in such programs as children. As expected, the relationship was positive. Of the adults who said that they had participated in "nature or environmental education in elementary school," 69% also said that trees had spiritual meaning, while only 54% of those who reported not participating in these programs felt that trees had such value (Figure 31.3A). This response demonstrated the potential of childhood gardening programs to influence adult attitudes positively.

Participation in environmental education in elementary school also was associated with an increased likelihood of taking gardening classes as an adult (Figure 31.3B). Of the adults who participated in environmental education in elementary school, 13% reported taking a gardening class in the last year, while only 8% of those who did not participate in these programs during childhood reported taking a gardening class.

## Childhood Experiences: Outdoor Activities

Visiting or Playing in Local Parks

We wondered if adult attitudes or actions might be affected by whether the interaction with plants during childhood was active or passive. We asked adults how often they

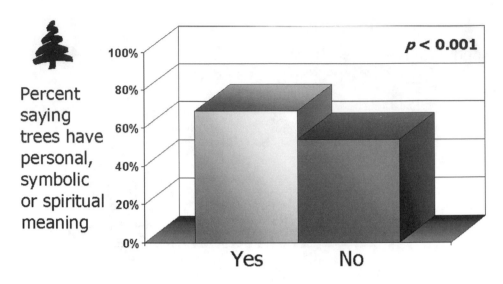

**A**

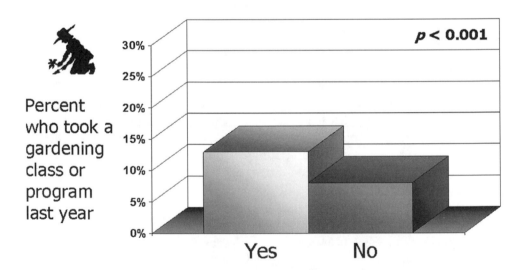

**B**

Figure 31.3.   Effect of nature education in elementary school on adult attitudes and actions.

visited or played in local parks as a child, a relatively passive interaction with nature. Frequency of park visits had a positive influence on people's feelings about trees (Figure 31.4). For example, 63% of adults who often visited parks as a child felt that trees had spiritual meaning, while only 53% of those who rarely visited parks felt this way about trees. Park visits during childhood did not affect adult participation in gardening classes (data not shown). For these questions, passive childhood nature interactions had a positive effect on adult attitudes, but not on adult actions.

## Picking Flowers, Fruits, or Vegetables

To examine active involvement with plants during childhood, we asked how often people picked "flowers, fruits, or vegetables from a garden" as a child. As expected, people who regularly participated in these activities as children were more likely to report that trees had spiritual meaning than were people who picked flowers, fruits, or vegetables less often (Figure 31.5A). This active involvement also affected adults' likelihood of taking gardening classes (Figure 31.5B).

Responses to questions about active or passive interactions with nature support the contention that active involvement with nature does have a stronger influence than passive involvement. These responses also clearly showed that any involvement with nature during childhood, whether active or passive, has positive value.

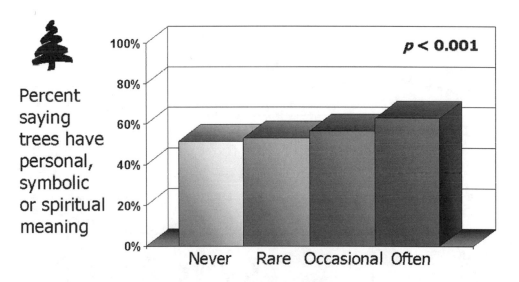

Figure 31.4.   Effect of visiting or playing in local parks on adult attitudes.

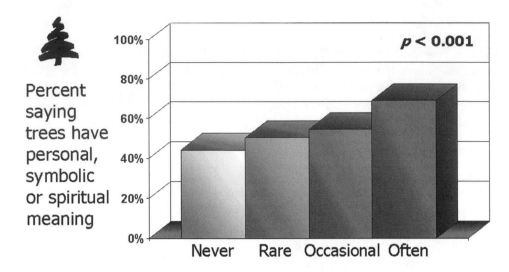

Percent
saying
trees have
personal,
symbolic
or spiritual
meaning

**Did you pick flowers, fruits, or
vegetables from a garden as a child?**

**A**

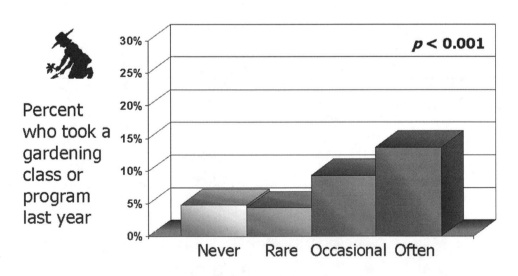

Percent
who took a
gardening
class or
program
last year

**Did you pick flowers, fruits, or
vegetables from a garden as a child?**

**B**

Figure 31.5.   Effect of picking flowers, fruits, or vegetables on adult attitudes and actions.

## CONCLUSIONS

These results indicate that childhood experiences with nature influence adult sensitivities to trees and that the influence is very strong. By understanding the relationships between childhood experiences and current attitudes toward trees in urban areas, we can better understand the influence of childhood participation in tree planting and gardening programs on the perspectives of those same children as adults. This information could allow us to tailor children's environmental and gardening activities more effectively to engender appreciation for nearby nature in our adult citizens.

This research only begins to address these important relationships; it was based on *memories* of childhood nature contact, rather than on accurately documented and quantified contact. It was designed to point the way to promising further studies, such as more resource-intensive longitudinal ones. The raw data will be made available to other researchers for this purpose. More information on this project, its results, and the raw data are available at http://www.wsu.edu/~lohr/hih/nucfac.

## LITERATURE CITED

Bullock, J. R. 1994. Helping children value and appreciate nature. *Day Care and Early Educ.* 21:4-8.

Cooper Marcus, C. 1992. Environmental memories. In *Place Attachment,* ed. I. Altman and S. M. Low, Plenum Press: New York. pp. 87-112.

Dwyer, J. F., E. G. McPherson, H. W. Schroeder, and R. A. Rowntree. 1992. Assessing the benefits and costs of the urban forest. *J. Arbor.* 18:227-234.

Dwyer, J., H. Schroeder, and P. Gobster. 1991. The significance of urban trees and forests: Towards a deeper understanding of values. *J. Arbor.* 17:276-284.

Groves, R., and M. Couper. 1998. *Nonresponse in household interview surveys.* John Wiley and Sons: New York.

Hull, R. B. 1992. How the public values urban forests. *J. Arbor.* 18:98-101.

Jaus, H. H. 1984. The development and retention of environmental attitudes in elementary school children. *J. Environ. Educ.* 15:33-36.

Lewis, C. A. 1996. Participation with green nature: Gardening. In *Green nature/human nature: The meaning of plants in our lives.* University of Illinois Press: Urbana.

Skelly, S. M., and J. M. Zajicek. 1998. The effect of an interdisciplinary garden program on the environmental attitudes of elementary school students. *HortTechnology* 8(4):579-583.

Sommer, R., F. Learey, J. Summit, and M. Tirrell. 1994. The social benefits of resident involvement in tree planting. *J. Arbor.* 20:170-175.

U.S. Census Bureau. 2000. Metropolitan area population estimates for July 1, 1999 and population change for April 1, 1990 to July 1, 1999 (includes April 1, 1990 population estimates base). <http://www.census.gov/population/estimates/metro-city/ma99-01.txt> (12 Dec. 2001).

Westphal, L. M. 1992. Why trees? Urban forestry volunteers values and motivations. In *Managing Urban and High-Use Recreation Settings,* ed. P. H. Gobster, USDA Forest Service: Chicago. pp. 19-23.

Wilson, E. O. 1993. Biophilia and the conservation ethic. In The Biophilia Hypothesis, ed. S. R. Kellert and E. O. Wilson, Island Press: Washington DC. 31-41.

# 32

# Exhilarating Fragrances in the Indo-Islamic Garden

S. Ali Akbar Husain

When the Turco Iranian and Afghan peoples swept into the subcontinent of India and Pakistan (from the 11th to the 16th centuries), they took with them not only their swords and horses, but also their carpets and attars and their immense love of gardens and plants. Plants had been subjects of royal interest in the centuries after the rise of Islam, in Baghdad, Damascus, Cairo, Cordoba, and other centers (Watson 1983), and this interest was pursued wherever the Muslims went. Although there is considerable evidence of gardens built in the subcontinent (India and Pakistan) from the 16th to the 18th centuries, little is known about how these gardens were planted except what may be gleaned from contemporary works on horticulture, medical botany, and garden poetry—and the (miniature) paintings that the Muslim rulers commissioned (Pasha 1911; Ghani 1926; Schimmel 1973; Zebrowski 1983).

Paintings were commonly commissioned in the subcontinent to illustrate the memoirs of the rulers as well as popular poetical works, particularly romances. Gardens often figured in such illustrations, frequently as settings for events the rulers wanted commemorated. Plants were examined in more detail in natural history paintings, as these paintings are now known. The subject of these natural history paintings tended to be flowers, and flowers also figured in portraiture paintings. As elsewhere in medieval Islam, the sultans of the subcontinent enjoyed representations of themselves, whether they were the Mughals who traced their ancestry to the Mongols and dominated northern India and Pakistan from the 16th to the 18th centuries or the Qutb Shahs and the Adil Shahs who had migrated to the subcontinent from the Iranian plateau and ruled over the Deccan in the 16th and 17th centuries (Spabe and Learmouth 1954).

Even a random selection of Mughal and Deccani portrait paintings serves to highlight the plant qualities that were valued. Whether Mughal or Deccani, the subjects of such paintings are almost invariably shown contemplating a spray of flowers or a solitary flower, often held up close enough to seemingly savor the perfume. Sometimes, it is the "hundred-petalled rose" (*Rosa damascena*) that is the object of admiration, and frequently too the Arabian jasmine (*Jasminum sambac*). But other flowers were also portrayed: the tuberose (*Polianthes tuberosa*) and the fragrant screwpine

279

(*Pandanus odoratissimus*), the pink lotus and the white night-opening water lily (*Nelumbium speciosum* and *Nymphaea alba*), the chrysanthemum (*Chrysanthemum chinensis*) and the carnation (*Dianthus caryopyllata*), the champa (*Michelia champaca*), and the mulsary (*Mimusops elengi*). The question that might be asked is why were fragrant flowers such as these so precious to the Islamic cultures of medieval times?

## SIGNIFICANCE OF FRAGRANCE

The significance of fragrance in Islam is suggested in a tradition associated with the Prophet of Islam. The Prophet Muhammad is believed to have expressed his fondness for perfumes by declaring that "Scent is the food of the spirit and the spirit is the vehicle for the faculties of man" (Elgood 1951). To Muslims, as well as to the Greeks whose humoral theory and materia medica the Muslims elaborated, the heart was the seat of human emotion (its monitor, that is), the "fountain of blood" within the body whose composition and flow affected the health and mood of a person. The heart, the Muslim physicians declared unanimously, needed to be fortified with joy continuously, with music and pleasant company, while its essential nourishment was the "rectified air (a person) inhaled from without" (Bar-sela et al. 1964, p. 27). Scents, whether incense and perfume compositions or the "vapors" of scented plants, were thought to purify the air that was breathed in and to strengthen the sensory powers thereby. Since pleasure was considered to be a function of perception, it followed that fragrances contributed to the experience of pleasure or, as the physicians reasoned, "When the sensory powers are strengthened—the sense of taste satisfied with sweet substances and the sense of smell with aroma—the sensation of pleasure is experienced" (Gruner 1983). Conversely, the failure to be happy, it was argued, resulted from a lack of perceiving power. "Dullness of understanding, failure of intelligence, and defects of memory" were said to result from breathing polluted air (Bar-sela et al. 1964).

Of the physician/philosophers of classical Islam, the most popular throughout the medieval Muslim world, from Spain to India, was Ibn Sina (Avicenna in the west, b. 980 AD, d. 1037 AD). Avicenna was widely acclaimed for the emphasis his thesis placed on the heart. To Avicenna, the "spirit" resident in the heart was an "air-and-blood" composition whose "temperament" (or quality) was manifest in heart and pulse rate, muscle tension, skin characteristics, and in the abundance and vigor of a person's breath and its quality (whether "refined" or "coarse"). By purifying the air that a person inhaled, Avicenna explained, fragrances contributed to the substance of the "spirit" in fact, enabling it to circulate freely through the body (presumably with blood), and to carry messages from one body organ to another. In this way, the sensation of pleasure that emanated in the brain was said to be transmitted to the heart, and the "spirit" therefore came to be "the vehicle for the faculties of man" (Hamid 1983).

In a treatise on cardiac drugs, Avicenna listed 63 aromatic simples, including 43 botanically based substances from diverse parts of the known world, as preventives

Table 32.1. List of plants reported by Avicenna in 10__ AD as preventives and tonics for the heart

| Scientific Name (Common Name) | Name in Text |
|---|---|
| *Abies alba* (Silver fir) | Zarnab |
| *Amomum subulatum* (Greater cardamum) | Qaqla |
| *Aquillaria agallocha* (Aloes wood) | 'Ud |
| *Asparagus racemosus* | Shaqaqul |
| *Bambusa arundinaceae* (Bamboo manna) | Tabashir |
| *Borago officinalis* (Bugloss) | Lisan al-thawr |
| *Boswellia glabra* (Frankincense) | Kundur |
| *Cumphora officinarum* (Camphor) | Kafur |
| *Centaurea behen* | Bihman |
| *Cichorium intybus* (Chicory) | Talkshaquq |
| *Cinnamomum zeylanicum* (Cinnamon) | Sadhij |
| *Citrus medica* (Citron) | Utrujj |
| *Coriandrum sativum* (Coriander) | Kuzbarah |
| *Crocus sativus* (Saffron) | Za'fran |
| *Cyperus rotundus* | Sa'd |
| *Delphinium denudatum* | Jadwar |
| *Doronicum hookeri* | Darunaj |
| *Elettaria cardamomum* (Smaller cardamum) | Khayr buwwa |
| *Emblica officinalis* (Emblic myrobalan) | Amlaj |
| *Helianthus annus* (Sunflower) | Azarbuyya |
| *Iris florentina* (Iris) | Sawsan azad |
| *Lavandula stoechas* (Lavender) | Ustukhuddus |
| *Melissa officinalis* (Balm) | Badranjbuyya |
| *Mentha arvensis* (Mint) | Na'na |
| *Myrtus communis* (Myrtle) | As |
| *Nardostachys jatamansi* (Nard) | Sunbul |
| *Nymphaea lotus* (Water lily) | Nilufar |
| *Ocimum basilicum* (Basil) | Badruj |
| *Ocimum gratissimum* (Basil) | Firanjmushk |
| *Paeonia officinalis* (Peony) | Fawania |
| *Pandanus odoratissimus* (Fragrant screwpine) | Armak |
| *Parmelia perlata* (Rock moss) | Ushna |
| *Pistacia terebinthus* (Pistachio) | Fistuq |
| *Punica granatum* (Pomegranate) | Rumman |
| *Pyrus communis* (Pear) | Kummathra |
| *Pyrus malus syn.* Malus domestica (Apple) | Tuffah |
| *Rheum emodi* | Ribas |
| *Rosa damascena* (Rose) | Ward |
| *Santalum album* (Sandalwood) | Sandal |
| *Tamarindus indica* (Tamarind) | Tamar-i Hindi |
| *Terminalia chebula* (Chebulic myrobolan) | Halilaj |
| *Thymus serphyllum* (Wild thyme) | Kahruba |
| *Zingiber zerumbet* (Zedoary) | Zaranbad |

Source: Hamid 1983.

and tonics for the heart (see Table 32.1). He called these aromatic substances "exhilarants." Exhilarants, he explained, functioned by virtue of their "intrinsic properties" and their "specific actions." For instance, lavender was exhilarating because it stimulated the respiratory tract, camphor and sandalwood cooled "the heat of the spirit," and musk was antidotal. Most floral smells, he declared, were "mild," "smoothly blended," and "refined," and "naturally" agreeable to the "spirit." Therefore, they fortified and refreshed the "spirit." It may be recalled that today, too, the oxidation of volatile oils in contact with air is related with the accompanying sense of refreshment, perhaps following the marginal lowering of air temperature which results (Piesse 1891). The degree to which this knowledge influences plant selection in contemporary landscape design is not known, of course. But a thousand years ago, floral fragrances were found refreshing and therapeutic, and a garden environment, it was presumed, facilitated the recovery from psychosomatic maladies and helped to control the symptoms of heart disease, together with compound drugs, aromatic baths and massage, and necessary dietary measures.

## THE USE OF FRAGRANCE IN THE ISLAMIC GARDEN

The use of exhilarating fragrances in an Islamic garden would seem implicit in the presumed therapeutic virtues of plant fragrances detailed in medicinal texts. Undeniably, the association of fragrances with pleasure-giving sensations has been conditioned by centuries of use. Therefore, it might seem that it was the idea of certain fragrances as pleasure-giving and pleasure-awakening (rather than their presumed physiological function), which was the greatest stimulus to the use of scented plants in Islamic gardens.

In horticultural texts compiled by Muslim writers over the centuries (Nadwi 1933; Pasha 1911), floral fragrances have been deemed to be among the chief ornaments of gardens, perhaps in continuity of older pre-Islamic traditions, but also because perfume plants characterize the arid and semiarid regions where Islamic gardens were largely built. Horticultural texts list all such garden ornaments, suggesting how such scent patterns are to be dispersed about the garden relative to walks, pools, and other built features. Such texts also elaborate how plants yielding scentless flowers could be caused to produce scented ones through appropriate breeding or by treating seeds and roots with odorous material. The former practice, it needs to be remembered, has been clearly understood since the time of Mendel (mid-19th century), but modern science has largely discounted the latter.

That fragrant plants should have been a prized ornament of Islamic gardens is understandable when it is recalled that such gardens were usually contained entities, walled, and, in the case of the larger gardens outside the citadel, roofed by a continuous canopy of trees. But, whether "attached as outdoor rooms" to the quarters of the royal household, in palace forts, or detached and built outside the city walls along perennial streams, or around reservoirs and storage "tanks," the Islamic gardens were essentially irrigated outdoor enclosures.

Enclosure afforded the means to awaken the senses and, indeed, to revel in this awakening. As might be thought, perfume played a significant role in these enclosed outdoor spaces, together with other sensory characteristics—whether perfume was contained and held within a palace court by the surrounding walls or, as in the case of the larger gardens, wafted along air tunnels along a canal to an airy pavilion on a platform.

The Islamic gardens of the subcontinent bore a strong "family likeness" to those on the Iranian plateau. Like the latter, they were "quartered in plan and sub-quartered," with a principal axis defined by raised canals and adjacent tree-shaded walks, along which the gardens typically stepped up in terraces. Typically, too, a reservoir with a pavilion centered within marked the intersection of the two plan axes, serving lines of fountain jets set in the canals following the quartered plan, the whole simulating a chessboard with trees, shrubs, and annuals filling in the compartments and pools centered within these (Byrom 1984). Vegetation defined and enclosed the water bodies. Trees and shrubs lined along, or around, a water body reduced evaporative losses, and created suitable growing conditions for moisture and shade-loving plants. As both garden poetry and horticultural texts suggest (Husain 2000), the pleasures of such planting—the glow of flower, fruit, and foliage—were multiplied in the water, and perfume too was more perceptible in the calm, humid conditions around a reservoir, or pool, or within a pavilion in its midst.

The large royal gardens in the Indo-Pakistan subcontinent provided the opportunity to witness the change of the seasons, particularly to celebrate the onset and duration of the monsoon rains and the first harvest of the year. These were occasions when visits to gardens were planned, and this occasional or seasonal use, entailing the enjoyment of the flowering peaks at these times of the year in the tropics and subtropics, is itself suggestive of the life of the garden, particularly the significance of floral values in planting.

To the Turco-Iranian Muslims gardening near the tropics, an Indian evening garden, prolonging the garden's enjoyment by day, was an additional page in the book of the garden. By day, the "flowers of the sun" (the annuals of the rains, or of the cooler season) were enjoyed for their brilliant colors. At night, pleasure was sought in the fragrances diffused by forest (and naturalized) trees and shrubs whose flowers opened in the evening for pollination—white and cream-colored bells, salvers, and tambourines that stood out against the dark foliage of the mango and the mulsary.

## DISCUSSION AND CONCLUSION

As stated earlier, there is little of such planting to be seen in the gardens of the subcontinent since no attempt has been made to restore the planting character of these gardens. As one observer commented, most gardens are "set in a sea of dhoob grass, British-inspired, and dotted about with inconsequential planting" (Byrom 1984). In the Deccan, in southern India, the gardens at the Qutb Shahi necropolis are so efficiently managed by the Department of Archaeology that few trees remain to detract from the architecture of the mausoleums, which, according to the officials, is what people come to see.

A comparison of perfume plants listed in Indo-Islamic perfumery texts with those recorded in poetical narratives (see Tables 32.2 and 32.3) would suggest that the bouquets that permeated the spaces of the royal chambers were also part of the garden's spaces—its arbors, walks, and pools. The essential beauty of this bouquet, it would seem, lay in its power to exhilarate the heart, conditioning it to the "Breath of the Merciful" (as the Sufis of Islam may have thought) or, equally, in an Indian context, to Kamadeva, the Indian god of Love.

Table 32.2. List of scented plants compiled from a 19__ and Indo-Islamic manuscript concerning recipes for perfume composition

| Scientific Name | Name in Text |
|---|---|
| *Acacia catechu* | Khayr |
| *Acorus calamus* | Bach |
| *Aloe vera* | Sugandh sa'iri |
| *Alpina officinarum* | Qulanjan |
| *Amomum subulatum* | 'Ilai'chi kalan |
| *Andropogon muricatus,* syn. *Cymbopogon m.,* *Vetiveria zizaniodes* | Khass |
| *Andropogon schoenanthus,* syn. *Cymbopogon s.* | Azkhar |
| *Anethum sowa* | Shabath |
| *Aquillaria agallocha* | Agar; 'Ud |
| *Aristolochia longa* | Bag nakh |
| *Artemesia pallens* and other spp. | Dawna; Birinjasaf |
| *Asarum europaeum* | Asarun |
| *Astralagus lusitanicus* | Baram |
| *Azadirakhta indica* | Nim |
| *Boswellia serrata* | Luban |
| *Carthamus tinctorius* | Asfar |
| *Cedrus deodora* | Diudar |
| *Centaurea behen* | Bihman |
| *Centaurea moschata* | Nafarmushk |
| Chrysanthemum chinensis | Da'udi |
| *Cinnamomum camphora* | Kafur |
| *Cinnamomum tamala* | Tejpat |
| *Cinnamomum zeylanicum* | Taj |
| *Cistus creticus* | 'Ambar; Ladan |
| *Citrus acida* | China patti |
| *Citrus aurantium* | (Gul-i)Bahar; Narangi |
| *Commiphora mukul,* syn. *C. guggula* | Sugandh guggula |
| *Crocus sativus* | Za'fran |
| *Cucumis sativus* | Khiyyar |
| *Cupressus sempervirens* | Sarw |
| *Curcuma caesia* | Narkachur |
| *Curcuma longa* | Dar-i Hald |
| *Curcuma zedoaria* | Dunki kachur |

(*Table 32.2 continued*)

| Scientific Name | Name in Text |
| --- | --- |
| *Eleagnus* sp. | Sinjid |
| *Elettaria cardamomum* | 'Ila'ichi khard |
| *Eugenia caryophyllata* | Qarnphul; Lawng |
| *Eugenia jambolana,* syn. *Syzygium j.* | Jamun |
| *Ficus carica* | Anjir |
| *Ficus glomerata* | Gular |
| *Hedychium spicatum* | Kapur kachri |
| *Hibiscus abelmoschus* | Mushk dana |
| *Ilicium verum* | Badiyyan khitai |
| *Inula helenium* | Rasan |
| *Ipomoea muricatus* | Nil dana |
| *Iris germanica* | Bikh banafsha |
| *Jasminum* sp. | Niwali |
| *Jasminum auriculatum* | Juhi |
| *Jasminum grandiflorum* | Chambell; Yasmin |
| *Jasminum sambac* | Motiyya |
| *Juglans regia* | Jawz |
| *Lawsonia inermis* | Hinna |
| *Lilium* sp. | Sawsan; Zambac |
| *Liquidambar orientalis* | Silaras |
| *Lisea* sp. | Mayda lakri |
| *Magnifera indica* | Amba |
| *Malus* sp. | Sib |
| *Matricaria chamomila* | Babuna |
| *Mentha* sp. | Pudina; Na'na |
| *Mesua ferea* | Narmushk |
| *Michelia champaca* | Champa |
| *Mimusops elengi* | Mulsary |
| *Myristica fragrans* | Jawz buyya |
| *Narcissus* sp. | Narjis bala; Nargis |
| *Nardostachys jatamansi,* syn. *Valeriana j.* | Sunbul al-Tibb |
| *Nigella sativa* | Shuniz |
| *Nyctanthes arbor-tristis* | Har singhar |
| *Nymphaea* sp. | Nilufar |
| *Ocimum basilicum* | Sabza; Rihan |
| *Ocimum gratissimum* | Firinjmushk; Rayhansiyyah |
| *Ocimum sanctum* | Tulsi |
| *Origanum vulgare/marw* | Marzanjush; Marwa |
| *Pandanus odoratissimus* | Kewra; Ketki |
| *Parmelia perlata* | Chharila |
| *Pavonia odorata* | Bala |
| *Piper betle* | Ghatuna; Tanbul |
| *Piper cubeba* | Kababa |
| *Pistacia integerrima* | Kakra singhi |
| *Pistacia lentiscus* | Mastaki |
| *Plantago ovata* | Bartang |

(*continued*)

*(Table 32.2 continued)*

| Scientific Name | Name in Text |
|---|---|
| *Prunus mahaleb* | Khila khili |
| *Psoralia corylifolia* | Bawanchi; Babchi |
| *Rosa brunonii* | Sewti; Nasrin |
| *Rosa damascena* | Gul-i surkh; Gulal |
| *Rosa lyelrii* | Kuza |
| *Salix caprea* | Bid mushk |
| *Santalum album* | Sandal |
| *Saussurea lappa* | Qust shirin |
| *Sida* sp. | Bala |
| *Symplocos racemosa* | Ludh |
| *Taxus baccata* | Barmi |
| *Valeriana wallichi* | Tagar |
| *Vateria indica* | Ral |
| *Vitis vinifera* | Angur |
| *Withania somnifera* | Asgandh |
| *Wrightia tinctoria* | Indarjau shirin |

Source: S. Ali Akbar Husain.

Table 32.3. List of scented plants compiled from descriptions of gardens in 17__ and Indo-Islamic poetry

| Trees and Shrubs | |
|---|---|
| Scientific Name | Name in Texts |
| *Aganosma caryophyllata* | Malati; Madmalati |
| *Aleurites moluccana* | Akhrut |
| *Amygdalus communis* | Badam |
| *Areca catechu* | Fawfal |
| *Artabotrys odoratissimus* | Madan mast; Madan an |
| *Artocarpus integrifolia* | Phannas |
| *Averrhoa karambola* | Kamrakh |
| *Borassus flabelliformis* | Tar |
| *Boswellia* sp. | Agar; Luban |
| *Butea* sp. | Kesu |
| *Calophyllum inophyllum* | Surpan |
| *Carissa* sp. (possibly) | Tindu |
| *Caryota urens* | Mar |
| *Cercis siliquastrum* | Arghwan |
| *Cinnamomum zeylanicum* | Dar sini |
| *Citrus chinensis,* syn. *C. aurantium* | Narangi |
| *Citrus medica* | Jhamberi; Kawnla |
| *Citrus medica* var. *acida* | Nibu |
| *Citrus maxima* | Zanbu |
| *Cocos nucifera* | Nalir; Narjil |

*(Table 32.3 continued)*

| Scientific Name | Name in Texts |
| --- | --- |
| *Cupressus sempervirens* | Sarw |
| *Cydonia* sp. | Safarjal |
| *Datura* sp. | Dhatura |
| *Dendrocalamus* sp. | Bans |
| *Eugenia caryophyllata* | Gul-i qalanfar; Qarnphul |
| *Eugenia jambolana,* syn. *Syzygium j.* | Jamun |
| *Eugenia jambos,* syn. *Syzygium j.* | Jam |
| *Ficus carica* | Anjir |
| *Jasminum auriculatum* | Juhi |
| *Jasminum grandiflorum* | Yasman; Chambeli; Saman |
| *Jasminum humile* | Jiyyu |
| *Jasminum sambac* | Mugra; Bat mogra; Bela; Motiyya |
| *Juglans regia* | Akhrut |
| *Lawsonia* sp. | Hinna |
| *Magnifera indica* | Amb; Naghzak |
| *Malus* sp. | Sib |
| *Mesua ferea* | Sankesar |
| *Michelia champaca* | Champa; Rai champa |
| *Mimusops elengi* | Mulsary; Gul-i mulsar |
| *Morus* sp. | Tut |
| *Myristica fragrans* | Jayphal |
| *Ochrocarpus* sp. | Surpan; Suringu |
| *Pandanus odoratissimus* | Kewra |
| *Phoenix dactylifera* | Khajur |
| *Phoenix sylvestris* | Sendhi |
| *Pinus gerardiana* | Chilghuza |
| *Pistacia vera* | Pista |
| *Platanus orientalis* | Chinar |
| *Plumeria* sp. | Gulchin |
| *Psidium guava* | Jam |
| *Punica granatum* | Anar |
| *Punica granatum* var. *nan* | Gulnar |
| *Quercus incana* | Maynphal |
| *Rosa damascena* and *R. bourboniana* | Gulal |
| *Rosa brunonii* | Sewti |
| *Rosa lyelii* | Kuza; Nastaran |
| *Salix caprea* | Bid mushk |
| *Santalum album* | Sandal; Chandan |
| *Sesbania aegytica* | Rawasin |
| *Sesbania grandiflora* | Jai |
| *Semecarpus marsupium* | Bhilawa |
| *Stereospermum* sp., syn. *Bignonia* sp. | Pa'idal |
| *Tamarindus indica* | Imbli |
| *Terminalia catappa* | Badam |
| *Terminalia chebula* | Harrah |
| *Vitis vinifera* | Angur; Dak |
| *Zizyphus* sp. | Bir |

*(continued)*

*(Table 32.3 continued)*

| Semishrubs, Perennials, Annuals | |
|---|---|
| Scientific Name | Name in Texts |
| *Abutilon* spp. | Sirna; Jhumka |
| *Ananassa sativus* | Annas |
| *Anethum sowa* | Suya |
| *Artemesia pallens* | Dawna |
| *Beta vulgaris* | Chuqandar |
| *Cassythia* spp. | Akas bel |
| *Celosia cristata* | Kalgha; Taj khurus |
| *Chrysanthemum chinensis* | Shewanti; Da'udi |
| *Citrullus vulgaris* | Kulangar |
| *Coriandrum sativum* | Kothmir |
| *Crocus sativus* | Za'fran |
| *Cuminum cyminum* | Kammun |
| *Gomphrena globosa* | Gul-i Awrang |
| *Helianthus annus* | Gul-i Sur |
| *Hyacinthus orientalis* | Sunbul |
| *Ipomoea* sp., syn. *Calonyction* sp. | Gul-i chand |
| *Iris* sp. | Sawsan |
| *Kaempferia rotundus* | Bhuin champa |
| *Lagenaria vulgaris* (Bottle gourd); also    *Benincasa cerifera* (White gourd) | Kaddu |
| *Lilium* sp. | Sawsan |
| *Luffa acutangula* | Tura'i |
| *Momordica charantia* | Karela |
| *Nelumbium speciossum* | Kanwal |
| *Nymphaea alba* | Kamudi kanwal |
| *Ocimum basilicum* | Rayhan; Sabza |
| *Ocimum gratissimum* | Gul-i dimaran |
| *Origanum* sp. | Marwa |
| *Orzya sativa* | Shalu |
| *Papaver rhoeas* | Lala |
| *Pavonia odorata* | Bala |
| *Physalis* sp. | Kakanad; Kakanaj |
| *Piper betle* | Pan ki bel |
| *Psoralia corylifolia* | Babchi |
| *Sachharum officinarum* | Nishkar |
| *Sida* sp. | Bala |
| *Tagetes erecta* | Gind makhmal |
| *Trichosanthes anguina* | Chachunda |
| *Trigonella foenum-graecum* | Methi |
| *Viola odorata* | Banafsha |
| *Withania somnifera* | Asgandh |

Source: S. Ali Akbar Husain.

## LITERATURE CITED

Bar-sela, A., H. E. Hoff, and E. Faris. 1964. transl. English, Fi Tadbir al Sihhah of Moses Maimanides as The Treatise sent to King Afdal son of Saladin concerning the Regimen of Health in *Transactions of American Philosophical Society* 54(4):16-64.

Byrom, J. B. 1984. Indian Gardens: An Introduction. In *University of Edinburgh: Landscape Occasional Papers,* Paper 3, March.

Elgood, C. 1951. *The Medical History of Persia,* Cambridge Press, Cambridge, England.

Ghani, Hakim M. M. 1926. *Khaza'in al Adwiya* (Treasury of Drugs), 7 volumes. Mian Abdul Majid Publishers, Lahore, Pakistan.

Gruner, O. C. 1983. The four emotions. In Hamid (ed).

Hamid, Hakim A., ed. 1983. transl. English, Al-Adwiyya al-qalbiyya of Ibn Sina in *Avicenna's Tract on Cardiac Drugs and Essays on Arab Cardiotherapy,* Karachi, Hamdard Foundation, Karachi, Pakistan.

Husain, S. Ali Akbar. 2000. *Scent in the Islamic Garden,* Oxford University Press, Karachi, Pakistan.

Nadwi, Hakim S. H. 1933. transl. Urdu, *Kitab al Falaha of Ibn al-Awwam,* Rampur.

Pasha, Kaniza Haji. 1911. *Zira'at-i-Asafiyya* (Agriculture under the Asifyyas). Burhan al Din Press, Hyderabad.

Piesse, C. H. Ed. 1891. *Piesse's Art of Perfumery,* Longmans and Green, London.

Schimmel, A. 1973. *The Islamic Literatures of India.* Wiesbaden, Harrassowitz.

Spabe, O. H. K., and A. T. A. Learmouth. 1954. *India and Pakistan: A Regional Geography.* Mathuen and Co. Ltd., London.

Watson, A. M. 1983. Agricultural Innovations in the Early Islamic World. Cambridge University Press, London.

Zebrowski, M. 1983. *Deccani painting.* University of California Press, Berkeley.

# 33

# Where the Lawn Mower Stops:
# The Social Construction of Alternative
# Front Yard Ideologies

Andrew J. Kaufman
Virginia I. Lohr

## INTRODUCTION

Visit just about any American neighborhood from coast to coast, and more often than not, you will see a unifying theme of front yards with green, well-maintained lawns. The lawn has truly become an American icon. In addition to being in almost every residential setting, lawns are found in business parks, shopping centers, public parks, and athletic facilities.

Lawns cover approximately 30 million acres in the United States (Jenkins 1994). In Iowa, the lawns of an estimated 870,878 single-family homes cover 592,000 acres, which equates to roughly 7,500 square feet of lawn per urban residence (Iowa Turfgrass Industry 2001). This patch of green carpet seems to be woven into not only the American psyche but also the American social fabric as a whole. When asked what percentage of homes in central Iowa have a front lawn, an industry representative replied, "There is no percentage, just about everyone does" (Iowa Turfgrass Industry 2001).

Having a front yard with a well-maintained lawn in the United States is the norm, yet not everyone goes along with it. What type of person would not have a lawn when almost everyone seems to want it? This study was designed to address this question.

## A BRIEF HISTORY OF THE LAWN

How did having a front yard landscape that includes a lawn become so popular? The American residential lawn started appearing in the 18th century when a few wealthy Americans, influenced by French and English aristocratic landscape architecture, began to adopt them. Indeed, Thomas Jefferson, the third president of the United States, has been credited with creating the first American lawn (Bormann et al. 1993). He established an English-style lawn at Monticello, his home in Virginia. Jefferson particularly admired the pastoral landscape quality that took form when the buildings were blanketed with green around their foundations.

In the mid-19th century, homeowners were being encouraged to cultivate their own "living green carpet," as popular garden magazines and garden writers of the time called it (Bormann et al. 1993; Jenkins 1994). Golf, with its great expanses of turf, was also growing in popularity. Even the United States Department of Agriculture was involved, conducting research on turf that could grow in all climates of the country. These influences suggested what a front yard "should" look like.

Frederick Law Olmsted, the father of modern American landscape architecture, addressed America's need for better living environments with countless projects, including his 1868 planned community of Riverside, Illinois (Tishler 1989). In Riverside, each lot had a lawn, and the houses were set back 30 feet from the street to give the entire development a park-like atmosphere. He believed that vast expanses of undulating lawn, incorporated with trees, would emulate a pastoral scene and give people a place to relieve the stress and toils of everyday life.

Originally, residential lawns were associated with upper class homes. They became a status symbol, and eventually status quo, for the middle class. With the invention of the lawn mower and development of lawn chemicals, maintaining a lawn became more feasible in terms of both time and expense. Another major factor that contributed to the booming development of residential lawns was federal funding for highways and veterans after World War II. Highways made access to the new suburban developments easier, and veterans could afford to purchase homes there (Bormann et al. 1993; Jenkins 1994).

The nearly universal appeal of lawns may be deeply rooted in the human subconscious. Balling and Falk (1982) found that people have an innate preference for savanna-like environments. They speculated that this preference arises from the evolution of humans on the savannas of East Africa. Characteristics of modern-day lawns, with their relatively smooth topography and color, can be likened to the setting of the savanna.

## SOCIAL AND ENVIRONMENTAL IMPACTS OF LAWNS

Lawns provide people with social and environmental benefits. For instance, lawns help replenish oxygen. A 50- by 50-foot lawn is purported to produce enough oxygen for a family of four (Professional Lawn Care Association of America [PLCAA] 2000; The Lawn Institute 2000). Lawns provide climate control by cooling neighborhoods. Lawns filter dust and pollen from the air, and they help prevent soil erosion by reducing runoff. They also improve water quality by filtering contaminants from rainwater.

Socially, grass, with its aesthetically pleasing color and uniform texture, fosters a sense of well-being. It provides a tough, yet soft, surface for recreation and sports. Often overlooked are the way lawns, which offer pleasant places for people to gather, contribute to people's emotional and sociological behavior (Eckbo 1950; Laurie 1979). For example, when people are in a beautifully designed vegetated space, their tendency is to become more at ease and more social with others (Kaplan and Kaplan 1982; Relf 1996; Ulrich 1985).

Although there are many social and environmental benefits associated with lawns, there are also potential negative impacts. Social downsides include receiving pressure from neighbors to conform to the societal norm and hearing gas-powered lawn mowers at 6 am on weekends. People's choices may also be restricted by city ordinances requiring that lawns be weed-free and maintained at certain heights.

Environmental issues include ground water and soil contamination from lawn chemicals. Concerns arise over the large quantities of potable water applied to lawns to keep them lush. A gasoline-powered lawn mower produces as much pollution in one hour as a new car does in 30 hours (Automobile Club of Southern California 1996). These problems are in addition to the associated economic costs of maintaining the aesthetic green carpet. The Iowa Turfgrass Industry (2001) estimates that Iowa residents pay $77,120,000 per year for professional lawn maintenance. Considering the benefits and concerns, the presentation of a lawn is a social statement with many societal ramifications.

## SOCIAL NORMS

A social norm is a process of mutual influences that results from similarities in the relationships and social interactions that occur among members of a group (Turner 1991). A feeling of "oughtness," which extends deeper than the notion of liking or disliking, develops (Turner 1991). It is a fundamental belief or moral obligation to adhere to something, even if one does not agree with it. Consequently, those who do not conform to the social actions of the group risk being penalized or ostracized. In this study, having a well-maintained lawn is considered the "normative" practice; lawn conformists perpetuate this dominant societal norm. Those who do not abide by this norm, such as someone with a lawn that is not well maintained, may be penalized by local ordinances or negative comments and actions from neighbors.

People who adhere to societal norms do not typically justify their actions (Mills 1972). Common, everyday occurrences, such as keeping your lawn maintained, are usually not questioned, because individuals simply accept them. Thus boasting about following a societal norm is not the same as justifying it (Mills 1972). When people boast about how green and weed-free their lawns are, they are not justifying the practice, since having a healthy green lawn is the norm. Someone who has a yellowish green lawn and describes it as "economical, because I don't waste money on fertilizer" would be socially justifying an alternate practice. A front yard with a lawn is the shared standard that almost everyone practices, so those who do not practice it probably have justified their actions.

## OBJECTIVE

Although most American homeowners have lawns, exceptions to this norm exist. These range from front yards with reduced areas of grass to yards with no grass at all. The goal of this case study was to typify the person who does not adhere to this norm.

The specific objective was to compare the attitudes of lawn conformists about front yards to the attitudes of lawn nonconformists. To achieve this, the primary investigator looked at homeowners through a social-psychological lens to reveal the characteristics of people who choose not to follow the societal norm and to see how they may differ from those who follow the norm. It evolved from principles of horticulture, landscape architecture, and environmental psychology.

## METHODOLOGY

"Lawn conformists" were operationally defined as people having a conventional landscape front yard consisting of more than 25% lawn. "Lawn nonconformists" were those having lawn grass in less than 25% of the front yard.

Primary information about residents with conventional and alternative front yard landscapes in central Iowa was obtained from Iowa State University faculty members, landscape architects, garden designers, landscape contractors, and garden centers. With this information, the primary investigator located potential participants living in single-family detached homes in Ames, Des Moines, and Gilbert, Iowa. A snowball method of asking participants about other potential participants was used to expand the sample size. After receiving approval by the Iowa State University Human Subjects Committee, a letter was mailed to potential participants outlining the project and asking respondents to participate (Dillman 1978). The final sample included six participants representing lawn conformity and 18 representing lawn nonconformity. More lawn nonconformists were selected for this study to investigate the different types of people who chose alternative landscapes.

A face-to-face interview with each participant was conducted. Interviews consisted of 21 open-ended questions about the participant's landscaping views and choices, such as *What is the function of your front yard?* and *How do you control weeds and pests in your yard?* The interview also included nine attitudinal questions, based on a seven-point scale, ranging from strongly disagree to strongly agree, and seven demographic questions. The interviews were recorded and later transcribed. Transcriptions were analyzed using content analysis methodology (Rels and Judd 2000).

To categorize the characteristic attitudes of lawn conformists and nonconformists, a typology, which is a classification based on shared characteristics, was developed. The criteria for the typologies were adopted from Roebuch and Frese (1976), who outline three sociological dimensions: (1) achieved and ascribed characteristics, (2) identities and perspectives, and (3) behavior on the scene. The categories were then broken down into subcategories to represent the main noncomformist themes.

## RESULTS AND DISCUSSION

### Demographics

Fifty-eight percent of the participants were male and 42% were female. They ranged in age from 28 to 74. On average, lawn conformists who participated in this study were younger (40 years) than nonconformists (55 years). Most of the lawn conformists' children were 14 or younger, while the children of nonconformists were mostly over 18. Only 40% of the lawn conformists grew up in a rural area, while 66% of the nonconformists did.

Participants' occupations ranged from retired telephone worker and auto mechanic to university professor and interior designer. Both groups had high educational attainment, with 83% of conformists and all of the lawn nonconformists being college graduates. There was a slight difference in their average household annual incomes ($40,000 for lawn conformists and $50,000 for nonconformists).

Although lawn nonconformists spent almost the same amount of time (4.2 hours per week) working on their yards as lawn conformists (4.5 hours per week), nonconformists were less likely to use a lawn care company to take care of their lawns (28%) than lawn conformists (67%). In addition, lawn nonconformists were more likely to belong to environmental or conservation groups (78%) than the lawn conformists (33%).

## Attitudes toward Neighbors

Four of the attitude questions were about the respondent's neighbors. Both lawn conformists and nonconformists believed that their neighbors liked their front yards (Table 33.1). Both groups also tended to agree that having a well-maintained lawn improves their relationship with neighbors. However, lawn nonconformists were less in agreement than lawn conformists that they have a close relationship with their neighbors. Furthermore, lawn nonconformists tended to disagree that their neighbors influence how they maintain their landscape, while lawn conformists tended to agree. This is consistent with the literature on social norms (Turner 1991).

## Social Norms Expressed

Content analysis of the statements of lawn conformists and nonconformists revealed characteristic attitudes toward front yard landscapes. Both groups shared some char-

Table 33.1. Attitudes* towards neighbors expressed by lawn conformists and lawn nonconformists in central Iowa single-family home neighborhoods

| Attitude | Lawn Conformists | Lawn Nonconformists |
|---|---|---|
| "My neighbors like my front yard." | 5.2 | 5.6 |
| "Having a well-maintained lawn improves my relationship with my neighbors." | 5.2 | 4.8 |
| "Would you say you have a close relationship with your neighbors?" | 5.5 | 4.4 |
| "My neighbors influence how I maintain my landscape." | 4.3 | 3.2 |

* Based on a 7-point scale from 1 (strongly disagree) to 7 (strongly agree).

acteristic attitudes. Notable differences in attitudes between lawn conformists and nonconformists were also evident (Table 33.2).

The influence of neighbors, which was documented through attitude questions (Table 33.1), also emerged in the statements of respondents. One lawn conformist, who voiced awareness of the opinions of neighbors, said, "The folks straight behind us obviously have a big vested interest in the way we keep our yard...Before we moved in, they asked our mutual friends—how well do they keep up their yard?" According to Turner (1991), a social norm is a process of mutual influences between people: lawn conformists appeared to be much more influenced by their neighbors than were nonconformists.

People who deviate from societal norms risk sanctions from the group (Turner 1991). People might express those sanctions in the form of negative opinions. This was reflected by a lawn conformist who said, "Their yards look like hell and it detracts from the whole house." Another noted, "I like the fact that all the neighbors keep up their yards and have a lot of green grass. I would be upset if someone let their yard go completely wild."

Goffman (1959) claims that people's day-to-day actions are similar to theatrical performances. These so-called "social performances" take place front stage and back-stage. The front yard resembles a front stage for both lawn conformists and noncon-formists, while the backstage could be likened to preparation work, such as consulting plant catalogs or seeking advice at a garden center. Both lawn conformists and nonconformists were aware of their "front stages." Lawn nonconformists, while expressing less concern for the opinions of neighbors than lawn conformists, still expressed a desire to get along with the community. One nonconformist said, "Even though we don't have grass—we do make an attempt to keep the yard well maintained

Table 33.2. Characteristic attitudes of residents of single-family homes in central Iowa towards front yard landscapes showing differences in attitudes between lawn con-formists and lawn nonconformists

| Lawn Conformists | Lawn Nonconformists |
| --- | --- |
| Neighbors have influence | Independent; not concerned about neighbors |
| Idea of low maintenance | Idea of low maintenance/environmentally concerned |
| Use of chemicals | Anti-chemical use |
| Feel the need to control nature | Feel a part of nature |
| Lawn must have good color and consistency | Use front yard to express creativity |
| Front yard landscape reflects the owner and house | Refer to native and natural qualities of the landscape |
| Yard care is work | Enjoy working with plants |
| Lawns are essential to the landscape | Lawns are negative |

so that the neighbors don't take offense at our yard." Another said, "I select plants that are already in the neighborhood, so there's a willingness to be part of the community."

## *Apparent Justifications*

The idea that everyday occurrences that follow social norms are usually not questioned and are not in need of justification (Mills 1972) was reflected in comments from lawn conformists. One lawn conformist said, "I enjoy the wide expanse of green grass. And I don't apologize for that at all." Another conformist stated, "I think there's a certain conception of what beauty is, the notion of a well-manicured lawn with grass that is green and mowed and shrubberies that have a sense of plan to them. Obviously, it's a fairly common conception, I guess what it says about me is I'm well socialized."

People who do not follow the social norm are more likely to justify their actions (Mills 1972). Lawn nonconformists expressed thoughts that could be considered justification for their actions. For example, one said, "A lot of people in this neighborhood walk by, so I planted a lot of those flowers so they would be able to enjoy the garden as they're walking." Another nonconformist said, "The neighbors are very precise, have very orderly yards. I love the freedom to do what I want to do, to plant what I want to plant." Another sign of justification came from some lawn nonconformists who voiced concerns about the costs of lawn care or the time involved in maintenance: "The plantings and the landscape are basically designed to remove as much of the yard as possible from mowing. . . . It's very low maintenance."

## *Purpose of a Front Yard Landscape*

Differences between lawn conformists and nonconformists emerged in their answers to questions about the purpose of their front yards. Lawn conformists were more likely to feel that their lawns were primarily for appearance, noting, "It doesn't get used, yeah you know, just curb appeal" and "So far, it's purely aesthetic." Lawn nonconformists often expressed additional purposes. Some were tangible: "I've also encouraged the neighbors to pick flowers if they want to." Other purposes were social: "I think of it as a public garden because the sidewalk goes through it. People can walk through it and enjoy the flowers and the plants." Some purposes were more personal. One lawn conformist said, "I just love working in the soil." Another noted, "In my front yard I feel creative. I'm out here almost every day doing something, but it's pleasure."

## *Chemicals for Lawn Care*

A major difference between lawn conformists and nonconformists arose regarding chemicals for lawn care. Lawn conformists used chemicals to obtain a lawn that fit their ideals. One remarked, "Ah, lawn chemicals. I do the full treatment. I know I probably apply twice as heavy as the bag says—it's an environmentalist's nightmare from that standpoint. I guess that I like a nice plush grass." Another stated, "I'm not great with messing with chemicals . . . I let a lawn company mess with it." Lawn non-conformists generally expressed concerns over the use of chemicals. One remarked, "Well, we were kind of worried that we're surrounded on two sides by chemical users." Another noted, "I have nothing against grass, I don't like the use of chemicals. . . . I guess my concern is there's enough groundwater problems in Iowa." One expressed grave concerns: "If there's somebody who has a monoculture lawn and maintains it that way, that would bother me more than anything else I guess. I just don't want the chemicals associated with it around."

## Apart from or a Part of Nature

Lawn conformists seemed to be apart from nature, almost trying to control it with their maintenance practices, whereas nonconformists seemed to be a part of nature. One nonconformist summed up the feeling of being a part of nature: "It's a dialogue between the owner and nature." Another nonconformist expressed a strong relation-ship with nature: "I need trees and plants. That's what feeds my soul." A desire to con-trol nature is evident in this conformist's words: "You know, weeds and all, com-pletely taken over by the creeping charlie, dandelions, and crab grass and not mowed very often—that would upset me." Another conformist said, "Somebody who never cuts the grass would annoy me, the grass is tall continuously or they got way too many trees—I don't mind trees, but I don't like to have a lot of trees."

## Learning Nonconformist Lawn Behavior

Some theorists suggest that nonconforming behavior is a learned process, influenced by intimate personal groups and, to a lesser extent, associations with media such as television and newspapers (Clinard and Meier 1995; Sutherland and Cressey 1974). This may have been the case for some lawn nonconformists. One noted the influence of a parent, saying, "My dad filled our yard with huge evergreens, so many trees, so there was very little grass, which can maybe be where my very little grass comes from." Parental influence was also expressed by another nonconformist who stated, "I inherited the love of flowers from my mother who was into gardening." Another non-conformist noted the role of grandparents: "My front yard looks very much like my grandparents' front yard."

Some lawn conformists also indicated that their values were learned. One said, "The lawn ethic is definitely from my Dad. He told me, 'you don't have weeds, you keep it watered, because that's important.' ... It's a reflection of—you know...being a responsible person."

## Types of Lawn Nonconformists

Lawn nonconformists were not all alike. They could be grouped into three subcate-gories. "Typical lawn nonconformists" (n=6) expressed all of the characteristic attitudes of nonconformists (Table 33.2). "Lawn conformist observers" (n=4) expressed many of the lawn nonconformist themes, but had a small portion of lawn that was well maintained. "Dandelion lovers" (n=8) expressed many of the lawn nonconformist themes, but had a small portion of lawn that was not well maintained. These subcategories of lawn nonconformists are not to be viewed as significant, but rather as illustrations of the levels of nonconformity observed in this study. Perhaps they are an indication of how strong the social pressures are to have a lawn. One lawn nonconformist commented, "This is my conversation to the neighbors. This is grass and is mowed. So, I maintain this strip here for them."

## CONCLUSION

Aldo Leopold once said, "a thing is right when it tends to preserve the integrity, stability, and the beauty of the biotic community. It is wrong when it does otherwise" (Leopold 1966, p. 262). The American lawn, with its chemical, water, and labor-dependent nature, along with its associated economic costs, may not be right according to Leopold's standard. What this study calls lawn nonconformity might really be conformity to American society's norm of individuality, distinctiveness, and originality.

For some homeowners in central Iowa who chose not to have front yards dominated by grass, the choice was driven by concerns over the time and expense of maintaining a lawn. Others chose their alternatives for environmental reasons, and many had a strong antichemical view. In fact, most of the lawn nonconformists belonged to environmental groups. Many also felt that their landscapes gave them a place to be part of nature. For some, the alternative landscape was a creative outlet.

People with alternate forms of front yards also held some views in common with people with traditional lawns. Both lawn conformists and nonconformists wanted their yards to be liked by their neighbors. Both believed a well-maintained lawn could improve relationships with their neighbors. Both spent similar amounts of time maintaining their yards. Lawn conformists and nonconformists were also very passionate about their front yard landscapes. Even though lawn conformists and nonconformists held different paradigms for the front yard, it was evident that, for both, "green nature is really a part of human nature" (Lewis 1996).

## ACKNOWLEDGMENT

This study was conducted under the supervision of Dr. Wendy Harrod, Department of Sociology, Iowa State University. The primary author wishes to thank Dr. Harrod for her support.

## LITERATURE CITED

Automobile Club of Southern California. 1996. *Avenues* 65(5):10. Los Angeles: Automobile Club of Southern California.

Balling, J. D., and J. H. Falk. 1982. Development of visual preference for natural environments. *Environment and Behavior* 14:5-28.

Bormann, F. H., D. Balmori, and G. T. Geballe. 1993. *Redesigning the American lawn.* New Haven: Yale University Press.

Clinard, M., and R. Meier. 1995. *Sociology of deviant behavior.* Fort Worth, TX: Harcourt Brace College Publishers.

Dillman, D. 1978. *Mail and telephone surveys: The total design method.* New York: John Wiley.

Eckbo, G. 1950. *Landscape for living.* New York: F. W. Dodge Corporation.

Goffman, E. 1959. *The presentation of self in everyday life.* New York: Anchor Books,

Iowa Turfgrass Industry. 2001. Personal phone conversation: December 26. (515) 232-8222. Iowa Turf Office. 17017 US Hwy 69, Ames, IA 50010.

Jenkins, V. S. 1994. *The lawn: A history of an American obsession.* Washington DC: Smithsonian Institution Press.

Kaplan, S., and R. Kaplan. 1982. *Humanscape.* Ann Arbor, MI: Ulrich's Bookstore.

Laurie, I. C., ed. 1979. *Nature in cities: The natural environment in the design and development of urban green space.* New York: John Wiley & Sons.

Lawn Institute, The. 2000. <http://www.lawninstitute.com>. Viewed March 23, 2000.

Leopold, A. 1966. *A Sand County almanac.* New York: Oxford University Press.

Lewis, C. A. 1996. *Green nature/human nature: The meaning of plants in our lives.* Champaign: University of Illinois Press.

Mills, C. W. 1972. Situated actions and vocabularies of motive. In *Symbolic interaction: A reader in social psychology,* 2nd Edition (pp. 393-404), eds. Manis, J. G., and B. N. Meltzer. Boston: Allyn & Bacon.

PLCAA. 2000. The Professional Lawn Care Association of America. <http://www.plcaa.org/index.cfm>. Viewed March 23, 2000.

Relf, D. 1996. The psycho-social benefits of greenspace. *Grounds Maintenance,* March 1996, 33-38.

Rels, H. T., and C. M. Judd. 2000. *Handbook of research methods in social and personality psychology.* New York: Cambridge University Press.

Roebuch, J. B., and W. Frese. 1976. *The Rendezvous: A case study of an after-hours club.* New York: Free Press.

Sutherland, E. H., and D. R. Cressey. 1974. *Criminology,* Ninth Edition. Philadelphia: J.B. Lippincott Company.

Tishler, W. H. 1989. *American landscape architecture: Designers and places.* Washington DC: The Preservation Press.

Turner, J. C. 1991. *Social influence.* Pacific Grove: Brooks Cole.

Ulrich, R. S. 1985. Human responses to vegetation and landscapes. *Landscape and Urban Planning* J. 13:29-44.

# V

# Abstracts

# 34

# Abstracts: Communications

### RESTORATIVE GARDENS: NATURE'S THERAPEUTIC COMPLEMENT TO HEALTHCARE ENVIRONMENTS

**David Kamp**

In this time of rapid change in health care, with emerging interests in complementary medicine, gardens can become a vital component in a comprehensive therapeutic environment. Successful gardens require a design team approach and effective communication to reach their full potential. This presentation featured several restorative gardens, each with challenging settings, budgets, and programs, and discussed how their design teams, incorporating designers, therapists, medical practitioners, and administrators, helped create responsive, cost-effective gardens.

### THE BANNING OF FLOWER CULTIVATION IN JAPAN NEAR THE END OF WORLD WAR II

**Haruo Konoshima**

By the Edo era, Japan had become famous throughout the world for its cultural tradition of flowers and ornamental plants. However, some 50 years later, near the close of World War II, the cultivation of flowers was prohibited in Japan, and those who raised such plants came to be ridiculed as enemies of the state. The facts regarding the enforcement of this ban near the end of the War in the Pacific have been investigated using the records kept by Chiba Prefecture, Hyogo Prefecture, Osaka Prefecture, and Kyoto Prefecture. In Chiba, there was a complete ban enforced, while in Hyogo and Kyoto Prefectures, depending on the wishes of the military and government, cultivation was allowed in certain cases. It has been found that, although the extent of the measures taken varied from place to place throughout the country, beginning in 1940, in the name of increasing food production, increasingly strict orders restricting cultivation of such plants were handed down one after another. However, the largest cause of the food shortage was the lack of a labor force, and thus the establishment of restrictions on planting (i.e., the ban on growing flowers and ornamental plants) did not result in an increase in food production. In this hostile environment, those who carried on the ornamental plant tradition, those who were ready to face the ridicule of being labeled enemies of the state, left for us the spirit of this cultural heritage. That they were able to do this can in fact be thought of as an expression of the universal value of all human life and mankind itself—mankind, the cultivator of flowers.

## ASSESSING THE IMPACT OF URBAN FORESTS ON ELDERLY PEOPLE IN LONG-TERM CARE SETTINGS: TOWARD A MULTICULTURAL FRAMEWORK

**Gowri Betrabet**
**Susana Alves**

As we face the growing demands placed on national resources by an increasingly multicultural aging population, we are compelled to discover new ways of bringing health benefits to people in long-term care settings. One of the potential interventions relates to the use of urban forests and their elements to ameliorate health problems. Although there are numerous studies in human-nature interactions, which demonstrate people's preferences for natural settings and stress-alleviating benefits, research studies focusing on influences of nature on elderly people in long-term care settings, especially health-related benefits, are scarce. There is a need for an integrative conceptual framework that reflects the special needs of the elderly, such as cultural diversity and how it might reflect upon human-nature interactions. Drawing from literature on natural environments, ethnicity in long-term care settings, and studies that focus on the role of culture, we propose a theoretical framework to map the findings and direct and guide future research.

## JAPANESE AND EUROPEAN NAMES OF COLORS ORIGINATING FROM PLANT NAMES IN JAPAN

**Eisuke Matsuo**
**Kinuko Masuda**
**H. Kweon**
**K. Mekaru**
**F. A. Miyake**

We have a close relationship to plants as they provide food, drink, medicine, cloth, dye, etc. This relationship is also evident in many names of colors, which are derived from the names of plants and their use. Thus, colors originating from plants are important considerations in people-plant relationships. This research was carried out using 12 Japanese pictorial books that were published from 1931 to 1996. Of 1,984 colors described, 176 colors were mentioned that originate from plant species, comprising 84 species used for food or drink, and their color names have been mostly adopted from flower or fruit colors. This presentation dealt with the increase in number of colors in Japanese usage originating from plant names, and the difference in the characteristics of the plants used in Japan and Europe.

# Present Status of Horticultural Therapy and Human Issues in Horticulture in Korea

Hye Ran Kwack
Paula Diane Relf

As the level of urbanization increases and the healing effect of plants become known, many people in Korea have begun to recognize this important aspect of horticulture. Today, many Koreans try to solve the difficulties of life and improve the quality of life through horticultural activities. As a result, the importance of therapeutic and well-being issues in horticulture are now being emphasized in various settings. Elementary, middle, and high schools have begun to operate garden-based programs. Eight universities include horticulture elements in their regular graduate programs. A few general and psychiatric hospitals have been applying horticulture as a means of work therapy. Two academic societies were founded to focus on Horticulture Therapy (HT) in 1997 and People-Plant Interaction in 1998. They have held several symposia and conducted various kinds of interdisciplinary studies. Many papers on HT and related issues have been published since these academic societies were founded. These papers deal with the basics of HT, experiments for the verification of the HT effects, and landscaping including healing garden effects and construction, school landscaping, and visual or psychological effects of landscape.

# Infection Control and Therapeutic Gardens: A Survey of Policies and Practice

Nancy J. Gerlach-Spriggs

An increased interest in the integration of gardens and horticultural therapy in mainstream American health care introduces potential risks. Design considerations for patients with cognitive and physical impairments are well known and routinely incorporated into the design process, but the risk of contracting infectious disease (Legionnaires, aspergillosis, and other opportunistic organisms) is rarely considered. This study surveyed a sample of diverse institutional types for their policies and practices regarding gardens and plant materials. Institutions were evaluated using a structured survey instrument and telephone interviews of infection control personnel, facilities management, and patient care staff. Preliminary data from nine healthcare institutions (long-term care facilities, community hospitals, and tertiary care centers) suggest there is wide variation in policies, procedures, and practices pertaining to patients' exposure to natural elements in healthcare settings. Some institutions ban all plants, including cut flowers, while others engage AIDS patients in horticultural therapy activities. Staff awareness of regulations, policies, and procedures varies as well.

Survey results from additional institutions were presented and compared to recommendations of Joint Commission on Accreditation of Healthcare Organizations, Centers for Disease Control and Prevention, and other regulatory authorities. Preliminary results indicate a need for clarification of health risks posed by therapeutic gardens and for education of health facility staff and designers about the safe implementation of plants in healthcare settings.

# 35

# Abstracts: Design

## A UNIVERSAL DESIGN APPROACH TO RECREATION AND LEARNING IN THE LANDSCAPE

**Susan Goltsman**

The outdoor environment is a rich setting for play and learning. A universal design approach to the development of outdoor settings permits a wide range of people with a variety of abilities to participate. Through case examples of built work, the principles of universal design will be demonstrated. Case examples will include children's gardens, zoos, parks, and schoolyards.

## RESTORATIVE GARDENS—METAPHORICALLY TRANSCENDING THE HUMAN EXPERIENCE OF LIFE

**Scott C. Scarfone**

Interpretation of restorative garden forms and themes has historically been hampered by the lack of direct evidence as to their meaning. Individual garden components serve as mental guideposts leading the spectator through a metaphysical journey whereby symbolic life-sustaining forces are subconsciously perceived. What are these mental guideposts and what are their meanings? Although the garden journey applies to all human beings, it becomes even more applicable to the sick and elderly. The restorative garden in this light becomes the pure embodiment of ongoing life. The renewed emphasis on restorative or healing gardens demands that they be examined under this pretense. This poster presentation explored the meaning of six different garden types, and dissected the circumstances that led a garden to be qualified as restorative and the net effect it had on the human experience. Restorative gardens— either contrived or natural—provide for and give life to a wounded or dying soul through their ability to invigorate mental stimulus and one's connection to nature. This iconographical analysis and documentation of physical form may provide a new direction in current thought and may ultimately influence the physical profile of future healing gardens.

## Therapeutic Garden Design in a Pediatric Healthcare Setting: A Developmental Approach

Roberta Hursthouse

Therapeutic garden design in a pediatric healthcare setting provides a sensory enriched, natural environment where children's developmental needs are met and nurtured. Through both passive and active involvement in a therapeutic garden, children and their families are given a powerful tool for coping with the stress and trauma of hospitalization. This presentation focused on the developmental needs of the hospitalized child and the development of a therapeutic garden that meets these needs in a developmentally appropriate manner. The "Garden Play" program of Children's Memorial Hospital, Chicago, Illinois, was used to illustrate innovative design and programming. This unique program was initiated in 1983 as a cooperative effort between the Chicago Botanic Garden's Horticultural Therapy Services Department and the Hospital's Child Life Department and has been implemented since 1985 by Roberta Hursthouse, H.T.R. A slide presentation featured adaptations and precautions for the pediatric gardener; "The Magic House," an inclusive dramatic play structure with outdoor "rooms"; and an intensive seasonal color rotation program for year-round visual impact. The Garden Court and "Garden Play" have been included in a chapter on "Healing Gardens for Children" in the recently published book *Healing Gardens* (Cooper Marcus and Barnes, 1999).

# 36

# Abstracts: Therapeutic Application

## HORTICULTURAL THERAPY IN REHABILITATION

**Katherine A. Feuillan**
**Alicia Gaca**

Horticultural therapy is a highly beneficial treatment modality for persons with physical and/or cognitive disabilities in a rehabilitation program. When using an interdisciplinary approach, horticultural therapy serves to augment and reinforce the entire treatment plan, while providing the participant with a concrete sense of warmth and an outlet for creative expression and, for some, an avocation that allows the person the opportunity to participate as a vital member of his or her community. Its value is cross-generational, and is one of the most requested activities among participants in a variety of treatment programs. Integrating horticultural activities into a client's individualized treatment plan can provide a variety of challenges to the individual, including but not limited to eye-hand coordination, gross and fine motor skill development, outlets for creative expression, stress reduction, and socialization. Other goals of the program are to assist the person in returning to a previous leisure interest, to teach a new leisure skill, or to educate the person on adaptive gardening techniques. Working with plants can give a person a sense of responsibility and being needed. This can encourage motivation for the future. Learning adaptive gardening techniques also encourages independence, which can lead to active living in the community.

## HORTICULTURAL THERAPY AT THE ROYAL NATIONAL ORTHOPEDIC HOSPITAL NHS TRUST

**Linda Exley**

Horticultural therapy has been an important part of the rehabilitation service offered by the Royal National Orthopedic Hospital (RNOH) occupational therapy department for the past 14 years. This hospital is the largest postgraduate hospital in the UK and treats patients whose healthcare needs range from the most acute spinal injury to chronic back pain. Within the extensive grounds, there is a garden with a level paved area for easy access for wheelchair users. Patients are provided with gardening therapy to build up their physical strength, increase movement, and help maintain balance, posture, and coordination. They are introduced to lightweight tools and shown techniques that can make projects achievable. Patients have remarked that horticultural therapy improves their physical well-being, and they speak of improved concentration and increased confidence helping them feel more motivated and able to communicate with others.

## SEED BALLS

Keiko Murayama
Ritsko Yasue

The purpose of this poster was to present an alternative method of seed propagation ideally suited for diverse client populations. The Gaia seed method is an alternative method of seed propagation gaining new popularity in Japan. Seeds are sown in a clay mixture that is then shaped into pea-sized balls. All the elements needed for seed germination are contained in the mixture, hence the name Gaia, which means "The Mother Earth" in Japanese. Since seed balls are self-sustaining, they do not require in-ground planting, and make this a far less labor-intensive method of planting than other traditional methods. This activity is easily adapted to meet the needs of diverse populations. It is a natural vehicle for promoting social interaction when used as a group activity. The repetitive and tactile aspects of the activity make it an excellent mechanism for stress reduction. Although excellent for incorporating psychosocial goals, this activity lends itself equally well to an array of other client goals including language development (receptive and expressive), sequencing, decision making, and fine motor development. Most importantly though, the Gaia seed method presents a fresh approach to connecting people with nature and creating a bond among all those sharing in the process. In Japan, Gaia seeds were introduced to the residents of an alcohol recovery program. Thirty residents worked together in making 12,000 seed balls, which they then shared with a nearby residential facility for children with severe developmental disabilities. Together these two groups "planted" the Gaia seed and in doing so created a relationship that has continued to blossom and be an inspiration to all involved. It is our hope that through the use of seed balls that similar symbiotic relationships can be replicated among multigenerational groups and people of all abilities.

## INTERACTIONS BETWEEN ELDERLY ADULTS AND PRESCHOOL CHILDREN IN A HORTICULTURAL THERAPY RESEARCH PROGRAM

Mary L. Predny
Paula Diane Relf

A study to examine the behavior of elderly adults and preschool children during horticultural therapy activities and to determine if combining intergenerational groups would complement or detract from the horticultural therapy goals for each group separately was conducted. During a 10-week observation period, data were collected on video, documenting attendance, participation time, and pattern during separate age groups and intergenerational activities. These data were used to determine if interactions changed over time or in response to different activities. Participation appeared to be affected by activity design, difficulty level, individual ability, and availability of assistance. Children's intergenerational participation scores appeared to show an increase in the category of "working with direct assistance," while elderly adults' intergenerational scores appeared to show an increase in the categories of "no participation" and "independent participation." In part, the intergenerational change appeared to be due to a decrease in assistance available from volunteers for each individual. For some individuals, the introduction of intergenerational groups appeared to detract from personal participation in horticultural activities. If the goal of the horticultural therapist is directly related to the individual's activity in horticulture (i.e., increase self-esteem from successfully designing and building a terrarium), the intergenerational element appears to reduce the potential for that benefit. The percentage of total social interaction time between the generations during activities increased over time. The intergenerational activities involving plant-based activities seemed to be more successful at increasing intergenerational exchange than the craft-type activities. Therefore, horticulture may be a useful activity for programs with a goal of increased intergenerational interaction.

## THE THERAPEUTIC ROLE OF HORTICULTURE IN EDUCATIONAL, RECREATIONAL, THERAPEUTIC, AND VOCATIONAL PROGRAMS

**Lori Keltner**

The Horticulture Connection is a contractual fee-based program that develops programs for schools, healthcare agencies, social service agencies, retirement facilities, community gardens, community organizations, and corporations. While serving these groups, plant-based activities are introduced through group interaction to foster self-esteem, the sense of belonging, the capacity to care for something, and a sense of trust. The Horticulture Connection provides educational, recreational, vocational, and therapeutic interventions by using plants and the environment to reach physical, social, emotional, and cognitive well-being. This poster presentation addressed how therapy and involvement in horticulture can improve a participant's way of living. The information explored the differences between educational, recreational, vocational, and therapeutic horticulture. Examples of programs and definitions of programs offered the conference participant a chance to explore how the effects of horticultural therapy benefit a variety of ages and populations.

## ENGAGING DISABLED STUDENTS IN DEVELOPING A FLOWER GARDEN AS A RECREATIONAL THERAPY

Karen Midden
Michelle McLernon

The extensive benefits of horticulture therapy for individuals with disabilities provided the impetus for the 1999 inception of the "Dig It!" program at Southern Illinois University Carbondale (SIUC). Results of an interest survey of participants in the Disabled Student Recreation program revealed high interest in horticulture activities, yet concerns of physical barriers and lack of experience prohibited their involvement. Efforts were made to eliminate perceived barriers and provide new opportunities for this group. Through cooperation of the Plant, Soil, and General Agriculture and Therapeutic Recreation departments and assistance from outside donors, resources were accessed to create this program. The 1999 program actively engaged the participants in the design, planting, and maintenance of an SIUC flower garden located in an approximate 375-square-foot existing brick planter. Although involvement in activities was subject to each participant's individual abilities, the diverse range of horticulture activities offered opportunities for everyone involved. However, modifications and nontraditional methods were used to meet the specific needs of this group. The success of the 1999 program, based on participants' verbal feedback and interest, has encouraged efforts to continue a horticulture therapy program and to develop a formal means of assessment.

## EVERY COMMUNITY NEEDS A PROVIDENCE FARM

Christine Pollard

Providing services for persons with barriers to employment and education can be daunting in a rural area, where resources are few. Providence Farm has, over a 20-year period, built a community within a community, becoming a multiuse facility where seniors, persons with developmental disabilities, persons with mental health issues, and persons with brain injury, learning disabilities, or drug and alcohol issues can participate in a horticulture/agriculture atmosphere. Providence Farm is an integrated therapeutic community where persons can participate in horticulture for therapeutic, employment, and social or recreational purposes. Our inclusive community works on the premise that everyone has a talent or ability to offer back to their community. Our programs and facilities are designed around the needs of the individual, and then incorporated into the larger nursery, greenhouse, market garden, livestock, orchard, or landscape projects. The results of having such a resource in our community has benefited the individuals, their families, and friends we work for in that these persons have a place in the community where they can contribute and be effective. The community-at-large benefits from our produce, our nursery stock, our value-added products, and our livestock products; and from a healthy community where there is more understanding of persons with disabilities and the contribution they can make. Our health and social systems benefit by having a healthier community, where preventative management can be practiced as opposed to crisis management. Any community can design and implement a Providence Farm model in their community. It will not be exactly like ours, but it will meet the needs of that community.

## INCORPORATING HORTICULTURE THERAPY WITH PSYCHOSOCIAL AND PHARMACOLOGICAL INTERVENTIONS OF MENTAL ILLNESS IN CHILDREN AND ADOLESCENTS

**Catherine Trapani**

A project incorporating horticulture therapy with traditional treatment paradigms used in addressing mental illness, including psychotherapy and pharmacology, was presented. The current research literature on psychopathology in childhood, with an emphasis on Major Depressive Disorder, was reviewed. Information included operational definitions, diagnostic criteria, and epidemiology. Current treatment approaches were presented. The inclusion of horticulture therapy in a milieu treatment setting was the focus of the session. A complete description of the program including treatment goals, program design, and therapeutic outcomes was provided. Qualitative clinical findings included testimonials from staff and student participants. The Horticulture Therapy project is entering its third season. In addition to contributing to the improvement of the environment of the school, students have experienced numerous positive treatment outcomes including increased prosocial behaviors, cooperative learning, and heightened awareness of self.

# 37

# Abstracts: Research

## EFFECTS OF HORTICULTURAL ACTIVITY ON THE LEVEL OF A STRESS HORMONE AND SUSCEPTIBILITY TO UPPER RESPIRATORY INFECTION

**Hyejin Cho**
**Richard H. Mattson**

Moderate physical activities improve health; reduce the risk of various disease, depression, and anxiety; and promote psychological well-being for a wide range of people. Effects of horticultural activity on physical and mental health have been reported, but the restorative experiences of horticulture are a unique benefit. Susceptibility to upper respiratory tract infection (URTI) mediated by immune function is influenced by biological and behavioral changes through various life events. Emotional stressors influence not only a hormonal pathway, but also duration and severity of infection. URTI is a relatively mild illness, but it has significant impact on medical practice and economic consequences due to work loss. In particular, mucosa immunity to URTI is an important host defense. Therefore, we hypothesize that horticultural activities will reduce the overall level of stress and susceptibility to URTI by means of providing moderate physical activity and restorative experiences over an extended time. This preliminary study was conducted to measure horticultural activity influences on human susceptibility to URTI. Salivary cortisol and secretory immunoglobulin A were measured to indicate the stress level and the mucosa immunity of college students while in sedentary and physically active horticulture laboratories. Students were asked to complete several questionnaires concerning general physical health, affective mood, and group environment.

## PERSONAL, SOCIAL, AND CULTURAL MEANINGS OF A COMMUNITY GARDEN FOR STUDENTS: AN IN-DEPTH CASE STUDY AND EVALUATION

**Catherine A. Bylinowski**
**Richard H. Mattson**

A case study was conducted of a 31-year-old community garden program organized at an on-campus university student-housing complex. In-depth interviews were conducted with the student gardeners from China, Korea, Germany, Pakistan, and the United States concerning personal, social, and cultural meaning of gardening, motivations for gardening, and how the garden functions as a social space. Preliminary interview results indicate that the inherently transitory nature of their participation does not diminish the value of the garden experience. Participant observation data have observed the number of people visiting the garden, garden activities and methods used, and the types of social interactions taking place. Stated benefits have included fulfilling a desire to be closer to nature, growing vegetables from one's country of origin, and making friends. The community garden has the potential to enhance the students' educational experience and serve to enhance multicultural awareness through increased interaction within the garden space and the wider university community.

## Physical and Psychological Responses of Patients to the Hospital Landscape

Chun-Yen Chang
Tzu-Hui Tseng

The purpose of this study was to test the relationship among the hospital landscape, the patients' experiences of the landscape, disease types, and the patients' physical and psychological responses. Renal stone and peptic ulcer patients at Chino Medical College Hospital and Chung-Shaw Medical and Dental College Hospital completed questionnaires on each of the first 3 days of their hospital stay. Major findings were less feelings of pain for patients with the greatest hospital landscape experience; disease type and the patient's psychological response to the hospital landscape were not related; and greatest hospital satisfaction was experienced by patients with the longer outdoor hospital landscape experience. The results of this research could assist landscape planners, designers, and hospital managers to understand the importance of the hospital landscape environment. Furthermore, the results should encourage medical experts and psychologists to use horticultural therapy as a component of their treatment plans.

## Plant Successions and Urban Successions: How Human Beings Have Affected Them

Hideki Hirano

When a city is formed and developed, the surrounding forest is required not only to supply materials for buildings and fire but also to conserve soils and to reserve water. As the city grows, the forest is not always able to continue to respond to these demands. Why? When trees are cut or fall down in a plant community of stable successions, this plant community becomes unstable and will impact the surrounding areas, including outside of the forest, i.e., urban space inhabited by human beings. In other words, destruction of the forest habitat continues to affect human community space known as a "city." This poster analyzed the relationships of urban populations, urban expansion, loss of original-growth forest (large-tree growth) area, age of groves of large trees within original-growth woodlands, and relation of water demand and supply, inspecting the connection between the human community and plant community.

## AIR CONDITIONING AND NOISE CONTROL USING VEGETATION

**Peter Costa**

Vegetation is increasingly being shown to perform by natural means functions that were previously the domain of mechanical engineering (for example, air and water filtration, heavy metals and toxic chemicals extraction, and now cooling, humidification, and room acoustics). "Air Conditioning and Noise Control Using Vegetation" is about two processes that have so far not been considered as engineering attributes of plants, namely cooling and humidification. The new, simple models proposed to quantify these processes in and on buildings are validated using quality data collected from diverse sources. The equations are based on the extension of established parameters. With a mean outdoor air temperature of 19°C the cooling equations predicts a Sol-air temperature of the order of 18°C for a vegetated surface compared to 67°C for a dark man-made surface, or 44°C for a light surface. This means that solar gain on that surface could be ignored. The humidification equation predicts the humidifying power of vegetation. Used on the outside of the building, evapotranspiration by vegetation permits the wet bulb depression associated with adiabatic humidification to be exploited thereby achieving sensible cooling of the building envelope. Used inside the building, evaporative cooling and humidification are achieved albeit at a reduced rate dependent on solar performance of glazing. These functions are all fully automatic and dependent on solar gain. Consequently, the operation of air conditioning systems could be reduced, and a reduction in energy consumption would result. This yields a reduction in carbon dioxide production at the power station. The new work presented at this conference was the validation of the two models that have been developed, which lends weight to the practical application of vegetation for another two mechanical engineering functions, namely cooling and humidification. Acoustic absorption by indoor plants was also presented.

## PSYCHOPHYSIOLOGICAL AND EMOTIONAL INFLUENCES OF FLORAL AROMA ON HUMAN STRESS

**Mingwang Liu**
**Richard H. Mattson**

Stress of modern life may deteriorate human health psychologically, physiologically, and emotionally. Flowers, gardens, and natural landscapes reduce life stress and improve human wellness. Olfaction is a critical human perception for appreciation of natural beauty. The sense of smell evokes powerful memories and changes human perception and behavior. Persuasive scientific evidence on the benefits of aromatic flowers to human wellness is needed for widespread use of aromatic plants in floral markets and landscape design. With the help of a computerized biomonitoring system, brainwaves, skin conductance, and skin temperature of participants exposed to lavender stimuli were measured and recorded simultaneously. The participants recorded their emotional responses to indicate their feelings on a five-point ZIPERS test. Through analysis of these physiological and emotional responses, mechanisms were explored on how floral aroma-related products reduce human stress, improve human wellness, and influence behavior of consumers. Based on research results, informative recommendations will be given to consumers of floral products and the floral industries.

# VI

# Appendix

# Appendix

## Symposium Overview

Interaction by Design: Bringing People and Plants Together for Health and
    Well Being
July 20–22, 2000
Chicago Botanic Garden
Glencoe, Illinois

*Sponsors:*
    American Horticultural Therapy Association
    People-Plant Council
    School of the Chicago Botanic Garden

*Symposium Chair:*
    Candice A. Shoemaker, Kansas State University

*Symposium Organizing Committee:*
    Tina Cade, Southwest Texas State University
    Jack Carman, Design for Generations
    Frank Clements, Wolff Clements & Associates, Ltd.
    Holly Estal, Chicago Botanic Garden
    Patricia Hill, Misericordia's Greco Gardens
    Elizabeth Messer Diehl, American Horticultural Therapy Association
    Paula Diane Relf, Virginia Tech University
    Gene Rothert, Chicago Botanic Garden
    Martha Tyson, Douglas Hills Associates
    Charles Waldheim, University of Illinois—Chicago
    Joanne Westphal, Michigan State University

*Funded in Part by:*
    Chicago Botanic Garden
    The Retirement Research Foundation

*Endorsed by:*

Alzheimer's Association
American Association of Botanic Gardens and Arboreta
American Federation for the Blind
American Horticultural Society
American Institute of Architects
American Occupational Therapy Association
American Planning Association
American Society for Horticultural Science
American Society of Interior Designers
American Society of Landscape Architects Therapeutic Garden Committee
Association of Collegiate Schools of Architecture
Association of Professional Landscape Designers
Canadian Horticultural Therapy Association
Council of Educators in Landscape Architecture
National Federation for the Blind
National Gardening Association

# Index

Academicians in the horticulture community, 6–9
Accessibility
    barriers and, 158–159
    postoccupancy evaluations (POE), 223
Acoustics, 119–120
Age
    of horticultural therapists, 44
    research of impact of cut roses in restaurants and,
        247
    urban forestry programs and, 268–277
Alexander, Christopher, 145
Alterra Clare Bridge, Oklahoma City, 113
Alva and Bernard F. Gimbel Garden, 152
Alzheimer's facilities
    activities and, 138–139
    American Society of Landscape Architects
        (ASLA) and, 99
    Garden Project of ASLA and, 111–114
    Memory Gardens (Alzheimer's disease),
        Portland, Oregon and, 99–110
    plant therapy/programs and, 25
Alzheimer's Memory Garden, Macon, Georgia, 113
Ambulatory settings, 25
American Community Gardening Association, 40
American Horticultural Therapy Association
    (AHTA)
    computer use survey, 46–50
    computer use survey results, 41–45
    designing of therapeutic environments and, 149
    as symposium sponsors, 321
American Society of Horticultural Sciences, 43
American Society of Landscape Architects (ASLA)
ANOVA test, 248
Alzheimer's facilities and, 99
Alzheimer's Garden Project and, 111–114
Americans with Disabilities Act
    horticulture customer base and, 15
Arboreta staff as educators, 9–10

Architecture
    hospital design (1900s) and, 143–144
    incorporation of stairs in, 146–147
    incorporation of towers in, 146
    integration of natural elements in, 145–147
    quadrangles design and, 117
    research on plant decorations and, 261–265
    tropical conservatory greenhouses, 176
Architecture of the Well-tempered Environment, 144
Art therapists/programs
    bonsai as life enhancement and, 175–178
    examples from, 21–23
    horticulture communication and, 13
    Northwestern Memorial Hospital (NMH) and, 24
ASLA/National Alzheimer's Association Garden
        Project, 111–114
Autobiographical Memory Coding Tool, 166
Avicenna, physician/philosopher, 280–282
Awaji Landscape Planning and Horticulture
        Academy (ALPHA), Hyogo, Japan, 57–60


Back of the Yards Council, 35–36
Banham, Reyner, 144
Barriers, 124–125, 158. See also Privacy
"Be Yourself, Say Hello! Communicating with
        People Who Have Disabilities" guide, 69
Beuhler Enabling Garden, Chicago Botanic Garden,
        152–153
Beyond Prozac, 146
Biofeedback, 139
Biophilia Hypothesis, 112
Biotechnology and horticulture, 7
Birds/butterflies/wildlife attraction
    displays at Buehler Enabling Garden and, 71
    Eden Alternative and, 188
    Enid A. Haupt Glass Garden and, 153
    mental health facilities and, 120

Birds/butterflies/wildlife attraction (*Continued*)
   planning for involvement and, 161–162
   small group process facilitation and, 80–81
Bonsai and elderly, 175–178
Botanic gardens staff as educators, 9–10. *See also*
      Staff in healthcare
Botanica, The Wichita Gardens, 203–208
British Columbia Professional Fire Fighters' Burns
      and Plastic Surgery Unit of Vancouver General
      Hospital, 91–97
Buehler Enabling Garden
   interpretations/exhibits of, 67–69
   marketing/publicity/opening celebrations, 69–71
   planning/design of, 65–67
   plant displays at, 71
   programming of plant displays at, 72
   universal design and, 61
   volunteers and, 69
Butterflies. *See* Birds/butterflies/wildlife attraction


Cancer, 141–142
Car barriers and children, 124–125
Cardiac tonics/preventatives, 280–282
Chicago Botanic Garden (CBG). *See also* Buehler
      Enabling Garden
   Beuhler Enabling Garden of, 152–153
   Chicago's School Garden Initiative (SGI) and, 38
   Garden Play program of Children's Memorial
      Hospital and, 308
   symposium sponsor/funding and, 321
   volunteers and, 69
Chicago Neighborhood Tours/City of Chicago
      Department of Cultural Affairs, 40
Chicago's 13th Street Garden, Chicago, Illinois, 37
Chicago's School Garden Initiative (SGI), Chicago,
      Illinois, 38
Children
   Eden Alternative and, 188
   elderly interaction with, 310–311
   horticultural therapy and depression in, 313
   rehabilitation facilities for, 124–125
   research on nature contact by, 267–277
   Rusk PlayGarden for, 124–129
   sensory garden tours and, 195–200
   therapeutic garden design for, 308
   urban forestry programs and, 267–277
Children's Memorial Hospital, Chicago, 308
City Farms (UK), 161–162
City managers as horticulture clients, 16–17
Clark, George, 37–38
Cognitive impairment
   bonsai as life enhancement and, 175
   reminiscence therapies and, 165–174
Collaboration in research, 234. *See also* Research
Color

Japanese plant names and, 304
lawns of America and, 292
mental health facilities and, 119
reminiscence therapies and, 167–168
research of effect of flower color, 253–259
Communication
   *"Be Yourself, Say Hello! Communicating with
      People Who Have Disabilities"* and, 69
   with healthcare community, 24–25
   horticulture research and, 234
   stress on staff and, 27, 29
   therapeutic garden abstracts, 303–305
Community gardeners
   cultural understanding and, 34–35
   as horticulture educators, 11
Community gardens
   Chicago's 13th Street Garden, Chicago, Illinois,
      37
   Chicago's School Garden Initiative (SGI),
      Chicago, Illinois, 38
   effect of childhood park experiences, 273–277
   Garfield Neighborhood Garden, Chicago, Illinois,
      37–38
   healing landscape and, 56
   Mary's Kids Gardens, Chicago, Illinois, 35–36
   Memory Gardens (Alzheimer's disease),
      Portland, Oregon and, 99–110
   Pennsylvania Horticulture Society's (PHS)
      Greene Countrie Towne Program,
      Philadelphia, 35
   personal/social/cultural meanings of, 315–316
   Providence Farm and, 312
   Sheriff's Garden, Chicago, Illinois, 36–37
Confidence. *See* Self-esteem
Connections with the past, 162
Consumer types, 16–18
*Contemplative Gardens*, 84
Cooperative Extension agents as educators, 9–10
Cooperative interactions and mental health
      facilities, 118
Corporations
   community gardens and, 34
   as horticulture clients, 16
Correctional facilities, 36–37
*Creating People-Friendly Parks*, 53
Critical care units, 25–26
Cultural awareness, 33–35, 39


DaySpring Assisted Living Residence, Muskegon
      County, Michigan, 113
Dementia patients. *See also* Alzheimer's facilities
   cooperative research project and, 7
Demographics
   horticulturalist therapists and, 41–42, 44–45
   Providence Farm and, 312

urban forestry programs and, 271–277
Denver Botanic Gardens, 195–201
Depression/anxiety. *See also* Stress
  emotions/perception and, 136–137
  horticultural therapy and, 313
  importance of lighting and, 143–144
  Japanese cut flowers effects and, 237–242
Design. *See* Planning
Disabled people
  accessibility and, 158–159
  Awaji Landscape Planning and Horticulture
    Academy (ALPHA), Hyogo, Japan and,
    57–60
  *"Be Yourself, Say Hello! Communicating with
    People Who Have Disabilities"* and, 69
  Buehler Enabling Garden and, 66–68
  flower garden recreational therapy and, 311–312
  importance of universal design in Japan and, 61
  mental health facilities and, 120
  PlayGardens and, 123, 129–130
  publicity directed to, 70
  Sensory Garden (Fureai-no-niwa), Osaka, Japan
    and, 53–56
  small group process facilitation and, 76–77
  as volunteers, 68
Drug use in healthcare
  previous plant therapy research and, 27
Dunn, Ken, 37, 39
Durfee Conservatory and Gardens, 180–181

Economics
  building practices and, 144
  community gardens and, 36
  of Eden Alternative, 190
  health outcomes research and, 20, 28
  importance of planning for, 61–62, 67
  initiating horticultural therapy programs and,
    207–208
  lawns of America and, 292–293, 299
  marketing at Buehler Enabling Garden, 69–71
  market size/net profit and, 14–15
  plant therapy/programs and, 25
  of research, 29–30
  salaries and horticulture, 5–6
  urban forestry programs and, 268
Eden Alternative, 187–193
Education
  experiential, 267–268
  of Healing Gardeners, 60
  horticultural intervention for stress in, 179–185
  of horticultural therapists, 44
  Japanese institutions and, 214
  Kolb's Theory of Learning Styles and, 79
  Medical Center Gardens Project (Georgetown
    University) and, 83–89

regarding Alzheimer's disease, 99–101
  therapeutic application abstracts and, 311
  urban forestry programs and, 270, 273–275
  for volunteers, 69, 208
  World Wide Web and, 10
Educators. *See also* Academicians
  types of, 9–11
80th Street Residence, New York City, 113
Elderly. *See also* Alzheimer's facilities
  Awaji Landscape Planning and Horticulture
    Academy (ALPHA), Hyogo, Japan and,
    57–60
  bonsai and, 175–178
  Eden Alternative and, 187–193
  Garden Project of ASLA and, 111–114
  hospitality and, 59–60
  importance of universal design in Japan and, 61
  plant therapy/programs and, 25
  preschool children interaction with, 310–311
  publicity directed to, 70
  reminiscence therapies and, 165–174
  Sensory Garden (Fureai-no-niwa), Osaka, Japan
    and, 53–56
  sensory garden tours and, 195–200
  urban forestry programs and, 304
  as volunteers, 68
Electrodermal activity (EDA), 237
Electroencephalograms (EEG), 237
Electromyography (EMG) use, 237–240, 253,
  256–259
Enid A. Haupt Glass Garden (Rusk Institute),
  152–153
Environmental issues. *See also* Nature
  air conditioning/noise control via vegetation and,
    317
  cooperative research project and, 7
  design for, 138–139
  land management and, 36, 38
  lawns of America and, 292–293
  planning for involvement and, 161–162
  research at Botanica, The Wichita Gardens and,
    204
Ethics and health outcomes, 21
Evaluations
  mental health facilities and, 121
  postoccupancy, 73
Exercise
  postoccupancy evaluations (POE) and, 222

Fight/flight reaction, 137–139
Fragrance, 279–288, 317–318. *See also* Senses
Frisbee Park of Georgetown University Medical
  Center Gardens Project, 88
Funding
  horticulture industry and, 6–9, 14–16

Funding (*Continued*)
  Medical Center Gardens Project (Georgetown
      University) and, 85
  Memory Gardens (Alzheimer's disease),
      Portland, Oregon and, 106–107
  Fureai-no-niwa (Sensory Garden), Osaka, Japan,
      53–56

"Garden in Every School", 38
Garden Project of ASLA, 111–114
Gardeners. *See* Community gardeners
*Gardener's Delight: Gardening Books from 1560-
    1960*, 187
Gardens. *See also* Community gardens
  Awaji Landscape Planning and Horticulture
      Academy (ALPHA), Hyogo, Japan, 57–60
  effect of childhood park experiences and,
      273–277
  healing landscape and, 56–57, 59–60
  horticulture customer base and, 15–16
  Indo-Islamic garden, 279–288
  influence/role of, 4–5
  Medical Center Gardens Project (Georgetown
      University) and, 83–89
  programming of, 72
  Sensory Garden (Fureai-no-niwa), Osaka, Japan,
      53–56
  small group process facilitation and, 80–81
Garfield Neighborhood Garden, Chicago, Illinois,
    37–38
Gender
  of horticulturalist therapists, 44
  research of impact of cut roses in restaurants and,
      247–248
Genetically modified food. *See* Biotechnology
Georgetown University Medical Center Gardens
    Project, 83–89
Gonzales, German, 35–36
Gonzales, Mary, 35–36
Government agencies
  community gardens and, 35
Grants. *See* Funding
Gratitude Hoop Work, 76
Greenfield Senior Center, Greenfield,
    Massachusetts, 177
Greenlining, 33–35. *See also* Community gardens
GreenNet: Chicago's Greening Network, 40
Group therapy programs
  designing of therapeutic environments and, 151
  Enid A. Haupt Glass Garden and, 153

Handicapped people. *See* Disabled people
Hathorn, Kathy, 24

"Healing and the Mind", 84
Healing Gardeners, 59–60
*Healing Gardens: A Natural Haven of Emotional
    and Physical Well-Being*, 84, 92, 308
Healing landscape, 56–57
Health outcomes
  control groups and, 30
  patient surveys and, 23–24
  plant therapy and, 20–21
  previous plant therapy research and, 27–28
  research methods and, 28–30
Healthcare
  consumer-/service-oriented trend of, 23–24
  demand/utilization patterns trend of, 24–25
  importance of garden planning and, 62, 72–73
  stress on staff in, 26–27, 29
Healthcare industry members
  health outcomes research and, 20
  horticulture communication and, 13
  Medical Center Gardens Project (Georgetown
      University) and, 83–89
Holistic healing
  healing landscape and, 56–57
  Medical Center Gardens Project (Georgetown
      University) and, 83–89
  small group process facilitation, 76–81
  spiritual aspect of, 75–76
Homeowners as horticulture clients, 16
Horticultural intervention and stress, 179–185
Horticultural Therapist Registered (HTR), 41,
    44–45
Horticultural Therapist Technician (HTT), 41,
    44–45
Horticultural therapy
  bonsai as life enhancement and, 176
  community gardens and, 315–316
  depression and, 313
  effect on stress/infections, 315
  healing landscape and, 56
  hospital patient responses to, 316
  Japanese institutions and, 214
  Kolb's Theory of Learning Styles, 79
  in Korea, 305
  mental health facilities and, 118
  plant displays/programming and, 71–72
  preschool children interaction with elderly and,
      310–311
  research on therapeutic effects of, 230–235
  storytelling and, 79
  therapeutic application abstracts and, 309–313
  therapists as educators and, 11
Horticultural vocational rehabilitation specialists, 11
Horticulture
  definition, 5
Horticulture community

academicians in, 6–9
components of, 4
importance of recognition of, 12–13
increasing students for, 8–9
industry and, 14–16
money and, 3
*HortScience*, 8
Hospitality, 59–60
Hospitals. *See also* Healthcare industry members
design of (1900s), 143–144
horticultural therapy and, 316
importance of settings in, 19–20
Kyushu, Japan and, 211–215
plant therapy/programs and, 25
postoccupancy evaluations (POE), 222–224
survey services from, 29–30
with therapeutic gardens (U.S.), 149
Hoyles, Martin, 187
Human Issues in Horticulture
horticultural therapists and, 43
research and, 227–228
Hypertension, 204–207. *See also* Stress

Ikebana, 176
Independent living settings, 25
Indo-Islamic garden, 279–288
Islamic gardens, 117

Japanese gardens
Awaji Landscape Planning and Horticulture
Academy (ALPHA), 57–60
banning of flower cultivation in, 303
cut flowers effects and, 237–242
Gaia seed ball method and, 310
importance of universal design and, 61
Kyushu Island and, 211–215
names of colors originating in, 304
plant decorations by houses/gates and, 261–265
Sensory Garden (Fureai-no-niwa), Osaka, Japan,
53–56
yugen and, 120
Johansson & Walcavage (Johansson Design
Collaborative, Inc.), 124
Joint Commission on Accreditation of Healthcare
Organizations proposed environmental
standards by, 28
*Journal of Environmental Horticulture, The*, 8
*Journal of Extension*, 8
*Journal of Vocational Technical Education*, 8

Kansas State University, 43
Kolb's Theory of Learning Styles, 79

Korean horticultural therapy, 305
Kyushu, Japan, 211–215

Land management, 36, 38
Landscape designers. *See also* Planning
horticulture customer base and, 15–16
as horticulture educators, 11
Olmsted, Frederick Law, 292
Lawns of America, 291–299
Learning styles, 79
Legacy Good Samaritan Hospital, Portland, Oregon,
26
Lighting
importance of, 141–147
postoccupancy evaluations (POE), 223–224
sensitivity to, 151
Long-term care units
Eden Alternative and, 187–193
plant therapy/programs and, 25
small group process facilitation and, 80–81
urban forestry programs and elderly in, 304

Making Connections study, 157
Marketing. *See* Economics
Mary's Kids Gardens, Chicago, Illinois, 35–36
Master Gardeners, 9–10, 13
Master Horticultural Therapist (HTM), 41, 44–45
Mattson, Richard, 50
*Meaning of Gardens*, 84
Media. *See* Marketing; Publicity
Medical Center Gardens Project (Georgetown
University), 83–89
Medicine Wheel Work, 76
Meditation Garden of Georgetown University
Medical Center Gardens Project, 88
Memory Gardens (Alzheimer's disease), Portland,
Oregon
as ASLA project, 113
objectives for, 99–100
publicity/fund raising and, 106–107
site selection and, 100–106
Mental health facilities
further research and, 121
horticultural therapy and, 313
nature interaction/description of, 117–120
security and, 117
Meyers-Thomas, Judith, 187
Mini Mental State Exam (MMSE) scores. *See*
Reminiscence therapies
Minton, Lea, 50
Monastery gardens, 117
Monroe Community Hospital, Rochester, New
York, 113

Morrison Horticultural Demonstration Center, Denver Botanic Gardens, 195–201
Music therapy, 124

National Alzheimer's Association
  Garden Project of ASLA and, 111–114
  Memory Gardens (Alzheimer's disease), Portland, Oregon and, 99–110
National Gardening Association, 40
Nature. *See also* Birds/butterflies/wildlife attraction
  bonsai as life enhancement, 175–178
  designing of therapeutic environments and, 149–153
  interactive PlayGardens and, 129–133
  lawns of America and, 291–299
  prisons and, 230
  research on children's contact with, 267–277
Nightingale, Florence, 143
Norden, Michael J., 146
Northwestern Memorial Hospital (NMH)
  art program of, 24
*Notes on Nursing*, 143
Nursing homes. *See* Elderly; Long-term care units

Oberlander, Cornelia, 97
Occupational therapists, 13
Olfaction. *See* Fragrance
Olmsted, Frederick Law, 292
Open Space Technology, 78
Out-patient settings and plant therapy, 25

Pain
  plant therapy/programs and, 25
  previous plant therapy research and, 27
Patient surveys/interviews. *See also* Planning
  art therapy and, 21–23
  economic acquisition of data via, 29–30
  participatory design (Vancouver General Hospital) and, 91–95
  physical environment and, 23–24
  small group process facilitation and, 80–81
*Pattern Language*, 145
Pavement vehicles, 159
Pennsylvania Horticulture Society's (PHS) Greene Countrie Towne Program, Philadelphia, Pennsylvania, 35
People-Plant Council (PPC)
  designing of therapeutic environments and, 149
  research and, 227–228
  as symposium sponsors, 321
Perfume, 284–288
Pert, Candace, 84
Pharmaceutical Garden of Georgetown University

Medical Center Gardens Project, 87, 89
Photosensitivity, 119
Planning. *See also* Universal design of outdoor spaces
  connections with the past and, 162
  for contact with outside world, 159–160
  designing of therapeutic environments and, 149–153, 162
  importance of, 61–62, 65–67, 72–73
  for involvement, 161–162
  Medical Center Gardens Project (Georgetown University) and, 84–86
  Memory Gardens (Alzheimer's disease), Portland, Oregon and, 99–106
  mental health facilities and, 121
  participatory design (Vancouver General Hospital) and, 91–95
  physiological reactions to environment and, 137–138
  rehabilitation facilities for children and, 124–125
  for relaxation, 160
  reminiscence therapies (case studies) and, 168–172
  role of perception in, 135–140
  small group process facilitation and, 76–78, 80–81
  for social contact, 161
  therapeutic garden abstracts, 307–309
Plants
  air conditioning/noise control via, 317
  banning of flower cultivation in Japan, 303
  for cardiac tonics/preventative, 280–282
  displays at Buehler Enabling Garden, 71
  horticultural intervention for stress and, 181
  Indo-Islamic garden and, 279–288
  Japanese cut flowers effects, 237–242
  Kyushu, Japan and, 213–214
  lawns of America, 291–299
  reminiscence therapies and, 166–167
  research of effect of flower color, 253–259
  research of impact of cut roses in restaurants, 245–251
  sensory stimulation in children and, 129–131
  urban settings and surrounding forests, 316–317
Plants for Clean Air Council, 17
PlayGardens, 123
Plaza Garden of Georgetown University Medical Center Gardens Project, 88
Pollution. *See also* Environmental issues
  air conditioning/noise control via vegetation and, 317
  lawns of America and, 292–293, 298
Portland Memory Garden. *See* Memory Gardens (Alzheimer's disease), Portland, Oregon
Postoccupancy evaluations (POE), 219–224
Privacy

need for, 95–96
plant decorations by houses/gates and, 261–265
postoccupancy evaluations (POE), 222
rehabilitation facilities for children and, 124
Wright, Frank Lloyd and, 146
Process facilitation, 76–81
Professional associations as educators, 9–11
Profile of Mood States (POMS), 237, 239
Programming of display gardens, 72
Providence Farm, 312
Public parks. *See* Community gardens
Publications
    peer-reviewed, 13
    research, 8
Publicity
    Buehler Enabling Garden and, 69–71
    Memory Gardens (Alzheimer's disease),
        Portland, Oregon and, 106–109


Quadrangle design, 117
Qualitative research, 233–234. *See also* Research
Quality of life and horticulture, 16–18
Quantitative research, 232–233. *See also* Research


Rainbow Hoop Work, 76
Real estate developers as horticulture clients, 16–17
Recreation
    therapeutic application abstracts and, 311
Recreational therapists, 13
Regina Medical Center, Hastings, Minnesota, 113
Reh, 29
Rehabilitation facilities
    active/passive concepts and, 151
    plant therapy/programs and, 25, 29
    Rusk Institute of Rehabilitation Medicine, New
        York University Medical Center, 124–125
Relaxation responses
    designing for, 139–140, 160
    horticultural intervention for stress and, 182–184
Reminiscence therapies, 162, 165–174
Research
    ASLA/National Alzheimer's Association Garden
        Project and, 114
    banning of flower cultivation in Japan, 303
    Botanica, The Wichita Gardens and, 203–208
    community gardens and, 39
    control groups and, 30
    effect of flower color and, 253–259
    funding and, 7–8
    health outcomes research and, 20
    horticultural computer use and, 41–45
    horticultural computer use (survey questions)
        and, 46–50
    horticultural intervention for stress in, 179–185

horticultural therapy and, 315–318
of human responses to horticulture, 228–235
impact of cut roses in restaurants and, 245–251
improved client service and, 11–12
infection control in therapeutic gardens and, 305
Japanese cut flowers effects and, 237–242
Korean horticultural therapy and, 305
mental health facilities and, 114
methods for strengthening, 28–30
names of colors originating in Japan, 304
nature and health outcomes, 27–28
on nature contact by children, 267–277
need for, 25, 157–158
on plant decorations by houses/gates and,
    261–265
postoccupancy evaluations (POE), 219–224
quality of life and, 16–17
urban forestry programs and elderly in long-term
    care, 304
Resources
    *"Be Yourself, Say Hello! Communicating with
        People Who Have Disabilities"* and, 69
    format availability of, 69
    horticultural computer use and, 43–44
    urban community gardens and, 40
Restaurant industry
    impact of cut roses on, 245–251
Restorative setting garden abstracts, 307–308
Royal National Orthopedic Hospital (RNOH) NHS
    Trust, 309–310
Rusk Institute of Rehabilitation Medicine, New
    York University Medical Center, 124, 152–153


Salaries, 5–6
Sally Stone Sensory Garden, 207
Sandboxes, 128, 130
School gardens, 11, 38, 56
Sculptures
    Medical Center Gardens Project (Georgetown
        University) and, 87, 89
    mental health facilities and, 119–121
Seasonal Affective Disorder (SAD), 143–144, 147
Seating arrangements in mental health facilities,
    118, 121
Security/safety issues
    postoccupancy evaluations (POE), 223
    quadrangles and, 117–118
    rehabilitation facilities for children and, 124–129
Self-esteem
    reminiscence therapies (case studies) and, 173
Senses/sensory gardens
    Awaji Landscape Planning and Horticulture
        Academy (ALPHA), Hyogo, Japan and,
        57–60
    Denver Botanic Gardens and, 195–201

Senses/sensory gardens (*Continued*)
   designing for, 160–161
   healing landscape and, 56–57
   horticultural intervention for stress and, 179–185
   Indo-Islamic garden and, 279–288
   perception in planning and, 135, 140
   PlayGardens and, 123, 129–130
   Sally Stone Sensory Garden, 207
   Sensory Garden (Fureai-no-niwa), Osaka, Japan,
      53–56
   small group process facilitation and, 76–77
Sheriff's Garden, Cook County Correctional
   Facility, Chicago, Illinois, 36–37
Signage
   Buehler Enabling Garden and, 68
   Medical Center Gardens Project (Georgetown
      University) and, 85
Sister Jean Bridgeman Memorial Garden of
   Georgetown University Medical Center
   Gardens Project, 87–88
Small group process facilitation, 76–81
*Small Group Process Facilitation Manual*, 77–78
Social and Economic Sciences Research Center
   (SESRC), 268
Social impacts of lawns in America, 292–299
Social skills
   designing for, 161
   interactive PlayGardens and, 132
   postoccupancy evaluations (POE), 222–223
Solheim Lutheran Home, Los Angeles, California,
   113
Solitude importance in mental health facilities, 120
Sound pollution, 224
Spirituality. *See* Holistic healing
Staff in healthcare
   mental health facilities and, 121
   plant therapy/programs and, 26–27, 29
   sensory garden tours and, 200
Stairs (incorporation of), 146–147
Storytelling, 79
Stress
   bonsai as life enhancement and, 176
   fragrances and, 317–318
   horticultural intervention and, 179–185
   horticultural therapy and, 315
   Japanese cut flowers effects and, 237–242
   lawns of America and, 292
   nature views by prisoners and, 230
   postoccupancy evaluations (POE), 222–224
   research at Botanica, The Wichita Gardens and,
      204–207
   "temporary escape" from, 92
Students
   horticultural intervention and, 179–185
   planning for involvement and, 162
   youth offenders and cooperative research

     projects, 7
Surveys/interviews. *See also* Patient
   surveys/interviews
   Eden Alternative and, 192
   horticultural computer use and, 41–45
   horticultural computer use (survey questions)
      and, 46–50
   Kyushu, Japan and, 211–215
   lawns of America and, 294
   mood assessment, 181
   urban forestry programs and, 268–277
Symposium overview, 321–322

Technology
   horticultural computer use and, 41–45
   horticultural computer use (survey questions)
      and, 46–50
Therapeutic Design Committee of the American
   Society of Landscape Architects (ASLA),
   149
Therapeutic environments
   Beuhler Enabling Garden and, 152–153
   designing for, 162
   designing of, 149–153
   Enid A. Haupt Glass Garden and, 152–153
   infection control and, 305
   therapeutic application abstracts and, 309–313
Thomas, William, 187–188
Time flow and mental health facilities, 120
*Timeless Healing*, 84–85
Tools
   enabling, 68
   horticulture customer base and, 15
Tourist-site managers
   community gardens and, 34
   as horticulture clients, 16–17
Towers (incorporation of), 146
Trade associations, 13
Tree planting programs, 267–277

Ulrich, Roger, 228–230
United States Department of Agriculture (USDA)
   Chicago's School Garden Initiative (SGI) and, 38
   funding and, 6
Universal design of outdoor spaces
   Awaji Landscape Planning and Horticulture
      Academy (ALPHA), Hyogo, Japan, 57–60
   Buehler Enabling Garden and, 63, 68–69
   healing landscape and, 56–57
   hospitality and, 59–60
   Medical Center Gardens Project (Georgetown
      University) and, 85
   Sensory Garden (Fureai-no-niwa), Osaka, Japan
      and, 53–61

Universal Garden Society (UGS), 91
Universities
    horticultural intervention for stress and,
        179–185
    horticulturalist therapists and, 43–44
University of Massachusetts (UMass/Amherst),
    179–185
Urban forestry programs
    effect on surrounding forests, 316–317
    elderly and, 304
    influence/role of, 267–277


Vancouver General Hospital, 91–97
Virginia Tech University, 43
Vocational rehabilitation, 13
Volunteers
    Botanica, The Wichita Gardens and, 208
    Denver Botanic Gardens and, 196–197
    disabled/elderly, 68–69
    Eden Alternative and, 191
    hospitality and, 59–60
    Medical Center Gardens Project (Georgetown
        University) and, 84

Memory Gardens (Alzheimer's disease),
    Portland, Oregon and, 99, 107


Water features. *See also* Sculptures
    Beuhler Enabling Garden and, 153
    interactive PlayGardens and, 131–132
    lawns of America and, 292
    mental health facilities and, 119, 121
Watts, Elizabeth, 95
Wildlife. *See* Birds/butterflies/wildlife attraction
Williams, Patrick, 29
Women Infant and Children Center (WIC), 37
World Wide Web and horticulture education, 10
Wright, Frank Lloyd
    integration of natural elements and, 143, 146
    mental health facilities and, 117


Youth offenders, 7
Yugen and mental health facilities, 120


ZIPERS test, 317